LABORATORY MANUAL

to accompany

Concepts of
BIOLOGY

Sylvia S. Mader

Connect
Learn
Succeed™

LABORATORY MANUAL TO ACCOMPANY CONCEPTS OF BIOLOGY

1 2 3 4 5 6 7 8 9 0 WDQ/WDQ 1 0 9 8 7 6 5 4 3 2 1 0

ISBN 978–0–07–729733–6
MHID 0–07–729733–4

Vice President, Editor-in-Chief: *Marty Lange*
Vice President, EDP: *Kimberly Meriwether David*
Vice-President New Product Launches: *Michael Lange*
Publisher: *Janice Roerig-Blong*
Executive Editor: *Michael S. Hackett*
Senior Developmental Editor: *Rose M. Koos*
Senior Marketing Manager: *Tamara Maury*
Project Coordinator: *Mary Jane Lampe*
Senior Buyer: *Sandy Ludovissy*
Senior Media Project Manager: *Jodi K. Banowetz*
Senior Designer: *Laurie B. Janssen*
Cover Image: *Bronzy Hermit (Glaucis aenea) hummingbird, feeding and pollinating a Perfumed Passion Flower (Passiflora vitifolia) flower, rainforest, Costa Rica.* © Michael & Patricia Fogden, Minden Pictures
Senior Photo Research Coordinator: *Lori Hancock*
Photo Research: *Connie Mueller*
Compositor: *Electronic Publishing Services Inc., NYC*
Typeface: *11/13 Slimbach*
Printer: *Quad/Graphics*

Contents

Preface   iv
Laboratory Safety   vii

*These laboratory sessions are especially recommended for single semester courses that want to minimize laboratory preparation and include the basics of biological concepts.

Preface

To the Instructor

The twenty-six laboratory sessions in this manual have been designed to introduce beginning students to the major concepts of biology while keeping in mind minimal preparation for sequential laboratory use. The laboratories are coordinated with *Concepts of Biology,* a general biology text that covers all fields of biology. This manual may also be used in coordination with other texts and can be adapted to a variety of course orientations and designs. Thirteen laboratories have been singled out in the Table of Contents as especially appropriate for a single semester course. Then, too, many activities may be performed as demonstrations, thereby shortening the time required to cover a particular concept.

The Exercises

All exercises have been tested for student interest, preparation time, estimated time of completion, and feasibility. The following features are particularly appreciated by adopters:

Integrated opening: Each laboratory begins with a list of Learning Outcomes organized according to the major sections of the laboratory. The major sections of the laboratory are numbered on the opening page and in the laboratory text material. This organization will help students better understand the goals of each laboratory session.

Self-contained content: Each laboratory contains all the background information necessary to understand the concepts being studied and to answer the questions asked. This feature will reduce student frustration and increase learning.

Scientific process: All laboratories stress the scientific process, and many opportunities are given for students to gain an appreciation of the scientific method. The first laboratory of this edition explicitly explains the steps of the scientific method and gives students an opportunity to use them. I particularly recommend this laboratory because it utilizes the pillbug, a living subject.

Student activities: A color bar is used to designate each student activity. Some student activities are Observations and some are Experimental Procedures. An icon appears whenever a procedure requires a period of time before results can be viewed. Sequentially numbered steps guide students as they perform each activity.

Live materials: Although students work with living material during some part of almost all laboratories, the exercises are designed to be completed within one laboratory session. This facilitates the use of the manual in multiple-section courses.

Laboratory safety: Laboratory safety is of prime importance, and the listing on page vii will assist instructors in making the laboratory experience a safe one.

Customized Editions

The 26 laboratories in this manual are now available as individual "lab separates," so instructors can custom-tailor the manual to their particular course needs.

Laboratory Resource Guide

The *Laboratory Resource Guide,* an essential aid for instructors and laboratory assistants, free to adopters of the *Laboratory Manual,* is online at www.mhhe.com/maderconcepts2. The answers to the Laboratory Review questions are in the *Resource Guide.*

To the Student

Special care has been taken in preparing the *Concepts of Biology Laboratory Manual* to enable you to *enjoy* the laboratory experience as you *learn* from it. The instructions and discussions are written clearly so that you can understand the material while working through it. Student aids are designed to help you focus on important aspects of each exercise.

Student Learning Aids

Student learning aids are carefully integrated throughout this manual. The Learning Outcomes set the goals of each laboratory session and help you review the material for a laboratory practical or any other kind of exam. In this edition, the major sections of each laboratory are numbered, and the Learning Outcomes are grouped according to these topics. This system allows you to study the chapter in terms of the outcomes presented.

The introduction reviews much of the necessary background information required for comprehending upcoming experiments. Color bars bring attention to exercises that require your active participation by highlighting Observations and Experimental Procedures, and an icon indicates a timed experiment. Throughout, space is provided for recording answers to questions and the results of investigations and experiments. Each laboratory ends with a set of review questions covering the day's work.

The appendix at the end of the book provides useful information on the metric system. Practical examination answer sheets are also provided.

Laboratory Preparation

Read each exercise before coming to the laboratory. *Study* the introductory material and the procedures. If necessary, to obtain a better understanding, read the corresponding chapter in your text. If your text is *Concepts of Biology*, by Sylvia S. Mader, see the "text chapter reference" column in the table of contents at the beginning of the *Laboratory Manual*.

Explanations and Conclusions

Throughout a laboratory, you are often asked to formulate explanations or conclusions. To do so, you will need to synthesize information from a variety of sources, including the following:

1. Your experimental results and/or the results of other groups in the class. If your data are different from those of other groups in your class, do not erase your answer; add the other answers in parentheses.
2. Your knowledge of underlying principles. Obtain this information from the laboratory Introduction or the appropriate section of the laboratory and from the corresponding chapter of your text.
3. Your understanding of how the experiment was conducted and/or the materials used. *Note:* Ingredients can be contaminated or procedures incorrectly followed, resulting in reactions that seem inappropriate. If this occurs, consult with other students and your instructor to see if you should repeat the experiment.

In the end, be sure you are truly writing an explanation or conclusion and not just restating the observations made.

Color Bar and Icon

Throughout each laboratory, a color bar designates either an Observation or an Experimental Procedure.

Observation: An activity in which models, slides, and preserved or live organisms are observed to achieve a Learning Outcome.

Experimental Procedure: An activity in which a series of steps uses laboratory equipment to gather data and come to a conclusion.

Time: An icon is used to designate when time is needed for an Experimental Procedure. Allow the designated amount of time for this activity. Start these activities at the beginning of the laboratory, proceed to other activities, and return to these when the designated time is up.

Laboratory Review

Each laboratory ends with a number of thought questions that will help you determine whether you have accomplished the outcomes for the laboratory.

Student Feedback

If you have any suggestions for how this laboratory manual could be improved, you can send your comments to:

The McGraw-Hill Companies
Product Development—General Biology
501 Bell Street
Dubuque, Iowa 52001

Laboratory Safety

The following is a list of practices required for safety purposes in the biology laboratory and in outdoor activities. Following rules of lab safety and using common sense throughout the course will enhance your learning experience by increasing your confidence in your ability to safely use chemicals and equipment. Pay particular attention to oral and written safety instructions given by the instructor. If you do not understand a procedure, ask the instructor, rather than a fellow student, for clarification. Be aware of your school's policy regarding accident liability and any medical care needed as a result of a laboratory or outdoor accident.

The following rules of laboratory safety should become a habit:

1. Wear safety glasses or goggles during exercises in which glassware and solutions are heated, or when dangerous fumes may be present, creating possible hazards to eyes or contact lenses.
2. Assume that all reagents are poisonous and act accordingly. Read the labels on chemical bottles for safety precautions and know the nature of the chemical you are using. If chemicals come into contact with skin, wash immediately with water.
3. **DO NOT**
   a. ingest any reagents.
   b. eat, drink, or smoke in the laboratory. Toxic material may be present, and some chemicals are flammable.
   c. carry reagent bottles around the room.
   d. pipette anything by mouth.
   e. put chemicals in the sink or trash unless instructed to do so.
   f. pour chemicals back into containers unless instructed to do so.
   g. operate any equipment until you are instructed in its use.
4. **DO**
   a. note the location of emergency equipment such as a first aid kit, eyewash bottle, fire extinguisher, switch for ceiling showers, fire blanket(s), sand bucket, and telephone (911).
   b. become familiar with the experiments you will be doing before coming to the laboratory. This will increase your understanding, enjoyment, and safety during exercises. Confusion is dangerous. Completely follow the procedure set forth by the instructor.
   c. keep your work area neat, clean, and organized. Before beginning, remove everything from your work area except the lab manual, pen, and equipment used for the experiment. Wash hands and desk area, including desk top and edge, before and after each experiment. Use clean glassware at the beginning of each exercise, and wash glassware at the end of each exercise or before leaving the laboratory.
   d. wear clothing that, if damaged, would not be a serious loss, or use aprons or laboratory coats, since chemicals may damage fabrics.
   e. wear shoes as protection against broken glass or spillage that may not have been adequately cleaned up.
   f. handle hot glassware with a test tube clamp or tongs. Use caution when using heat, especially when heating chemicals. Do not leave a flame unattended; do not light a Bunsen burner near a gas tank or cylinder; do not move a lit Bunsen burner; do keep long hair and loose clothing well away from the flame; do make certain gas jets are off when the Bunsen burner is not in use. Use proper ventilation and hoods when instructed.
   g. read chemical bottle labels; be aware of the hazards of all chemicals used. Know the safety precautions for each.
   h. stopper all reagent bottles when not in use. Immediately wash reagents off yourself and your clothing if they spill on you, and immediately inform the instructor. If you accidentally get any reagent in your mouth, rinse the mouth thoroughly, and immediately inform your instructor.
   i. use extra care and wear disposable gloves when working with glass tubing and when using dissection equipment (scalpels, knives, or razor blades), whether cutting or assisting.
   j. administer first aid immediately to clean, sterilize, and cover any scrapes, cuts, and burns where the skin is broken and/or where there may be bleeding. Wear bandages over open skin wounds.
   k. report all accidents to the instructor immediately, and ask your instructor for assistance in cleaning up broken glassware and spills.
   l. report to the instructor any condition that appears unsafe or hazardous.
   m. use caution during any outdoor activities. Watch for snakes, poisonous insects or spiders, stinging insects, poison oak, poison ivy, and so on. Be careful near water.

I understand the safety rules as presented. I agree to follow them and all other instructions given by the instructor.

Name: _____   Date: _____

Laboratory Class and Time: _____

1

Scientific Method

Learning Outcomes

Introduction

This laboratory will provide you with an opportunity to use the scientific method in the same manner as scientists. Scientists often begin by making observations about the subject of interest. Today our subject is the pillbug, *Armadillidium vulgare*, a type of crustacean that lives on land (Fig. 1.1).

Pillbugs have overlapping "armored" plates that make them look like little armadillos. A pillbug can roll up into such a tight ball that its legs and head are no longer visible, earning it the nickname "roly-poly." They are commonly found in damp leaf litter, under rocks, and in basements or crawl spaces under houses. Pillbugs breathe by gills located on the underside of their bodies. The gills must be kept slightly moist, and that is why they are usually found in damp places.

In winter, pillbugs are inactive, but when spring arrives they become active and mate. Females have a pouch on the underside of their body, where they can carry from 7 to 200 eggs. The eggs hatch several weeks after mating, and the young look like miniature adults. The young stay in the pouch another six weeks, and then they leave and begin to feed. They eat decaying plants and animals and some living plants. They can live up to three years.

Figure 1.1 Pillbugs (roly-poly bugs).
The pillbug, *A. vulgare,* lives on land. Pillbugs are wingless, and coloration varies from brown to slate grey. They breathe by gills, and they have seven pairs of jointed legs for locomotion on land. Pillbugs live in damp places, mostly under rocks, wood, and decaying leaves. They feed at night.

Pillbugs molt (shed the exoskeleton) four or five times. They have three body parts: head, thorax, and abdomen. Among crustaceans, pillbugs are classified as isopods because they are dorsoventrally flattened, lack a carapace, have compound eyes, and have two pairs of antennae.

Isopods are the only crustaceans that include forms adapted to living their entire life on land, although moisture is required. Currently, it is believed that they do not transmit diseases, nor do they bite or sting. Because they eat dead organic matter, such as leaves, they are easy to find and keep in a moist terrarium with leaf litter, rocks, and wood chips. You are encouraged to collect some for your experiment. Since they live in the same locations as snakes, be careful when collecting them.

First, become acquainted with your subject and how it normally moves. Then, you will use your knowledge of the pillbug to hypothesize whether it will be attracted to, repelled by, or indifferent to various substances of your choice. After you have tested your hypotheses, you will conclude whether they are supported or not. Finally, your conclusions may lead to other hypotheses, and if time permits, you may go ahead and test those also.

1.1 Using the Scientific Method

Scientists use the scientific method (Fig. 1.2) to come to a conclusion about the natural world. When a scientist begins a study, he or she uses preliminary observations and previous data to formulate a hypothesis. A **hypothesis** is a tentative explanation of observed phenomena.

After a hypothesis is formulated, it must be tested by doing new experiments and/or making new observations. Experiments are done and observations made in such a way that others can repeat them. Only repeatable observations and experiments are accepted as valid contributions to the field of science.

Data are any factual information that can be observed either independently of, or as a result of, experimentation. On the basis of the data, a scientist comes to a **conclusion**—whether the observations support the hypothesis or prove it false. Scientists are always aware that further observations and experiments could lead to a change in prior conclusions. Therefore, it is never said that the data prove a hypothesis true, but we could say that the data support the hypothesis. The arrow in Figure 1.2 indicates that research often enters a cycle of hypothesis—experiments and observations—conclusion—hypothesis. Explain this cycle. _____

_____

After many years of testing and study, the scientific community may develop a **scientific theory,** a concept that ties together many varied conclusions into a generalized statement. For example, after testing the cause of many individual diseases, the germ theory of disease was formulated. It states that infectious diseases are caused by pathogens (e.g., bacteria and viruses) that can be passed from one person to another. How is a scientific theory different from a conclusion?

_____

_____

Figure 1.2 **Flow diagram for the scientific method.**
On the basis of observations, a scientist formulates a hypothesis. The hypothesis is tested by further experiments and/or observations. The scientist then concludes whether the data support or do not support the hypothesis. The return arrow shows that if the data do not support the hypothesis, the hypothesis is rejected and the scientist starts again. If the hypothesis is supported, scientists continue to work and eventually may formulate a scientific theory.

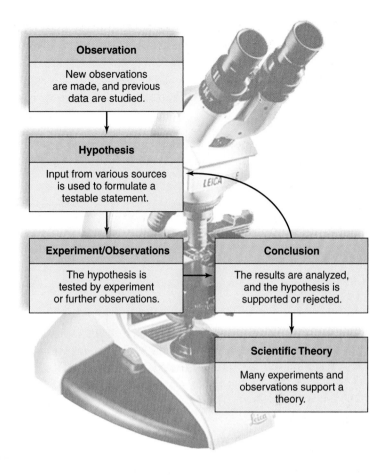

Observation
New observations are made, and previous data are studied.

Hypothesis
Input from various sources is used to formulate a testable statement.

Experiment/Observations
The hypothesis is tested by experiment or further observations.

Conclusion
The results are analyzed, and the hypothesis is supported or rejected.

Scientific Theory
Many experiments and observations support a theory.

1.2 Observing the Pillbug

Wash your hands before and after handling pillbugs. Please handle them carefully so they are not crushed. When touched, they roll up into a ball or "pill" shape as a defense mechanism. They will soon recover if left alone.

Observation: Pillbug's External Anatomy

1. Obtain a pillbug that has been numbered with white correction fluid or tape tags. First examine the shell and body with the unaided eye and then with a magnifying lens or dissecting microscope. Put the pillbug in a small glass or plastic dish to keep it contained.

2. Examine the shell shape, color, and texture. Note the number of legs and antennae and whether there are any posterior appendages, such as uropods (paired appendages at end of abdomen) or brood pouches. (Females have leaflike growths at the base of some legs where developing eggs and embryos are held in pouches.) Locate the eyes. Count the number of overlapping plates.

3. In the following space, draw a large outline of your pillbug (at least 10–12 cm across). Label the head, thorax, abdomen, antennae, eyes, uropods, and one of the seven pairs of legs.

4. Draw the pillbug rolled into a ball.

Observation: Pillbug's Motion

1. Watch a pillbug's underside as the pillbug moves up a transparent surface, such as the side of a graduated cylinder or beaker. Describe the action of the feet and any other motion you see.

    _____

2. As you watch the pillbug, identify behaviors that might

    a. protect it from predators _____

    b. help it acquire food _____

    c. protect it from the elements _____

    d. allow interaction with the environment _____

3. Allow a pillbug to crawl on your hand. Describe how it feels and how it acts.

    _____

4. Place the pillbug on a graduated cylinder. Experiment with the angle of the cylinder and the position of the pillbug to determine the pillbug's preferred direction of motion. For example, place the cylinder on end, and position the pillbug so that it can move up or down. Try other arrangements also. Repeat this procedure with three other pillbugs. Record the preferred direction of motion and other observations for each pillbug in Table 1.1.

Table 1.1	Preferred Direction of Motion	
Pillbug	**Direction Moved**	**Comments**
1		
2		
3		
4		

5. Measure the speed of the pillbug. Use what you learned about each pillbug's preferred direction of motion (see Table 1.1) to get maximum cooperation from each of the four pillbugs you worked with in step 4. Place each pillbug on a metric ruler, and use a stopwatch to measure the time it takes for the pillbug to move a certain number of centimeters. Record your results for each pillbug in Table 1.2. Calculate each pillbug's average speed in millimeters (mm) per second, and record your data in Table 1.2.

Table 1.2	Pillbug Speed		
Pillbug	**Millimeters Traveled**	**Time (sec)**	**Average Speed (mm/sec)**
1			
2			
3			
4			

1.3 Formulating Hypotheses

Hypotheses are often stated as "if-then" statements. For example, *if* the pillbug is exposed to _____, *then* it will be _____. You will be testing whether pillbugs are attracted to, repelled by, or unresponsive to particular substances. Pillbugs move away from a substance when they are repelled by it, and they move toward and eat a substance they are attracted to. If a pillbug simply rolls into a ball, nothing can be concluded, and you may wish to choose another pillbug or wait a minute or two to check for further response.

1. Choose
    a. two or three powders, such as flour, cornstarch, coffee creamer, baking soda, fine sand.
    b. two or three liquids, such as milk, orange juice, ketchup, applesauce, a carbonated beverage, water.

2. Hypothesize in Table 1.3 how you expect the pillbug to respond, and offer an explanation for your reasoning.

Table 1.3	Hypotheses About Pillbug's Reaction to Common Powders and Liquids	
Substance Tested	Hypothesis About How Pillbug Will Respond to Substance	Reasoning for Hypothesis
1		
2		
3		
4		
5		
6		

1.4 Performing an Experiment

Design an experiment to test the pillbug's reaction to the chosen substances. The pillbug must be treated humanely. No substance must be put directly on the pillbug, nor can the pillbug be placed directly onto the substance. Since pillbugs tend to walk around the edge of a petri dish, you could put the wet or dry substance around the edge of the dish. Or for wet substances, you could put liquid-soaked cotton in the pillbug's path.

A good experimental design contains a **control.** A control group goes through all the steps of an experiment but is not exposed to, or lacks, the factor being tested. If you are testing the pillbug's reaction to a liquid, water can be the control substance substituted for the test liquid. If you are testing the pillbug's reaction to a powder, substitute fine sand for the test powder.

Experimental Procedure: Pillbug's Reaction to Common Substances

1. What substances are you testing? Include in your list any controls, and complete the first column in Table 1.4.
2. Obtain a small beaker. Between procedures, rinse your pillbug over the beaker by spritzing it with distilled water from a spray bottle. Then put it on a paper towel to dry off.
3. Test the pillbug's reaction using the method described previously.

4. Watch the pillbug's reaction to each substance, and record it in Table 1.4.
5. Do your results support your hypotheses? Answer *yes* or *no* in the last column.

Table 1.4	Pillbug's Reaction to Common Substances	
Substance Tested	Pillbug's Reaction	Hypothesis Supported?
1		
2		
3		
4		
5		
6		

6. Compare your results with those of other students who tested the same substances. Complete Table 1.5.

Table 1.5	Class Results								
Group No.	Exp. 1 Direction	Exp. 2 Speed (mm/sec)	Exp. 3 Dry Control	Substances			Exp. 4 Wet Control	Substances	
1									
2									
3									
4									
5									
6									
7									
8									

Continuing the Experiment
7. Study your results and those of other students, and decide what factors may have caused the pillbug to be attracted to or repelled by a substance. _____

_____

On the basis of your decision, what are your new hypotheses? _____

_____

8. Test your hypotheses, and describe your results here. If possible, make a table to display your results.

9. Based on your new data, what is your conclusion?

Laboratory Review 1

1. What are the essential elements of the scientific method?
   _____
   _____

2. What is a hypothesis?
   _____

3. Is it sufficient to do a single experiment to test a hypothesis, why or why not?
   _____

4. What do you call a sample that goes through all the steps of an experiment but does not contain the factor being tested?
   _____

5. What part of a pillbug is for protection, and what does a pillbug do to protect itself?
   _____

6. State the data you used to formulate your hypotheses regarding pillbug reactions toward various substances.
   _____

7. Why is it important to test one substance at a time when doing an experiment?
   _____

Indicate whether statements 8 and 9 are hypotheses, conclusions, or theories.

8. The data show that vaccines protect people from disease. _____

9. All living things are made of cells. _____

2

Metric Measurement and Microscopy

Learning Outcomes

Introduction

This laboratory introduces you to the metric system, which biologists use to indicate the sizes of cells and cell structures. This laboratory also examines the features, functions, and use of the compound light microscope and the binocular dissecting microscope. Transmission and scanning electron microscopes are explained, and micrographs produced using these microscopes appear throughout this lab manual. The binocular dissecting microscope and the scanning electron microscope view the surface and/or the three-dimensional structure of an object. The compound light microscope and the transmission electron microscope can view only extremely thin sections of a specimen. If a subject was sectioned lengthwise for viewing, the interior of the projections at the top of the cell, called cilia, would appear in the micrograph (Fig. 2.1). A lengthwise cut through any type of specimen is called a **longitudinal section (l.s.).** On the other hand, if the subject in Figure 2.1 was sectioned crosswise below the area of the cilia, you would see other portions of the interior of the subject. A crosswise cut through any type of specimen is called a **cross section (c.s.).**

Figure 2.1 Longitudinal and cross sections.
a. Transparent view of a cell. b. A longitudinal section would show the cilia at the top of the cell. c. A cross section shows only the interior where the cut is made.

a. Cell   **b.** Longitudinal section   **c.** Cross section

2.1 The Metric System

The **metric system** is the standard system of measurement in the sciences, including biology, chemistry, and physics. It has tremendous advantages because all conversions, whether for volume, mass (weight), or length, can be in units of ten. Refer to the Appendix, page 362, for an in-depth look at the units of the metric system.

Length

Metric units of length measurement include the **meter (m), centimeter (cm), millimeter (mm), micrometer (μm),** and **nanometer (nm)** (Table 2.1). The prefixes milli- (10^{-3}), micro- (10^{-6}), and nano (10^{-9}) are used with length, weight, and volume.

Table 2.1	Metric Units of Length Measurement			
Unit	**Meters**	**Millimeters**	**Centimeters**	**Relative Size**
Meter (m)	1 m	1,000 mm	100 cm	Largest
Centimeter (cm)	0.01 (10^{-2}) m	10 mm	1 cm	
Millimeter (mm)	0.001 (10^{-3}) m	1.0 mm	0.1 cm	
Micrometer (μm)	0.000001 (10^{-6}) m	0.001 (10^{-3}) mm	0.0001 (10^{-4}) cm	
Nanometer (nm)	0.000000001 (10^{-9}) m	0.000001 (10^{-6}) mm	0.0000001 (10^{-7}) cm	Smallest

Experimental Procedure: Length

1. Obtain a small ruler marked in centimeters and millimeters. How many centimeters are represented? _____ One centimeter equals how many millimeters? _____ To express the size of small objects, such as cell contents, biologists use even smaller units of the metric system than those on the ruler. These units are the micrometer (μm) and the nanometer (nm). According to Table 2.1, 1 μm = _____ mm, and 1 nm = _____ mm.

   Therefore, 1 mm = _____ μm = _____ nm.

2. Measure the diameter of the circle shown below to the nearest millimeter. This circle is _____ mm = _____ μm = _____ nm.

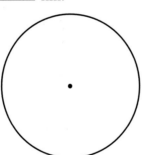

Volume

Two metric units of volume are the **liter (l)** and the **milliliter (ml).** One liter = 1,000 ml.

Experimental Procedure: Volume

1. Volume measurements can be related to those of length. For example, use a millimeter ruler to measure a wooden block to get its length, width, and depth.

   length = _____ cm; width = _____ cm; depth = _____ cm

The volume, or space, occupied by the wooden block can be expressed in cubic centimeters (cc or cm^3) by multiplying: length × width × depth = _____ cm^3. For purposes of this Experimental Procedure, 1 cubic centimeter equals 1 milliliter; therefore, the wooden block has a volume of _____ ml.

2. In the biology laboratory, volume is usually measured directly in liters or milliliters with appropriate measuring devices. For example, use a 50 ml graduated cylinder to add 20 ml of water to a test tube. First, fill the graduated cylinder to the 20 ml mark. To do this properly, you have to make sure that the lowest margin of the water level, or the **meniscus** (Fig. 2.2), is at the 20 ml mark. Place your eye directly parallel to the level of the meniscus, and add water until the meniscus is at the 20 ml mark. (Having a dropper bottle filled with water on hand can help you do this.) A large, blank, white index card held behind the cylinder can also help you see the scale more clearly. Now pour the 20 ml of water into the test tube.

3. What procedure would allow you to determine the total volume of the test tube? _____

_____

_____

What is the test tube's total volume? _____

4. Fill a 50 ml graduated cylinder with water to about the 20 ml mark. How could you use this setup to calculate the volume of an object? _____

_____

Now perform the operation you suggested. The object, _____, has a volume of _____ ml.

5. How could you determine how many drops from the pipette of the dropper bottle equal 1 ml?

_____

Now perform the operation you suggested. How many drops from the pipette of the dropper bottle equal 1 ml? _____ Some pipettes are graduated and can be filled to a certain level as a way to measure volume directly. Your instructor will demonstrate this. Are pipettes customarily used to measure large or small volumes? _____

Figure 2.2 Meniscus.
The proper way to view the meniscus.

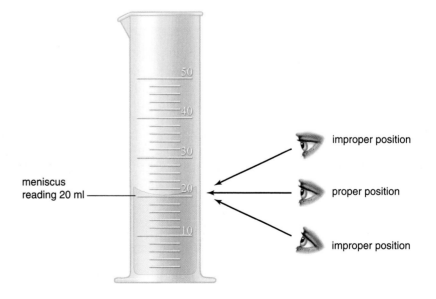

meniscus
reading 20 ml

improper position

proper position

improper position

2.2 Microscopy

Because biological objects can be very small, we often use a microscope to view them. Many kinds of instruments, ranging from the hand lens to the electron microscope, are effective magnifying devices. A short description of two kinds of light microscopes and two kinds of electron microscopes follows.

Light Microscopes

Light microscopes use light rays passing through lenses to magnify the object. The **binocular dissecting microscope** (stereomicroscope) is designed to study entire objects in three dimensions at low magnification. The **compound light microscope** is used for examining small or thinly sliced sections of objects under higher magnification than that of the binocular dissecting microscope. The term **compound** refers to the use of two sets of lenses: the ocular lenses located near the eyes and the objective lenses located near the object. Illumination is from below, and visible light passes through clear portions but does not pass through opaque portions. To improve contrast, the microscopist uses stains or dyes that bind to cellular structures and absorb light. Photomicrographs, also called light micrographs, are images produced by a compound light microscope (Fig. 2.3a).

Figure 2.3 Comparative micrographs.
Micrographs of a lymphocyte, a type of white blood cell. **a.** A photomicrograph (light micrograph) shows less detail than a **(b)** transmission electron micrograph (TEM). **c.** A scanning electron micrograph (SEM) shows the cell surface in three dimensions.

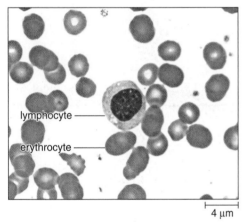

lymphocyte

erythrocyte

4 µm

a. Photomicrograph or light micrograph (LM)

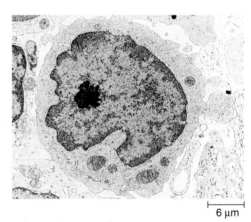

6 µm

b. Transmission electron micrograph (TEM)

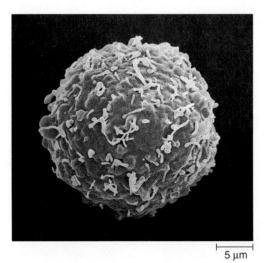

5 µm

c. Scanning electron micrograph (SEM)

Electron Microscopes

Electron microscopes use beams of electrons to magnify the object. The beams are focused on a photographic plate by means of electromagnets. The **transmission electron microscope** is analogous to the compound light microscope. The object is ultra-thinly sliced and treated with heavy metal salts to improve contrast. Figure 2.3*b* is a micrograph produced by this type of microscope. The **scanning electron microscope** is analogous to the dissecting light microscope. It gives an image of the surface and dimensions of an object, as is apparent from the scanning electron micrograph in Figure 2.3*c*.

The micrographs in Figure 2.3 demonstrate that an object is magnified more with an electron microscope than with a compound light microscope. The difference between these two types of microscopes, however, is not simply a matter of magnification; it is also the electron microscope's ability to show detail. The electron microscope has greater resolving power. **Resolution** is the minimum distance between two objects at which they can still be seen, or resolved, as two separate objects. The use of high-energy electrons rather than light gives electron microscopes a much greater resolving power since two objects that are much closer together can still be distinguished as separate points. Table 2.2 lists several other differences between the compound light microscope and the transmission electron microscope.

Table 2.2	Comparison of the Compound Light Microscope and the Transmission Electron Microscope	
Compound Light Microscope		**Transmission Electron Microscope**
1. Glass lenses		1. Electromagnetic lenses
2. Illumination by visible light		2. Illumination due to beam of electrons
3. Resolution $\cong$ 200 nm		3. Resolution $\cong$ 0.1 nm
4. Magnifies to 1,500×		4. Magnifies to 100,000×
5. Costs up to tens of thousands of dollars		5. Costs up to hundreds of thousands of dollars

Conclusions

- Which two types of microscopes view the surface of an object? _____
- Which two types of microscopes view objects that have been sliced and treated to improve contrast? _____
- Of the microscopes just mentioned, which one resolves the greater amount of detail?

_____

Rules for Microscope Use

Observe the following rules for using a microscope:

1. The lowest power objective (scanning or low) should be in position, both at the beginning and end of microscope use.
2. Use only lens paper for cleaning lenses.
3. Do not tilt the microscope as the eyepieces could fall out, or wet mounts could be ruined.
4. Keep the stage clean and dry to prevent rust and corrosion.
5. Do not remove parts of the microscope.
6. Keep the microscope dust-free by covering it after use.
7. Report any malfunctions.

2.3 Binocular Dissecting Microscope (Stereomicroscope)

The **binocular dissecting microscope** allows you to view objects in three dimensions at low magnifications. It is used to study entire small organisms, any object requiring lower magnification, and opaque objects that can be viewed only by reflected light. It is also called a stereomicroscope because it produces a three-dimensional image.

Identifying the Parts

After your instructor has explained how to carry a microscope, obtain a binocular dissecting microscope and a separate illuminator, if necessary, from the storage area. Place it securely on the table. Plug in the power cord, and turn on the illuminator. There is a wide variety of binocular dissecting microscope styles, and your instructor will discuss the specific style(s) available to you. Regardless of style, the following features should be present:

Figure 2.4 **Binocular dissecting microscope (stereomicroscope).**
Label this microscope with the help of the text material.

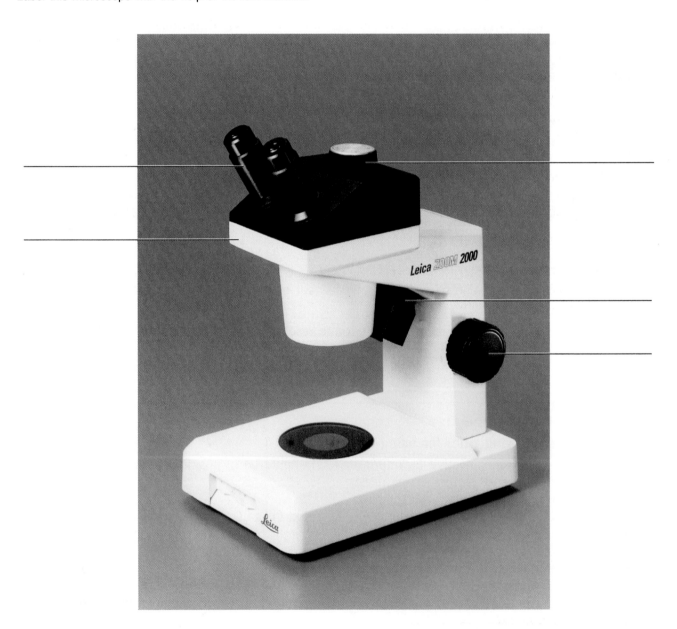

1. **Binocular head:** Holds two eyepiece lenses that move to accommodate for the various distances between different individuals' eyes.
2. **Eyepiece lenses:** The two lenses located on the binocular head. What is the magnification of your eyepieces? _____ Some models have one **independent focusing eyepiece** with a knurled knob to allow independent adjustment of each eye. The nonadjustable eyepiece is called the **fixed eyepiece.**
3. **Focusing knob:** A large, black or gray knob located on the arm; used for changing the focus of both eyepieces together.
4. **Magnification changing knob:** A knob, often built into the binocular head, used to change magnification in both eyepieces simultaneously. This may be a **zoom** mechanism or a **rotating lens** mechanism of different powers that clicks into place.
5. **Illuminator:** Used to illuminate an object from above; may be built into the microscope or separate.

Locate each of these parts on your binocular dissecting microscope, and label them on Figure 2.4.

Focusing the Binocular Dissecting Microscope

1. In the center of the stage, place a plastomount that contains small organisms.
2. Adjust the distance between the eyepieces on the binocular head so that they comfortably fit the distance between your eyes. You should be able to see the object with both eyes as one three-dimensional image.
3. Use the focusing knob to bring the object into focus.
4. Does your microscope have an independent focusing eyepiece? _____ If so, use the focusing knob to bring the image in the fixed eyepiece into focus, while keeping the eye at the independent focusing eyepiece closed. Then adjust the independent focusing eyepiece so that the image is clear, while keeping the other eye closed. Is the image inverted? _____
5. Turn the magnification changing knob, and determine the kind of mechanism on your microscope. A zoom mechanism allows continuous viewing while changing the magnification. A rotating lens mechanism blocks the view of the object as the new lenses are rotated. Be sure to click each lens firmly into place. If you do not, the field will be only partially visible. What kind of mechanism is on your microscope? _____
6. Set the magnification changing knob on the lowest magnification. Sketch the object in the following circle as though this represents your entire field of view:

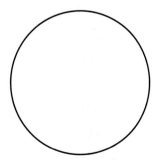

7. Rotate the magnification changing knob to the highest magnification. Draw another circle within the one provided to indicate the reduction of the field of view.
8. Experiment with various objects at various magnifications until you are comfortable with using the binocular dissecting microscope.
9. When you are finished, return your binocular dissecting microscope and illuminator to their correct storage areas.

2.4 Use of the Compound Light Microscope

As mentioned, the name **compound light microscope** indicates that it uses two sets of lenses and light to view an object. The two sets of lenses are the ocular lenses located near the eyes and the objective lenses located near the object. Illumination is from below, and the light passes through clear portions but does not pass through opaque portions. This microscope is used to examine small or thinly sliced sections of objects under higher magnification than would be possible with the binocular dissecting microscope.

Identifying the Parts

Obtain a compound light microscope from the storage area, and place it securely on the table. *Identify the following parts on your microscope, and label them in Figure 2.5.*

Figure 2.5   Compound light microscope.
Compound light microscope with binocular head and mechanical stage. Label this microscope with the help of the text material.

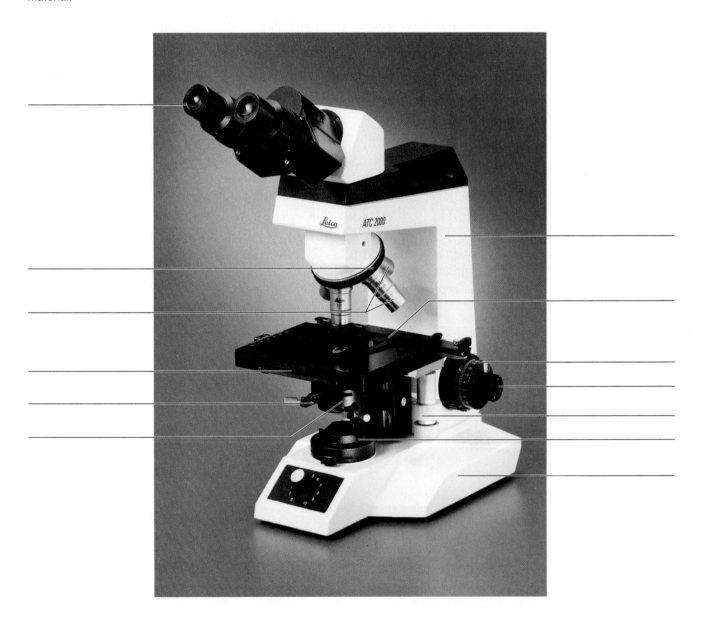

1. **Ocular lenses (eyepieces):** What is the magnifying power of the ocular lenses on your microscope? _____

2. **Body tube:** Holds nosepiece at one end and eyepiece at the other end; conducts light rays.

3. **Arm:** Supports upper parts and provides carrying handle.

4. **Nosepiece:** Revolving device that holds objectives.

5. **Objective lenses (objectives):**

   a. **Scanning objective:** This is the shortest of the objective lenses and is used to scan the whole slide. The magnification is stamped on the housing of the lens. It is a number followed by an ×. What is the magnifying power of the scanning objective lens on your microscope? _____

   b. **Low-power objective:** This lens is longer than the scanning objective lens and is used to view objects in greater detail. What is the magnifying power of the low-power objective lens on your microscope? _____

   c. **High-power objective:** If your microscope has three objective lenses, this lens will be the longest. It is used to view an object in even greater detail. What is the magnifying power of the high-power objective lens on your microscope? _____

   d. **Oil immersion objective:** (on microscopes with four objective lenses): Holds a 95× (to 100×) lens and is used in conjunction with immersion oil to view objects with the greatest magnification. Does your microscope have an oil immersion objective? _____ If this lens is available, your instructor will discuss its use when the lens is needed.

6. **Coarse-adjustment knob:** Knob used to bring object into approximate focus; used only with low-power objective.

7. **Fine-adjustment knob:** Knob used to bring object into final focus.

8. **Condenser:** Lens system below the stage used to focus the beam of light on the object being viewed.

9. **Diaphragm** or **diaphragm control lever:** Controls amount of illumination used to view the object.

10. **Light source:** An attached lamp that directs a beam of light up through the object.

11. **Base:** The flat surface of the microscope that rests on the table.

12. **Stage:** Holds and supports microscope slides.

13. **Stage clips:** Hold slides in place on the stage.

14. **Mechanical stage** (optional): A movable stage that aids in the accurate positioning of the slide.

    Does your microscope have a mechanical stage? _____

15. **Mechanical stage control knobs** (optional): Two knobs usually located below the stage. One knob controls forward/reverse movement, and the other controls right/left movement.

Focusing the Compound Light Microscope—Lowest Power

1. Turn the nosepiece so that the *lowest* power objective on your microscope is in straight alignment over the stage.

2. Always begin focusing with the *lowest* power objective on your microscope (4× [scanning] or 10× [low power]).

3. With the coarse-adjustment knob, lower the stage (or raise the objectives) until it stops.

4. Place a slide of the letter *e* on the stage, and stabilize it with the clips. (If your microscope has a mechanical stage, pinch the spring of the slide arms on the stage, and insert the slide.) Center the *e* as best you can on the stage or use the two control knobs located below the stage (if your microscope has a mechanical stage) to center the *e*.

5. Again, be sure that the lowest-power objective is in place. Then, as you look from the side, decrease the distance between the stage and the tip of the objective lens until the lens comes to an automatic stop or is no closer than 3 mm above the slide.

6. While looking into the eyepiece, rotate the diaphragm (or diaphragm control lever) to give the maximum amount of light.

7. Using the coarse-adjustment knob, slowly increase the distance between the stage and the objective lens until the object—in this case, the letter *e*—comes into view, or focus.

8. Once the object is seen, you may need to adjust the amount of light. To increase or decrease the contrast, rotate the diaphragm slightly.

9. Use the fine-adjustment knob to sharpen the focus if necessary.

10. Practice having both eyes open when looking through the eyepiece, as this greatly reduces eyestrain.

Inversion

Inversion refers to the fact that a microscopic image is upside down and reversed.

Observation: Inversion

1. In space 1 provided here, draw the letter *e* as it appears on the slide (with the unaided eye, not looking through the eyepiece).

1.	2.

2. In space 2, draw the letter *e* as it appears when you look through the eyepiece.

3. What differences do you notice? _____

4. Move the slide to the right. Which way does the image appear to move? _____

   Explain. _____

Focusing the Compound Light Microscope—Higher Powers

Compound light microscopes are **parfocal;** that is, once the object is in focus with the lowest power, it should also be almost in focus with the higher power.

1. Bring the object into focus under the lowest power by following the instructions in the previous section.

2. Make sure that the letter *e* is centered in the field of the lowest objective.

3. Move to the next higher objective (low power [10×] or high power [40×]) by turning the nosepiece until you hear it click into place. Do not change the focus; parfocal microscope objectives will not hit normal slides when changing the focus if the lowest objective is initially in focus. (If you are on low power [10×], proceed to high power [40×] before going on to step 4.)

4. If any adjustment is needed, use only the *fine*-adjustment knob. (*Note:* Always use only the fine-adjustment knob with high power.) On your drawing of the letter *e*, draw a circle around the portion of the letter that you are now seeing with high-power magnification.

5. When you have finished your observations of this slide (or any slide), rotate the nosepiece until the lowest-power objective clicks into place, and then remove the slide.

Total Magnification

Total magnification is calculated by multiplying the magnification of the ocular lens (eyepiece) by the magnification of the objective lens. The magnification of a lens is imprinted on the lens casing.

Observation: Total Magnification

Calculate total magnification figures for your microscope, and record your findings in Table 2.3.

Table 2.3	Total Magnification		
Objective	**Ocular Lens**	**Objective Lens**	**Total Magnification**
Scanning power (if present)			
Low power			
High power			
Oil immersion (if present)			

2.5 Microscopic Observations

When a specimen is prepared for observation, the object should always be viewed as a **wet mount.** A wet mount is prepared by placing a drop of liquid on a slide or, if the material is dry, by placing it directly on the slide and adding a drop of water or stain. The mount is then covered with a cover-slip, as illustrated in Figure 2.6. Dry the bottom of your slide before placing it on the stage.

Figure 2.6 Preparation of a wet mount.

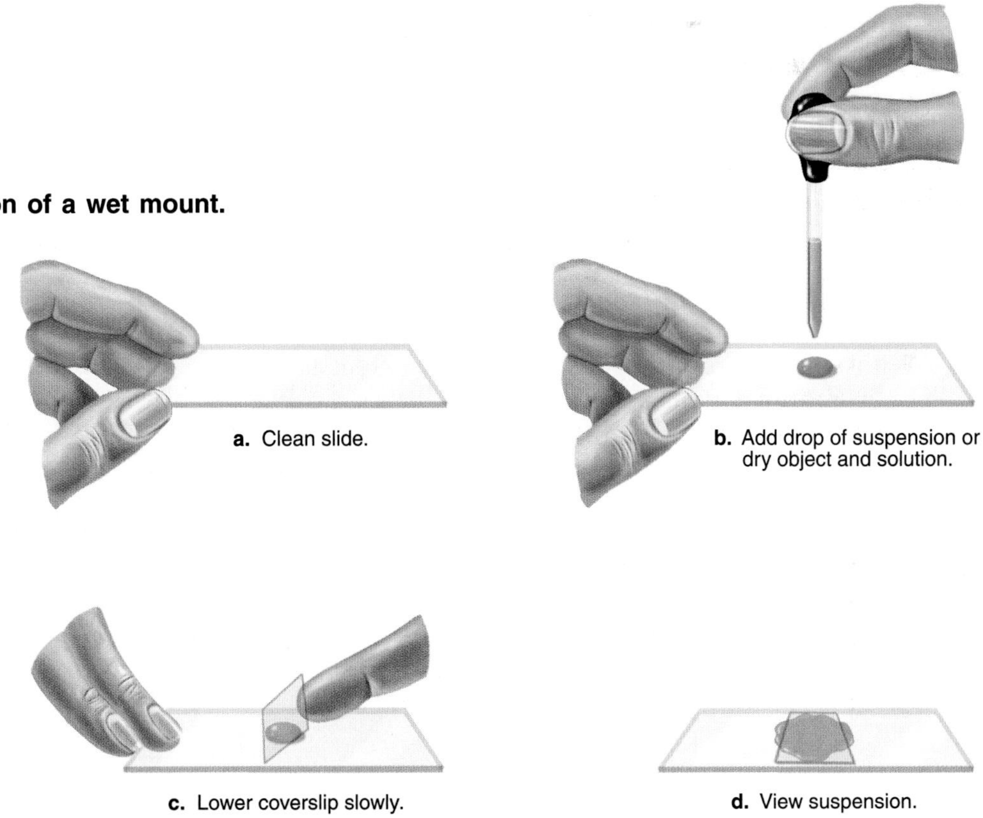

a. Clean slide.

b. Add drop of suspension or dry object and solution.

c. Lower coverslip slowly.

d. View suspension.

Human Epithelial Cells

Epithelial cells cover the body's surface and line its cavities.

> **Caution:** **Methylene blue** Avoid ingestion, inhalation, and contact with skin, eyes, and mucous membranes. Exercise care in using this chemical. If any should spill on your skin, wash the area with mild soap and water. Methylene blue will also stain clothing. Follow your instructor's directions for its disposal.

Observation: Human Epithelial Cells

1. Obtain a prepared slide, or make your own as follows:

   a. Obtain a prepackaged flat toothpick (or sanitize one with alcohol or alcohol swabs).

   b. Gently scrape the inside of your cheek with the toothpick, and place the scrapings on a clean, dry slide. Discard used toothpicks in the biohazard waste container provided.

   c. Add a drop of very weak *methylene blue* or *iodine solution*, and cover with a coverslip.

2. Observe under the microscope.
3. Locate the nucleus (the central, round body), the cytoplasm, and the plasma membrane (outer cell boundary). *Label Figure 2.7.*
4. Because your epithelial slides are biohazardous, they must be disposed of as indicated by your instructor.

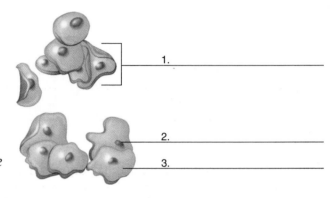

Figure 2.7 **Cheek epithelial cells.**
Label the nucleus, the cytoplasm, and the plasma membrane.

1. _____

2. _____

3. _____

Onion Epidermal Cells

Epidermal cells cover the surfaces of plant organs, such as leaves. The bulb of an onion is made up of fleshy leaves.

Observation: Onion Epidermal Cells

1. With a scalpel, strip a small, thin, transparent layer of cells from the inside of a fresh onion leaf.
2. Place it gently on a clean, dry slide, and add a drop of *iodine solution* (or *methylene blue*). Cover with a coverslip.
3. Observe under the microscope.
4. Locate the cell wall and the nucleus. *Label Figure 2.8.*

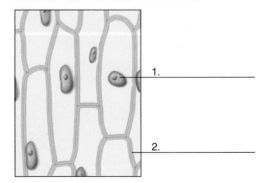

Figure 2.8 **Onion epidermal cells.**
Label the cell wall and the nucleus.

1. _____

2. _____

5. Note some obvious differences between the human cheek cells and the onion cells, and list them in Table 2.4.

Table 2.4	Differences Between Human Epithelial and Onion Epidermal Cells	
Differences	**Human Epithelial Cells (Cheek)**	**Onion Epidermal Cells**
Shape (sketch)		

Euglena

Examination of *Euglena* (a unicellular organism with a flagellum to facilitate movement) will test your ability to observe objects with the microscope and to control illumination to heighten contrast.

Observation: Euglena

1. Make a wet mount of *Euglena* by using a drop of a *Euglena* culture and adding a drop of Protoslo® (methyl cellulose solution) onto a slide. The Protoslo® slows the organism's swimming.
2. Mix thoroughly with a toothpick, and add a coverslip.
3. Scan the slide for *Euglena:* Start at the upper left-hand corner, and move the slide forward and back as you work across the slide from left to right. The *Euglena* may be at the edge of the slide because they show an aversion to Protoslo®. Use Figure 2.9 to help identify the internal appearance of *Euglena*.
4. Experiment by using scanning, low-power, and high-power objective lenses; by focusing up and down with the fine-adjustment knob; and by adjusting the light so that it is not too bright.
5. Compare your *Euglena* specimens with Figure 2.9. How do your specimens compare to Figure 2.9? _____

_____

Figure 2.9 *Euglena.*
Euglena is a unicellular, flagellated organism.

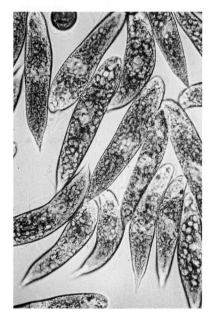

1. Make the following conversions:

    a. 1 mm = _____ µm = _____ cm

    b. 15 mm = _____ cm = _____ µm

    c. 50 ml = _____ liter

    d. 5 g = _____ mg

2. Explain the designation "compound light" microscope:

    a. compound _____

    b. light _____

3. What function is performed by the diaphragm of a microscope?

    _____

4. Briefly describe the necessary steps for observing a slide at low power under the compound light microscope.

    _____

    _____

5. Why is it helpful for a microscope to be parfocal?

    _____

6. Why is locating an object more difficult if you start with the high-power objective than with the low-power objective?

    _____

7. How much larger than normal does an object appear with a low-power objective? _____

    _____

8. A virus is 50 nm in size.

    a. Would you recommend using a binocular dissecting microscope, compound light microscope, or an electron microscope to see it? _____ Why? _____

    _____

    b. How many micrometers is the virus? _____ µm

9. What type of microscope, aside from the compound light microscope, might you use to observe the organisms found in pond water? _____

10. State two differences between onion epidermal cells and human epithelial cells.

    _____

    _____

3

Chemical Composition of Cells

Learning Outcomes

3.1 Proteins
- Cite examples of typical structural and functional proteins and describe their respective functions.   24
- Describe the relationship of amino acids to the composition of proteins (polypeptides) and the role that the peptide bond plays in their assembly.   24
- Identify the test for the presence of protein and explain why it is able to detect the presence of larger proteins or polypeptides.   24

3.2 Carbohydrates
- Explain how monosaccharides are related to disaccharides and polysaccharides.   26
- Distinguish between tests that identify starch from those that identify sugars.   27–28
- Explain why the Benedict's test might produce varied results.   28

3.3 Lipids
- Identify the structural composition of a common lipid such as fat.   30–31
- Describe a simple test for the detection of fat.   31

Introduction

All living things consist of basic units of matter called **atoms.** Molecules form when atoms bond with one another. Inorganic molecules are often associated with nonliving things, and organic molecules are associated with living organisms. Several classes of organic molecules have biological significance. In this laboratory, you will be studying **proteins, carbohydrates** (monosaccharides, disaccharides, polysaccharides), and **lipids** (i.e., fat).

Large organic molecules form during a dehydration reaction when smaller molecules bond as water is given off. During a hydrolysis reaction, bonds are broken as water is added. (The dehydration reaction and the hydrolysis reaction are illustrated in Figs. 3.1, 3.2, and 3.4). A fat contains one glycerol and three fatty acids. Proteins and some carbohydrates (called polysaccharides) are **polymers** because they are made up of smaller molecules called **monomers.** Proteins contain a large number of amino acids (the monomer) joined together by a peptide bond. A polysaccharide, such as starch, contains a large number of glucose molecules (the monomer) joined together.

Various chemicals will be used in this laboratory to test for the presence of specific organic molecules. Most often, you will be looking for a particular color change. If the change is observed, the test is said to be *positive* because it indicates that the molecule is present. If the color change is not observed, the test is said to be *negative* because it indicates that the molecule is not present.

A control should be included in each experiment for comparison. The **control** will go through all the steps of the experiment, but it will lack the factor being tested. This missing factor allows you to observe the difference between a positive result and a negative result. If the control sample tests positive, you know your test is invalid.

3.1 Proteins

Proteins have numerous functions in cells. Antibodies are proteins that combine with pathogens so that the pathogens are destroyed by the body. Transport proteins combine with and move substances from place to place. Hemoglobin transports oxygen throughout the body. Albumin is another transport protein in our blood. Regulatory proteins control cellular metabolism in some way. For example, the hormone insulin regulates the amount of glucose in blood so that cells have a ready supply. Structural proteins include keratin, found in hair, and myosin, found in muscle. **Enzymes** are proteins that speed chemical reactions. A reaction that could take days or weeks to complete can happen within an instant if the correct enzyme is present. Amylase is an enzyme that speeds the breakdown of starch in the mouth and small intestine.

Proteins are made up of **amino acids** joined together. About 20 different common amino acids are found in cells. All amino acids have an acidic group (—COOH) and an amino group ($H_2N—$). They differ by the **R group** (remainder group) attached to a carbon atom, as shown in Figure 3.1. The R groups have varying sizes, shapes, and chemical activities.

A chain of two or more amino acids is called a **peptide**, and the bond between the amino acids is called a **peptide bond.** A **polypeptide** is a very long chain of amino acids. A protein can contain one or more polypeptide chains. Insulin contains a single chain, while hemoglobin contains four polypeptides. A protein has a particular shape, which is important to its function. The shape comes about because the R groups of the polypeptide chain(s) can interact with one another in various ways.

Figure 3.1 Peptide bond.

Peptide bond formation between amino acids creates a dipeptide. This dehydration reaction involves the removal of one water molecule. During a hydrolysis reaction, water is added, and the peptide bond is broken. In a polypeptide, many amino acids are held together by multiple peptide bonds.

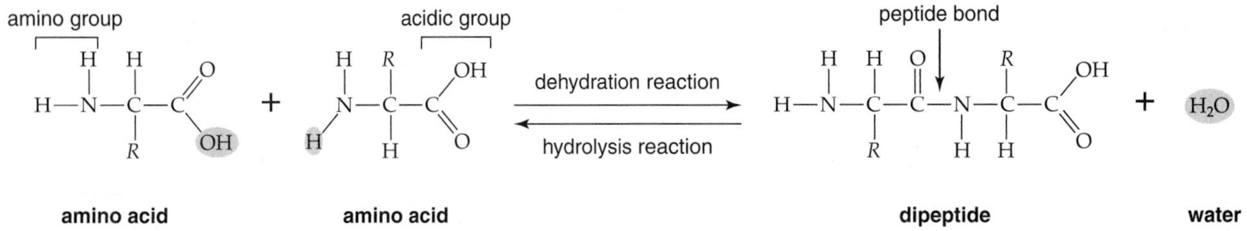

Test for Proteins

Biuret reagent (blue color) contains a strong solution of sodium or potassium hydroxide (NaOH or KOH) and a small amount of dilute copper sulfate ($CuSO_4$) solution. The reagent changes color in the presence of proteins or peptides because the peptide bonds of the protein or peptide chemically combine with the copper ions in biuret reagent (Table 3.1).

Table 3.1	Test for Protein	
	Protein	Peptides
Biuret reagent (blue)	Purple	Pinkish-purple

> Caution: **Biuret reagent**   Biuret reagent is highly corrosive. Exercise care in using this chemical. If any should spill on your skin, wash the area with mild soap and water. Follow your instructor's directions for its disposal.

With a millimeter ruler and a wax pencil, label and mark four clean test tubes at the 1 cm level. After filling a tube, cover it with Parafilm®, and swirl well to mix. (Do not turn upside down.) The reaction is almost immediate.

Tube 1   **1.** Fill to the mark with *distilled water,* and add about five drops of *biuret reagent.*
         2. Record the final color in Table 3.2.
Tube 2   **1.** Fill to the mark with *albumin solution,* and add about five drops of *biuret reagent.*
         2. Record the final color in Table 3.2.
Tube 3   **1.** Fill to the mark with *pepsin solution,* and add about five drops of *biuret reagent.*
         2. Record the final color in Table 3.2.
Tube 4   **1.** Fill to the mark with *starch solution,* and add about five drops of *biuret reagent.*
         2. Record the final color in Table 3.2.

Table 3.2	Biuret Test		
Tube	**Contents**	**Final Color**	**Conclusions**
1	Distilled water		
2	Albumin		
3	Pepsin		
4	Starch		

Conclusions

- From your test results, conclude what kind of chemical is present and why the results occurred. Enter your conclusions in Table 3.2.

- If your results are not as expected, offer an explanation. _____

  Then inform your instructor, who will advise you how to proceed.

- Which of the four tubes is the control sample? _____ Why? _____

  _____

  Why do experimental procedures include control samples? _____

  _____

- Pepsin is an enzyme. Enzymes are composed of what type of organic molecule? _____

- Did test tube 4 (starch) give a positive or a negative reaction? _____

3.2 Carbohydrates

Carbohydrates include sugars and molecules that are chains of sugars. **Glucose,** which has only one sugar unit, is a monosaccharide; **maltose**, which has two sugar units, is a disaccharide (Fig. 3.2). Glycogen, starch, and cellulose are polysaccharides, made up of chains of glucose units (Fig. 3.3).

Glucose is used by all organisms as an energy source. Energy is released when glucose is broken down to carbon dioxide and water. This energy is used by the organism to do work. Animals store glucose as glycogen and plants store glucose as starch. Plant cell walls are composed of cellulose.

Figure 3.2 **Formation of maltose, a disaccharide.**
During a dehydration reaction, a bond forms between the two glucose molecules, the components of water are removed, and maltose results. During a hydrolysis reaction, the components of water are added, and the bond is broken.

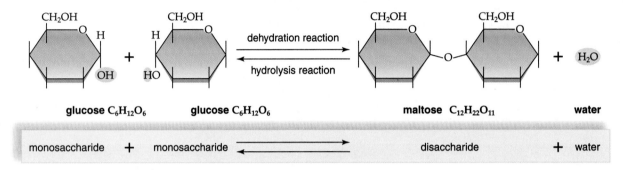

Figure 3.3 **Starch.**
Starch is a polysaccharide composed of many glucose units. **a.** Photomicrograph of starch granules in plant cells. **b.** Structure of starch. Starch consists of amylose that is nonbranched and amylopectin that is branched.

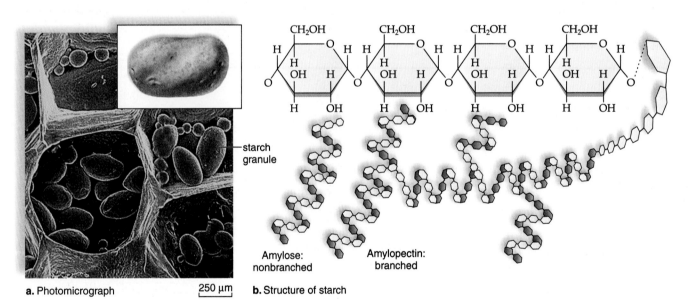

a. Photomicrograph   250 µm   b. Structure of starch

Test for Starch

In the presence of starch, iodine solution (yellowish-brown) reacts chemically with starch to form a blue-black color.

Experimental Procedure: Test for Starch

With a wax pencil, label and mark five clean test tubes at the 1 cm level.

Tube 1   1. Fill to the 1 cm mark with *water*, and add five drops of *iodine solution*.
            2. Note the final color change, and record your results in Table 3.3.

Tube 2   1. It is very important to shake the *starch suspension* well before taking your sample. After shaking, fill this tube to the 1 cm mark with the 1% *starch suspension*. Add five drops of *iodine solution*.
            2. Note the final color change, and record your results in Table 3.3.

Tube 3   1. Add a few drops of *onion juice* to the test tube. (Obtain the juice by adding water and crushing a small piece of onion with a mortar and pestle. Clean mortar and pestle after using.) Add five drops of *iodine solution*.
            2. Note the final color change, and record your results in Table 3.3.

Tube 4   1. Add a few drops of *potato juice* to the test tube. (Obtain the juice by adding water and crushing a small piece of potato with a mortar and pestle. Clean mortar and pestle after using.) Add five drops of *iodine solution*.
            2. Note the final color change, and record your results in Table 3.3.

Tube 5   1. Fill to the 1 cm mark with *glucose solution*, and add five drops of *iodine solution*.
            2. Note the final color change, and record your results in Table 3.3.

Table 3.3	Iodine (IKI) Tests for Starch		
Tube	**Contents**	**Color**	**Conclusions**
1	Water		
2	Starch suspension		
3	Onion juice		
4	Potato juice		
5	Glucose solution		

Conclusions

- From your test results, draw conclusions about what organic compound is present in each tube. Write these conclusions in Table 3.3.
- If your results are not as expected, offer an explanation. Then inform your instructor, who will advise you how to proceed.

Potato

1. With a sharp razor blade, slice a very thin piece of potato. Place it on a microscope slide, add a drop of *water* and a coverslip, and observe under low power with your compound light microscope. Compare your slide with the photomicrograph of starch granules (see Fig. 3.3*a*). Find the cell wall (large, geometric compartments) and the starch grains (numerous clear, oval-shaped objects).

2. Without removing the coverslip, place two drops of *iodine solution* onto the microscope slide so that the iodine touches the coverslip. Draw the iodine under the coverslip by placing a small piece of paper towel in contact with the water on the opposite side of the coverslip.

3. Microscopically examine the potato again on the side closest to where the iodine solution was applied.

   What is the color of the small, oval bodies? _____

   What is the chemical composition of these oval bodies? _____

Onion

1. Put a piece of onion skin on a microscope slide.
2. Add a large drop of *iodine solution*.
3. Does onion contain starch? _____

Test for Sugars

Sugars react with **Benedict's reagent** after being heated in a boiling water bath. Increasing concentrations of sugar give a continuum of colored products (Table 3.4). This experiment tests for the presence (or absence) of varying amounts of sugars in a variety of materials and chemicals.

Table 3.4	Benedict's Reagent (Some Typical Reactions)	
Chemical	**Chemical Category**	**Benedict's Reagent (After Heating)**
Water	Inorganic	Blue (no change)
Glucose	Monosaccharide (carbohydrate)	Varies with concentration: very low—green low—yellow moderate—yellow-orange high—orange very high—orange-red
Maltose	Disaccharide (carbohydrate)	Varies with concentration—see "Glucose"
Starch	Polysaccharide (carbohydrate)	Blue (no change)

> **Caution:** **Benedict's reagent** Benedict's reagent is highly corrosive. Exercise care in using this chemical. If any should spill on your skin, wash the area with mild soap and water. Follow your instructor's directions for disposal of this chemical.

With a wax pencil, label and mark five clean test tubes at the 1 cm level.

Tube 1   **1.** Fill to the 1 cm mark with *water,* then add about five drops of *Benedict's reagent.*
     2. Heat in a boiling water bath for 5 to 10 minutes, note any color change, and record in Table 3.5.

Tube 2   **1.** Fill to the 1 cm mark with *glucose solution,* then add about five drops of *Benedict's reagent.*
     2. Heat in a boiling water bath for 5 to 10 minutes, note any color change, and record in Table 3.5.

Tube 3   **1.** Place a few drops of *onion juice* in the test tube. (Obtain the juice by adding water and crushing a small piece of onion with a mortar and pestle. Clean mortar and pestle after using.)
     2. Fill to the 1 cm mark with *water,* then add about five drops of *Benedict's reagent.*
     3. Heat in a boiling water bath for 5 to 10 minutes, note any color change, and record in Table 3.5.

Tube 4   **1.** Place a few drops of *potato juice* in the test tube. (Obtain the juice by adding water and crushing a small piece of potato with a mortar and pestle.)
     2. Fill to the 1 cm mark with *water,* then add about five drops of *Benedict's reagent.*
     3. Heat in a boiling water bath for 5 to 10 minutes, note any color change, and record in Table 3.5.

Tube 5   **1.** Fill to the 1 cm mark with *starch suspension,* then add about five drops of *Benedict's reagent.*
     2. Heat in a boiling water bath for 5 to 10 minutes, note any color change, and record in Table 3.5.

Table 3.5	Benedict's Reagent Test		
Tube	Contents	Color (After Heating)	Conclusions
1	Water		
2	Glucose solution		
3	Onion juice		
4	Potato juice		
5	Starch suspension		

Conclusions

- From your test results, conclude what kind of chemical is present and why the results occurred. Enter your conclusions in Table 3.5. Which tube served as a control? _____
- Compare Table 3.3 with Table 3.5. Sugars are an immediate energy source in cells. In plant cells, glucose (a primary energy molecule) is often stored in the form of starch. Is glucose stored as starch in the potato? _____ Is glucose stored as starch in the onion? _____ Does this explain your results in Table 3.5? Why? _____

Starch Composition

To show that starch contains sugar, perform the following experiment involving the enzyme amylase:

$$\text{starch + water} \xrightarrow{\text{amylase}} \text{maltose}$$

Experimental Procedure: Starch Composition

With a wax pencil, label and mark two clean test tubes at the 2 cm, 4 cm, and 6 cm levels.

Tube 1
1. Fill to the 2 cm mark with *water* and to the 4 cm mark with 1% pancreatic *amylase*.
2. Shake and then wait for 30 minutes.
3. Add *Benedict's reagent* to the 6 cm level and heat in a boiling water bath.
4. Note any color change, and record your results in Table 3.6.

Tube 2
1. Fill to the 2 cm mark with *starch* and to the 4 cm mark with 1% pancreatic *amylase*.
2. Shake and then wait for 30 minutes.
3. Add *Benedict's reagent* and heat as before.
4. Note any color change, and record your results in Table 3.6.

Table 3.6	Starch Composition		
Tube	Contents	Color Change	Conclusions
1	Water Amylase		
2	Starch Amylase		

Conclusions

- From your test results, you may conclude that starch is composed of what kind of chemical?

  _____

- How do you know? _____

  Enter your conclusions in Table 3.6.

3.3 Lipids

Lipids are compounds that are insoluble in water and soluble in solvents, such as alcohol and ether. Lipids include fats, oils, phospholipids, steroids, and cholesterol. Typically, fats and oils are composed of three molecules of fatty acids bonded to one molecule of glycerol forming a triglyceride (Fig. 3.4). Phospholipids have the same structure as fats, except in place of the third fatty acid there is a phosphate group (a grouping that contains phosphate). Steroids are derived from cholesterol and, like this molecule, have skeletons of four fused rings of carbon atoms, but they differ by functional groups (attached side chains). Fat, as we know, is long-term stored energy in the human body. Phospholipids are found in the plasma membrane of cells. In recent years, cholesterol, a molecule transported in the blood, has been implicated in causing cardiovascular disease. Regardless, steroids are very important compounds in the body; for example, the sex hormones are steroids.

Figure 3.4 Formation of a fat.

A fat molecule forms when glycerol joins with three fatty acids as three water molecules are removed during a dehydration reaction. During a hydrolysis reaction, water is added, and the bonds are broken.

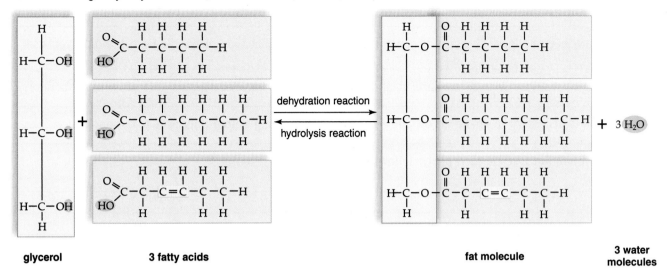

glycerol · · · 3 fatty acids · · · fat molecule · · · 3 water molecules

---

Test for Fat

Fats do not evaporate from brown paper; instead, they leave an oily spot.

Experimental Procedure: Test for Fat

1. Place a small drop of *water* on a square of brown paper. Describe the immediate effect. _____

   _____

2. Place a small drop of *vegetable oil* on a square of brown paper. Describe the immediate effect.

   _____

3. Wait at least 15 minutes for the paper to dry. Evaluate which substance penetrates the paper and which is subject to evaporation. Record your observations and conclusions in Table 3.7.

Table 3.7	Test for Fat	
Sample	**Observations**	**Conclusions**
Water spot		
Oil spot		

1. What macromolecules studied today are present in cells? _____

2. You have been assigned the task of constructing a protein. What type of building block would you use?
   _____

3. A digestive enzyme such as amylase breaks down starch to what disaccharide studied in this laboratory?
   _____

4. Why is it necessary to shake an oil and vinegar salad dressing before adding it to a salad? _____
   _____

5. How would you test for each of the following substances?
   a. Sugars _____
   b. Fat _____
   c. Starch _____
   d. Protein _____

6. Assume that you have tested an unknown sample with both biuret solution and Benedict's solution and that both tests result in a blue color. What have you learned? _____
   _____

7. What purpose is served when a test is done using water instead of a sample substance? _____
   _____

8. A test tube contains albumin. The test for protein is positive and the test for starch is negative. Explain.
   _____

9. A test tube contains starch and the enzyme amylase. After 30 minutes, the test for starch is negative and the test for sugar is positive. Explain.
   _____

LABORATORY

4

Cell Structure and Function

Learning Outcomes

4.1 Prokaryotic Versus Eukaryotic Cells
- Explain how to determine if a cell is prokaryotic and eukaryotic.   34

4.2 Animal Cell and Plant Cell Structure
- Identify the major organelles in photomicrographs of animal and plant cells and state their respective functions.   35–38
- Contrast the differences in structure and function between the two types of cells by comparing photomicrographs of plant and animal cells.   35–38

4.3 Diffusion
- Describe how the process of diffusion occurs and why it might occur differently in a living organism than it would in the physical environment.   39–41
- Predict which substances will or will not diffuse across a plasma membrane.   40–41

4.4 Osmosis
- Explain how the terms *isotonic, hypertonic,* and *hypotonic* relate to NaCl concentrations and how they influence osmosis.   42–45
- Predict the effect of different tonicities on animal (e.g., red blood) cells and on plant (e.g., *Elodea*) cells.   43–45

4.5 pH and Cells
- Predict the pH before and after the addition of an acid to nonbuffered and buffered solutions.   46–47

Introduction

The molecules we studied in the last laboratory are not alive—the basic units of life are cells. The **cell theory** states that all living things are composed of cells and that cells come only from other cells. Some organisms, such as *Euglena,* which you observed in Laboratory 2, are unicellular, but multicellular organisms are composed of many cells. While we are accustomed to considering the heart, the liver, or the intestines as enabling the human body to function, it is actually cells that do the work of these organs.

Figure 4.1 is a human cheek epithelial cell as viewed by an ordinary compound light microscope available in general biology laboratories. It shows that the content of a cell, called the **cytoplasm,** is bounded by a **plasma membrane.** The plasma membrane regulates the movement of molecules into and out of the cytoplasm. In this lab, we will study how the passage of water into a cell depends on the difference in concentration of solutes (particles) between the cytoplasm and the surrounding medium or solution. The well-being of cells also depends upon the pH of the solution surrounding them. We will see how a buffer can maintain the pH within a narrow range and how buffers within cells can protect them against damaging pH changes.

Because a photomicrograph shows only a minimal amount of detail, it is necessary to turn to the electron microscope to study the contents of a cell in greater depth. The models of plant and animal cells available in the laboratory today are based on electron micrographs.

Figure 4.1 Photomicrograph of an epithelial cell.
(Magnification 250x)

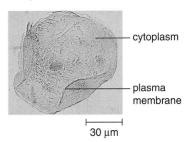

cytoplasm

plasma membrane

30 μm

4.1 Prokaryotic Versus Eukaryotic Cells

All living cells are classified as either prokaryotic or eukaryotic. One of the basic differences between the two types is that prokaryotic cells do not contain nuclei (*pro* means "before"; *karyote* means "nucleus"), while eukaryotic cells do contain nuclei (*eu* means "true"; *karyote* means "nucleus"). Only bacteria (including cyanobacteria) and archaea are prokaryotes; all other organisms are eukaryotes.

Prokaryotes also don't have the organelles found in eukaryotic cells (Fig. 4.2). **Organelles** are small, membranous bodies, each with a specific structure and function. Prokaryotes do have **cytoplasm,** the material bounded by a plasma membrane and cell wall. The cytoplasm contains ribosomes, small granules that coordinate the synthesis of proteins; thylakoids (only in cyanobacteria) that participate in photosynthesis; and innumerable enzymes. Prokaryotes also have a nucleoid, a region in the bacterial cell interior in which the DNA is physically organized but not enclosed by a membrane.

Figure 4.2 Prokaryotic cell.
Prokaryotic cells lack membrane-bounded organelles, as well as a nucleus. Their DNA is in a nucleoid region.

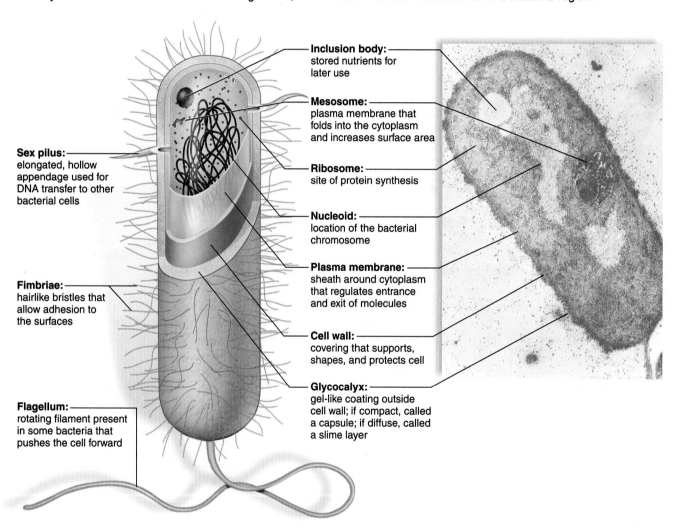

Observation: Prokaryotic/Eukaryotic Cells

Two microscope slides will show you the main difference between prokaryotic and eukaryotic cells.

1. Examine a prepared slide of a prokaryote. There are no nuclei in these cells. Sketch this cell.
2. Examine a prepared slide of cuboidal cells from a human kidney. You can make out a nucleus.

4.2 Animal Cell and Plant Cell Structure

Table 4.1 lists the structures found in animal and plant cells. The **nucleus** in a eukaryotic cell is bounded by a **nuclear envelope** and contains **nucleoplasm.** The *cytoplasm,* found between the plasma membrane and the nucleus, consists of a background fluid and the organelles. Many **organelles** are membranous, such as the nucleolus, endoplasmic reticulum, Golgi apparatus, vacuoles and vesicles, lysosomes, peroxisome, mitochondrion, and chloroplast.

Table 4.1	Eukaryotic Structures in Animal Cells and Plant Cells	
Name	**Composition**	**Function**
Cell wall*	Contains cellulose fibrils	Provides support and protection
Plasma membrane	Phospholipid bilayer with embedded proteins	Outer cell surface that regulates entrance and exit of molecules
Nucleus	Enclosed by nuclear envelope; contains chromatin (threads of DNA and protein) and nucleolus	Storage of genetic information; synthesis of DNA and RNA
Nucleolus	Concentrated area of chromatin, RNA, and proteins	Produces subunits of ribosomes
Ribosome	Protein and RNA in two subunits	Carries out protein synthesis
Endoplasmic reticulum (ER)	Membranous, flattened channels and tubular canals; rough ER and smooth ER	Synthesis and/or modification of proteins and other substances; transport by vesicle formation
Rough ER	Studded with ribosomes	Protein synthesis
Smooth ER	Lacks ribosomes	Synthesis of lipid molecules
Golgi apparatus	Stack of membranous saccules	Processes, packages, and distributes proteins and lipids
Vesicle	Membrane-bounded sac	Stores and transports substances
Lysosome	Vesicle containing hydrolytic enzymes	Digests macromolecules and cell parts
Peroxisome	Vesicle containing specific enzymes	Breaks down fatty acids and converts resulting hydrogen peroxide to water; various other functions
Mitochondrion	Membranous cristae bounded by an outer membrane	Carries out cellular respiration, producing ATP molecules
Chloroplast*	Membranous thylakoids bounded by two membranes	Carries out photosynthesis, producing sugars
Cytoskeleton	Microtubules, intermediate filaments, actin filaments	Maintains cell shape and assists movement of cell parts
Cilia and flagella	Contains microtubules	Movement of cell
Centrioles**	Contains microtubules	As a part of centrosome, may participate in cell division

*Plant cells only

**Animal cells only

Study Table 4.1 to determine structures that are unique to plant cells and unique to animal cells, and write them below the examples given.

	Plant Cells	**Animal Cells**
Unique structures:	1. Large central vacuole	Small vacuoles
	2. _____	_____
	3. _____	_____

Animal Cell Structure

With the help of Table 4.1, give a function for each of these structures, and label Figure 4.3. See also Figure 4.5a in the text.

Structure	Function
Plasma membrane	
Nucleus	
Nucleolus	
Ribosome	
Endoplasmic reticulum	
Rough ER	
Smooth ER	
Golgi apparatus	
Vesicles	
Lysosome	
Mitochondrion	
Centrioles in centrosome	
Cytoskeleton	

Figure 4.3 Animal cell structure.

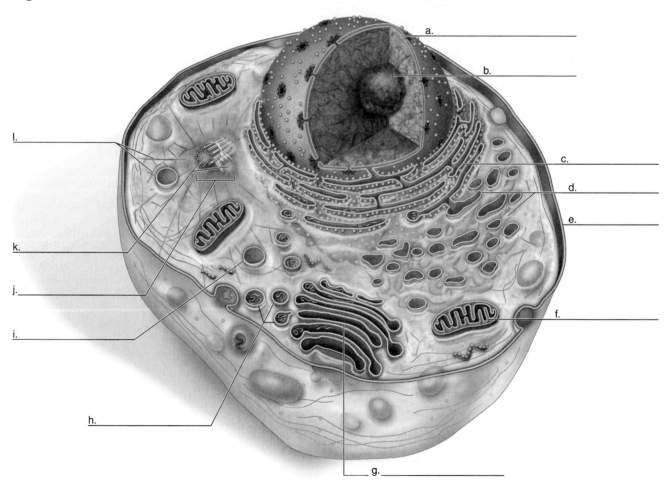

Plant Cell Structure

With the help of Table 4.1, give a function for these structures unique to plant cells, and label Figure 4.4. See also Figure 4.5*b* in the text.

Structure	Function
Cell wall	
Central vacuole, large	
Chloroplasts	

Figure 4.4 Plant cell structure.

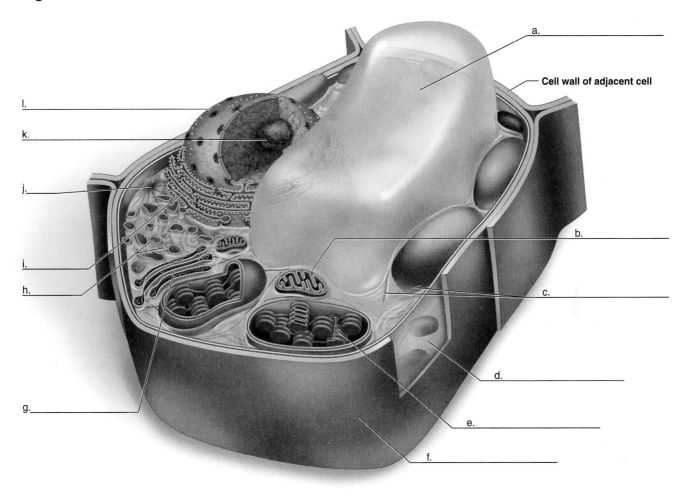

Cell wall of adjacent cell

Observation: Plant Cell Structure

1. Prepare a wet mount of a small piece of young *Elodea* leaf in fresh water. *Elodea* is a multi-cellular, eukaryotic plant found in freshwater ponds and lakes.
2. Have the drop of *water* ready on your slide so that the leaf does not dry out, even for a few seconds. Take care that the leaf is mounted with its top side up.
3. Examine the slide using low power, focusing sharply on the leaf surface.
4. Select a cell with numerous chloroplasts for further study, and switch to high power.
5. Carefully focus on the side and end walls of the cell. The chloroplasts appear to be only along the sides of the cell because the large, fluid-filled, membrane-bounded central vacuole pushes the cytoplasm against the cell walls (Fig. 4.5*a*). Then focus on the surface and notice an even distribution of chloroplasts (Fig. 4.5*b*).
6. Can you locate the cell nucleus? _____ It may be hidden by the chloroplasts, but when visible, it appears as a faint, grey lump on one side of the cell.
7. Can you detect movement of chloroplasts in this cell or any other cell? _____ The chloroplasts are not moving under their own power but are being carried by a streaming of the nearly invisible cytoplasm.
8. Save your slide for use later in this laboratory.

Figure 4.5 *Elodea* cell structure.

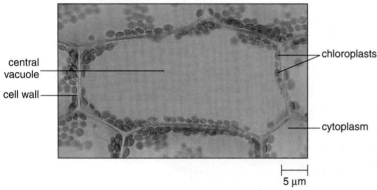

central vacuole
cell wall
chloroplasts
cytoplasm

5 µm

a. Middle of the cell. Due to the central vacuole, chloroplasts are visible around the perimeter and not in the center.

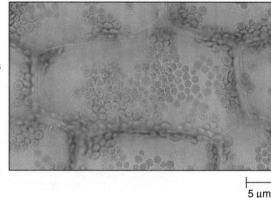

5 µm

b. Upper surface of cells. Chloroplasts are in the middle, as well as around the perimeter.

4.3 Diffusion

Diffusion is the movement of molecules from a higher to a lower concentration until equilibrium is achieved and the molecules are distributed equally (Fig. 4.6). At this point, molecules may still be moving back and forth, but there is no net movement in any one direction.

Figure 4.6 Process of diffusion.
Diffusion is apparent when dye molecules have equally dispersed.

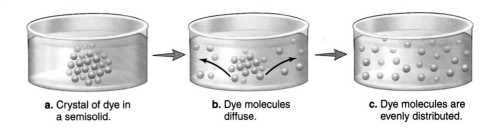

a. Crystal of dye in a semisolid.　　**b.** Dye molecules diffuse.　　**c.** Dye molecules are evenly distributed.

Diffusion is a general phenomenon in the environment. The speed of diffusion is dependent on such factors as the temperature, the size of the molecule, and the type of medium.

Experimental Procedure: Diffusion

> Caution: **Potassium permanganate ($KMnO_4$)** $KMnO_4$ is highly poisonous and is a strong oxidizer. Avoid contact with skin and eyes and with combustible materials. If spillage occurs, wash all surfaces thoroughly. $KMnO_4$ will also stain clothing.

Diffusion Through a Semisolid

1. Observe a petri dish containing 1.5% gelatin (or agar) to which potassium permanganate ($KMnO_4$) was added in the center depression at the beginning of the lab.
2. Obtain *time zero* from your instructor, and record *time zero* and the *final time* (now) in Table 4.2. Calculate the length of time in hours and minutes. Convert the time to hours: _____ hr.
3. Using a ruler placed over the petri dish, measure (in mm) the movement of color from the center of the depression outward in one direction: _____ mm.
4. Calculate the speed of diffusion: _____ mm/hr.
5. Record all data in Table 4.2.

Diffusion Through a Liquid

1. Add enough water to cover the bottom of a glass petri dish.
2. Place the petri dish over a thin, flat ruler.
3. With tweezers, add a crystal of potassium permanganate ($KMnO_4$) directly over a millimeter measurement line. Note the *time zero* in Table 4.2.
4. After 10 minutes, note the distance the color has moved. Record the *final time, length of time,* and *distance moved* in Table 4.2.
5. Multiply the length of time and the distance moved by 6 to calculate the *speed of diffusion:* _____ mm/hr. Record in Table 4.2.

Diffusion Through Air

1. Measure the distance from a spot designated by your instructor to your laboratory work area today. Record this distance in the fifth column of Table 4.2.
2. Record *time zero* in Table 4.2 when a perfume or similar substance is released into the air.

3. Note the time when you can smell the perfume. Record this as the *final time* in Table 4.2. Calculate the *length of time* since the perfume was released, and record it in Table 4.2.
4. Calculate the speed of diffusion: _____ mm/hr. Record in Table 4.2.

Table 4.2	Speed of Diffusion				
Medium	Time Zero	Final Time	Length of Time (hr)	Distance Moved (mm)	Speed of Diffusion (mm/hr)
Semisolid					
Liquid					
Air					

Conclusions

- In which experiment was diffusion the fastest? _____
- What accounts for the difference in speed? _____

Diffusion Across the Plasma Membrane

Some molecules can diffuse across a plasma membrane, and some cannot. In general, small, non-charged molecules can cross a membrane by diffusion, but large molecules cannot diffuse across a membrane. The dialysis tube membrane in the experimental procedure simulates a plasma membrane.

Experimental Procedure: Diffusion Across Plasma Membrane

At the start of the experiment,

1. Cut a piece of dialysis tubing approximately 15 cm (approximately 6 inches) long. Soak the tubing in water until it is soft and pliable.
2. Close one end of the dialysis tubing with two knots.
3. Fill the bag halfway with *glucose solution*.
4. Add 4 full droppers of *starch solution* to the bag.
5. Hold the open end while you mix the contents of the dialysis bag. Rinse off the outside of the bag with *distilled water*.
6. Fill a beaker 2/3 full with *distilled water*.

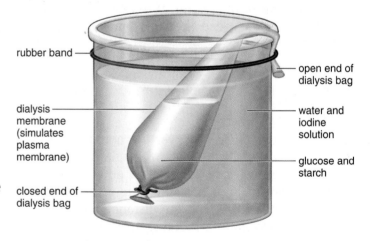

Figure 4.7 Placement of dialysis bag in water containing iodine.

rubber band
open end of dialysis bag
dialysis membrane (simulates plasma membrane)
water and iodine solution
glucose and starch
closed end of dialysis bag

7. Add droppers of *iodine solution* (IKI) to the water in the beaker until an amber (tealike) color is apparent.
8. Record the color of the solution in the beaker in Table 4.3.
9. Place the bag in the beaker with the open end hanging over the edge. Secure the open end of the bag to the beaker with a rubber band as shown (Fig. 4.7). Make sure the contents do not spill into the beaker.

After about 5 minutes, at the end of the experiment,

10. You will note a color change. Record the color of the bag contents in Table 4.3.
11. Mark off a test tube at 1 cm and 3 cm.

> **Caution:** **Benedict's reagent** Exercise care in using this chemical. It is highly corrosive. If any should spill on your skin, wash the area with mild soap and water. Follow your instructor's directions for its disposal.

12. Draw solution from near the bag and at the bottom of the beaker for testing with Benedict's reagent. Fill the test tube to the first mark with this solution. Add *Benedict's reagent* to the 3-cm mark. Heat in a boiling water bath for 5–10 minutes, observe any color change, and record your results as positive or negative in Table 4.3. (Optional use of glucose test strip: Dip glucose test strip into beaker. Compare stick with chart provided by instructor.)
13. Remove the dialysis bag from the beaker. Dispose of it and the used Benedict's reagent solution in the manner directed by your instructor.

Table 4.3	Diffusion Across Plasma Membrane				
	At Start of Experiment		*At End of Experiment*		
	Contents	Color	Color	Benedict's Test (+) or (−)	Conclusion
Bag	Glucose Starch	—		—	
Beaker	Water Iodine		—		

Conclusions

- Based on the color change noted in the bag, conclude what solute diffused across the dialysis membrane from the beaker to the bag, and record your conclusion in Table 4.3.
- From the results of the Benedict's test on the beaker contents, conclude what solute diffused across the dialysis membrane from the bag to the beaker, and record your conclusion in Table 4.3.
- Which solute did not diffuse across the dialysis membrane from the bag to the beaker? _____
  _____ Explain. _____

Laboratory Notes

_____

_____

_____

_____

_____

_____

_____

4.4 Osmosis

Osmosis is the diffusion of water across the plasma membrane of a cell. Just like any other molecule, water follows its concentration gradient and moves from the area of higher concentration to the area of lower concentration.

Experimental Procedure: Osmosis

To demonstrate osmosis, a thistle tube is covered with a membrane at its lower opening and partially filled with 50% corn syrup (a polysaccharide) or a similar substance. The whole apparatus is placed in a beaker containing distilled water (Fig. 4.8). The water concentration in the beaker is 100%. Water molecules can move freely between the thistle tube and the beaker.

1. Note the level of liquid in the thistle tube, and measure how far it travels in 10 minutes:
   _____ mm.

2. Calculate the speed of osmosis under these conditions: _____ mm/hr.

Conclusions

- In which direction was there a net movement of water? _____
  Explain what is meant by "net movement" after examining the arrows in Figure 4.8*b*.

  _____

- If the sugar molecules in corn syrup moved from the thistle tube to the beaker, would there
  have been a net movement of water into the thistle tube? _____ Why wouldn't large sugar
  molecules be able to move across the membrane from the thistle tube to the beaker?

  _____

- Explain why the water level in the thistle tube rose: In terms of solvent concentration, water
  moved from the area of _____ water concentration to the area of _____ water
  concentration across a differentially permeable membrane.

Figure 4.8 Osmosis demonstration.
a. A thistle tube, covered at the broad end by a differentially permeable membrane, contains a corn syrup solution. The beaker contains distilled water. **b.** The solute is unable to pass through the membrane, but the water (arrows) passes through in both directions. There is a net movement of water toward the inside of the thistle tube, where there is a lower percentage of water molecules. **c.** Due to the incoming water molecules, the level of the solution rises in the thistle tube.

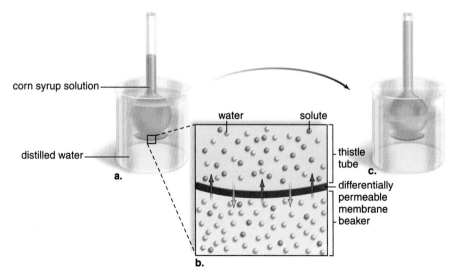

corn syrup solution

distilled water

water   solute

thistle tube

differentially permeable membrane

beaker

a.

b.

c.

Tonicity

Tonicity is the relative concentration of solute (particles), and therefore also of solvent (water), outside the cell compared with inside the cell.

- An **isotonic solution** has the same concentration of solute (and therefore of water) as the cell. When cells are placed in an isotonic solution, there is no net movement of water.
- A **hypertonic solution** has a higher solute (therefore, lower water) concentration than the cell. When cells are placed in a hypertonic solution, water moves out of the cell into the solution.
- A **hypotonic solution** has a lower solute (therefore, higher water) concentration than the cell. When cells are placed in a hypotonic solution, water moves from the solution into the cell.

Experimental Procedure: Potato Strips

1. Cut two strips of potato, each about 7 cm long and 1.5 cm wide.
2. Label two test tubes 1 and 2. Place one *potato strip* in each tube.
3. Fill tube 1 with *water* to cover the potato strip.
4. Fill tube 2 with 10% *sodium chloride* (NaCl) to cover the potato strip.
5. After 1 hour, remove the potato strips from the test tubes and place them on a paper towel. Observe each strip for limpness (water loss) or stiffness (water gain). Which tube has the limp potato strip? _____ Why did water diffuse out of the potato strip in this tube?

_____

Which tube has the stiff potato strip? _____ Why did water diffuse into the potato strip in this tube? _____

Red Blood Cells (Animal Cells)

A solution of 0.9% NaCl is isotonic to red blood cells. In such a solution, red blood cells maintain their normal appearance (Fig. 4.9a). A solution greater than 0.9% NaCl is hypertonic to red blood cells. In such a solution, the cells shrivel up, a process called **crenation** (Fig. 4.9b). A solution of less than 0.9% NaCl is hypotonic to red blood cells. In such a solution, the cells swell to bursting, a process called **hemolysis** (Fig. 4.9c).

Figure 4.9 Tonicity and red blood cells.

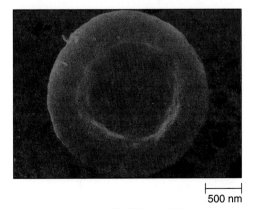

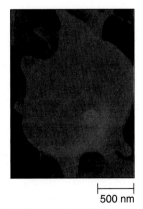

a. **Isotonic solution.** Red blood cell has normal appearance due to no net gain or loss of water.

b. **Hypertonic solution.** Red blood cell shrivels due to loss of water.

c. **Hypotonic solution.** Red blood cell fills to bursting due to gain of water.

> **Caution:** Do not remove the stoppers of test tubes during this procedure.

Three stoppered test tubes on display have the following contents:

Tube 1: 0.9% NaCl plus a few drops of whole sheep blood
Tube 2: 10% NaCl plus a few drops of whole sheep blood
Tube 3: 0.9% NaCl plus distilled water and a few drops of whole sheep blood

1. In the second column of Table 4.4, record the tonicity of each tube in relation to red blood cells.
2. Hold each tube in front of one of the pages of your lab manual. Determine whether you can see the print on the page through the tube. Record your findings in the third column of Table 4.4.

Table 4.4	Tonicity and Print Visibility		
Tube	**Tonicity**	**Print Visibility**	**Explanation**
1			
2			
3			

Conclusion

- Explain in the fourth column of Table 4.4 why you can or cannot see the print.

Elodea (Plant Cells)

When plant cells are in a hypotonic solution, the large central vacuole gains water and exerts pressure, called **turgor pressure.** The cytoplasm, including the chloroplasts, is pushed up against the cell wall (Fig. 4.10a).

When plant cells are in a hypertonic solution, the central vacuole loses water, and the cytoplasm, including the chloroplasts, pulls away from the cell wall. This is called **plasmolysis** (Fig. 4.10b).

Experimental Procedure: Elodea Cells

Hypotonic Solution

1. If possible, use the *Elodea* slide you prepared earlier in this laboratory. If not, prepare a new wet mount of a small *Elodea* leaf using fresh water.
2. After several minutes, focus on the surface of the cells, and compare your slide with Figure 4.10a.
3. Complete the portion of Table 4.5 that pertains to a hypotonic solution.

Hypertonic Solution

1. Prepare a new wet mount of a small *Elodea* leaf using a 10% NaCl solution.
2. After several minutes, focus on the surface of the cells, and compare your slide with Figure 4.10*b*.
3. Complete the portion of Table 4.5 that pertains to a hypertonic solution.

Table 4.5	Effect of Tonicity on *Elodea* Cells	
Tonicity	**Appearance of Cells**	**Due to (scientific term)**
Hypotonic		
Hypertonic		

Conclusions

- In a hypotonic solution, the large central vacuole of plant cells exerts _____ pressure, and the chloroplasts are seen _____ the cell wall.
- In a hypertonic solution, the central vacuole loses water, and the cytoplasm including the chloroplasts have _____ the cell wall.

Figure 4.10 *Elodea* cells.

a. Surface view of cells in a hypotonic solution (*above*) and longitudinal section diagram (*below*). The large central vacuole, filled with water, pushes the cytoplasm, including the chloroplasts, right up against the cell wall. **b.** Surface view of cells in a hypertonic solution (*above*) and longitudinal section diagram (*below*). When the central vacuole loses water, cytoplasm, including the chloroplasts, piles up in the center of the cell because the cytoplasm has pulled away from the cell wall. (*a:* Magnification 400×)

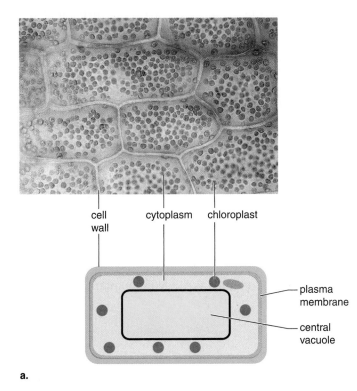

cell wall   cytoplasm   chloroplast

plasma membrane

central vacuole

a.

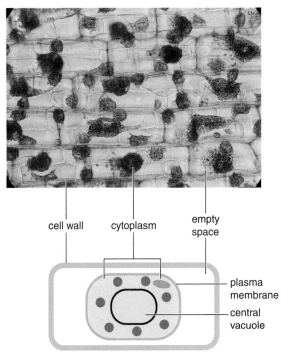

cell wall   cytoplasm   empty space

plasma membrane

central vacuole

b.

4.5 pH and Cells

The pH of a solution tells its hydrogen ion concentration [H^+]. The **pH scale** ranges from 0 to 14. A pH of 7 is neutral. A pH lower than 7 indicates that the solution is acidic (has more hydrogen ions than hydroxide ions), whereas a pH greater than 7 indicates that the solution is basic (has more hydroxide ions than hydrogen ions). A **buffer** is a system of chemicals that takes up excess hydrogen ions or hydroxide ions, as appropriate.

The concept of pH is important in biology because living organisms are very sensitive to hydrogen ion concentration. For example, in humans the pH of the blood must be maintained at about 7.4 or we become ill. All living things need to maintain the hydrogen ion concentration, or pH, at a constant level.

Why are cells and organisms buffered? _____

_____

Experimental Procedure: pH and Cells

> **Caution:** Hydrochloric acid (HCl) used to produce an acid pH is a strong, caustic acid. Exercise care in using this chemical. If any HCl spills on your skin, rinse immediately with clear water. Follow your instructor's directions for disposal of tubes that contain HCl.

1. Label three test tubes, and fill them to the halfway mark as follows: tube 1: *water;* tube 2: *buffer* (inorganic) solution; and tube 3: *simulated cytoplasm* (buffered protein solution).
2. Use pH paper to determine the pH of each tube. Dip the end of a stirring rod into the solution, and then touch the stirring rod to a 5-cm strip of pH paper. Read the current pH by matching the color observed with the color code on the pH paper package. Record your results in the "pH Before Acid" column in Table 4.6.
3. Add one drop of 0.1 N hydrochloric acid (HCl) to each tube. Shake or swirl. Use pH paper as in step 2 to determine the new pH of each solution. (Do as many pH determinations as possible on your pH paper before using a clean strip.) Record your results in the "pH After Acid" column in Table 4.6.

Table 4.6	pH and Cells			
Tube	**Contents**	**pH Before Acid**	**pH After Acid**	**Explanation**
1	Water			
2	Buffer			
3	Cytoplasm			

Conclusion

- Use the information in column two to explain your test results. Write your explanations in the last column of Table 4.6.

1. Pour the contents of tube 2 into a clean 50 ml beaker labeled "2." Pour the contents of tube 3 into a clean 50 ml beaker labeled "3."
2. Add hydrochloric acid in 5-drop increments to beakers 2 and 3. Mix and test the pH after every 5 drops, and record these values in Table 4.7.
3. Record the patterns of pH changes on the graphs provided. Use the top graph for the pattern observed with the inorganic buffer and the bottom graph for the pattern observed with simulated cytoplasm.

Table 4.7	Drops of HCl and pH Values	
# Drops	pH	pH
	Beaker 2 (Buffer)	Beaker 3 (Cytoplasm)
0		
5		
10		
15		
20		
25		
30		
35		
40		
45		
50		

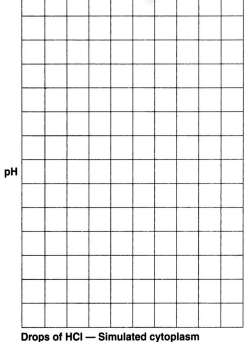

pH

Drops of HCl — Inorganic buffer

pH

Drops of HCl — Simulated cytoplasm

Conclusion

• Do the two graphs have a similar pattern?

_____

Explain.

_____

_____

_____

_____

1. What part of a prokaryotic cell contains their chromosomes? _____
   _____

2. What characteristics do all eukaryotic cells have in common? _____
   _____

3. Why would you predict that an animal cell, but not a plant cell, might burst when placed in a hypotonic solution? _____
   _____

4. Which of the cellular organelles would be included in a category called:

   a. Membranous canals and vacuoles? _____
   _____

   b. Energy-related organelles? _____
   _____

5. How do you distinguish between rough endoplasmic reticulum and smooth endoplasmic reticulum?

   a. Structure _____
   b. Function _____

6. If a dialysis bag filled with water is placed in a molasses solution, what do you predict will happen to the weight of the bag over time? _____
   Why? _____
   _____

7. What is the relationship between plant cell structure and the ability of plants to stand upright?
   _____
   _____

8. The police are trying to determine if material removed from the scene of a crime was plant matter. What would you suggest they look for? _____
   _____

9. A test tube contains red blood cells and a salt solution. When the tube is held up to a page, you cannot see the print. With reference to a concentration of 0.9% sodium chloride (NaCl), how concentrated is the salt solution? _____

10. Predict the microscopic appearance of cells in the leaf tissue of a wilted plant. _____
   _____

5

How Enzymes Function

Introduction

The cell carries out many chemical reactions. All the chemical reactions that occur in a cell are collectively called **metabolism.** A possible chemical reaction can be indicated like this:

$$A + B \longrightarrow C + D$$
$$\text{reactants} \qquad \text{products}$$

In all chemical reactions, the **reactants** are molecules that undergo a change, which results in the **products.** The arrow stands for the change that produced the product(s). The number of reactants and products can vary; in the one you are studying today, a single reactant breaks down to two products.

All the reactions that occur in a cell have an enzyme. **Enzymes** are organic catalysts that speed metabolic reactions. Because enzymes are specific and speed only one type of reaction, they are given names. In today's laboratory, you will be studying the action of the enzyme **catalase.** The reactants in an enzymatic chemical reaction are called **substrate(s).**

Enzymes are specific because they have a shape that accommodates the shape of their substrate. At one time it was thought that an enzyme and a substrate fit together like a key fits a lock, but now we know that the active site undergoes a slight change in shape to accommodate the substrate(s). This is called the **induced fit model** because the enzyme is induced to undergo a slight alteration to achieve optimum fit. The change in shape of the active site facilitates the reaction that now occurs. After the reaction has been completed, the product is released, and the active site returns to its original state, ready to bind to another substrate molecule. A cell needs only a small amount of enzyme because enzymes are not used up by the reaction. Enzymatic reactions can be indicated like this:

$$E + S \longrightarrow ES \longrightarrow E + P$$

In this reaction, E = enzyme, ES = enzyme-substrate complex, and P = product.

Figure 5.1 Enzymatic action.

The enzyme and substrate come together, and the reaction occurs on the surface of the enzyme at the active site. The product leaves the enzyme, and then the enzyme can be used again. **a.** During degradation, the substrate is broken down. **b.** During synthesis, substrates combine to produce the product.

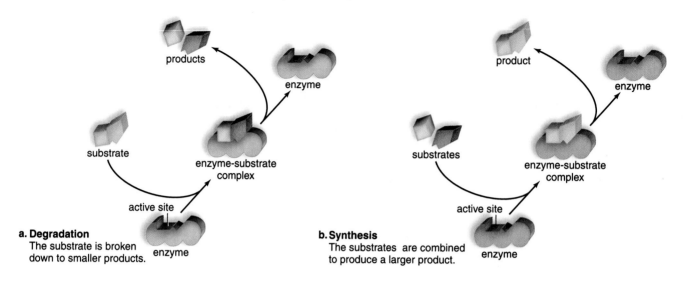

a. Degradation
The substrate is broken down to smaller products.

b. Synthesis
The substrates are combined to produce a larger product.

Two types of enzymatic reactions common to cells are shown in Figure 5.1. During degradation reactions, the substrate is broken down to the product(s), and during synthesis reactions, the substrates are joined to form a product. Notice in each case how the shape of the enzyme accommodates its substrate. The location where the enzyme and substrate form an enzyme-substrate complex is called the **active site** because the reaction occurs here.

At the end of the reaction, the product is released, and the enzyme can then combine with the substrate again. A cell needs only a small amount of an enzyme because enzymes are used over and over. Some enzymes have turnover rates well in excess of a million product molecules per minute.

All enzymes are complex proteins that generally act in an organism's closely controlled internal environment, where the temperature and pH remain within a rather narrow range. *Extremes* in pH or temperature may denature the enzyme by permanently altering its chemical structure. Even a small change in the protein's structure will change the enzyme's shape enough to prevent formation of the enzyme-substrate complex and thus keep the reaction from occurring. As more substrate molecules fill active sites, more product results per unit time. Therefore, in general, an increase in enzyme or substrate speeds enzymatic reactions. In this laboratory, you will test the effect of temperature, enzyme concentration, and pH on an enzymatic reaction.

5.1 Catalase Activity

In the Experimental Procedures that follow, you will be working with the enzyme catalase. Catalase is present in cells, where it speeds the breakdown of the toxic chemical hydrogen peroxide to water and oxygen:

$$2 \text{ H}_2\text{O}_2 \xrightarrow{\text{catalase}} 2 \text{ H}_2\text{O} + \text{O}_2$$

hydrogen peroxide   water   oxygen

What is the reactant in this reaction? _____ What is the substrate for catalase? _____

What are the products in this reaction? _____ and _____ Bubbling occurs as the reaction

proceeds. Why? _____

Experimental Procedure: Catalase Activity

With a wax pencil, label and mark three clean test tubes at the 1 cm and 5 cm levels.

Tube 1
1. Fill to the first mark with *catalase* buffered at pH 7.0, the optimum pH for catalase.
2. Fill to the second mark with *hydrogen peroxide.* Swirl well to mix, and wait at least 20 seconds for bubbling to develop.
3. Measure the height of the bubble column (in millimeters), and record your results in Table 5.1.

Tube 2
1. Fill to the first mark with *water.*
2. Fill to the second mark with *hydrogen peroxide.* Swirl well to mix, and wait at least 20 seconds.
3. Measure the height of the bubble column (in millimeters), and record your results in Table 5.1.

Tube 3
1. Fill to the first mark with *catalase.*
2. Fill to the second mark with *sucrose solution.* Swirl well to mix; wait 20 seconds.
3. Measure the height of the bubble column, and record your results in Table 5.1.

Table 5.1	Catalase Activity		
Tube	**Contents**	**Bubble Column Height**	**Explanation**
1	Catalase Hydrogen peroxide		
2	Water Hydrogen peroxide		
3	Catalase Sucrose solution		

Conclusions

- Which tube showed the bubbling you expected? _____ Conclude why this tube showed bubbling, and record your explanation in Table 5.1.
- Which tubes are a control? _____ If this tube showed bubbling, what could you conclude about your procedure? _____ Record your explanation in Table 5.1 for this tube.
- Enzymes are specific; they speed only a reaction that contains their substrate. Which tube exemplifies this characteristic of an enzyme? _____ Record your explanation in Table 5.1 for this tube.

5.2 Effect of Temperature on Enzyme Activity

In general, cold temperatures slow chemical reactions, and warm temperatures speed chemical reactions. Boiling, however, causes an enzyme to denature in a way that inactivates it.

Experimental Procedure: Effect of Temperature

With a wax pencil, label and mark three clean test tubes at the 1 cm and 5 cm levels.

1. Fill each tube to the first mark with *catalase* buffered at pH 7.0, the optimum pH for catalase.
2. Place tube 1 in a refrigerator or cold water bath, tube 2 in an incubator or warm water bath, and tube 3 in a boiling water bath. Complete the second column in Table 5.2. Wait 15 minutes.
3. As soon as you remove the tubes one at a time from the refrigerator, incubator, and boiling water, fill to the second mark with *hydrogen peroxide.*
4. Swirl well to mix, and wait 20 seconds.
5. Measure the height of the bubble column (in millimeters) in each tube, and record your results in Table 5.2.
6. Plot your results in Figure 5.2. Put temperature (°C) on the X-axis and bubble column height (mm) on the Y-axis.

Table 5.2	Effect of Temperature		
Tube	Temperature °C	Bubble Column Height (mm)	Explanation
1 Refrigerator			
2 Incubator			
3 Boiling water			

Conclusions

- The amount of bubbling corresponds to the degree of enzyme activity. Explain in Table 5.2 the degree of enzyme activity per tube.

- What is your conclusion concerning the effect of temperature on enzyme activity?

  _____
  _____
  _____
  _____
  _____

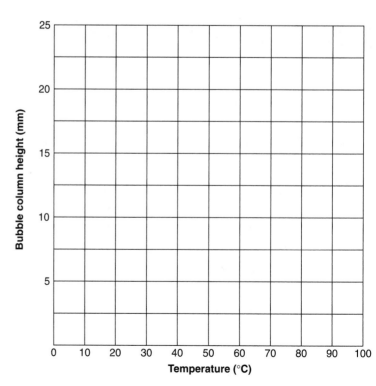

Figure 5.2 Effect of temperature on enzyme activity.

5.3 Effect of Concentration on Enzyme Activity

In general, a higher enzyme or substrate concentration results in faster enzyme activity—that is, the amount of product per unit time for any particular reaction will increase.

Experimental Procedure: Effect of Enzyme Concentration

With a wax pencil, label three clean test tubes.

Tube 1
1. Mark this tube at the 1 cm and 5 cm levels.
2. Fill to the first mark with buffered *catalase* and to the second mark with *hydrogen peroxide.*
3. Swirl well to mix, and wait 20 seconds.
4. Measure the height of the bubble column (in millimeters), and record your results in Table 5.3.

Tube 2
1. Mark this tube at the 2 cm and 6 cm levels.
2. Fill to the first mark with buffered *catalase* and to the second mark with *hydrogen peroxide.*
3. Swirl well to mix, and wait 20 seconds.
4. Measure the height of the bubble column (in millimeters), and record your results in Table 5.3.

Tube 3
1. Mark this tube at the 3 cm and 7 cm levels.
2. Fill to the first mark with buffered *catalase* and to the second mark with *hydrogen peroxide.*
3. Swirl well to mix, and wait 20 seconds.
4. Measure the height of the bubble column (in millimeters), and record your results in Table 5.3.

Table 5.3	Effect of Enzyme Concentration		
Tube	Amount of Enzyme	Bubble Column Height (mm)	Explanation
1	1 cm		
2	2 cm		
3	3 cm		

Conclusions

- The amount of bubbling corresponds to the degree of enzyme activity. Explain in Table 5.3 the degree of enzyme activity per tube.

- If unlimited time were allotted, would the results be the same in all tubes? _____
  Explain why or why not. _____

- Would you expect similar results if the substrate concentration were varied in the same manner as the enzyme concentration? _____ Why or why not? _____

5.4 Effect of pH on Enzyme Activity

Each enzyme has a pH at which the speed of the reaction is optimum (occurs best). Any higher or lower pH affects hydrogen bonding and the structure of the enzyme, leading to reduced activity.

Experimental Procedure: Effect of pH

> **Caution:** Hydrochloric acid (HCl) used to produce an acid pH is a strong, caustic acid, and sodium hydroxide (NaOH) used to produce a basic pH is a strong, caustic base. Wear eye protection and otherwise exercise care in using these chemicals. Follow your instructor's directions for disposal of tubes that contain these chemicals. If any acidic or basic solutions spill on your skin, rinse immediately with clear water.

With a wax pencil, label and mark three clean test tubes at the 1 cm, 3 cm, and 7 cm levels. Fill each tube to the 1 cm level with nonbuffered *catalase*.

Tube 1
1. Fill to the second mark with *water* adjusted to pH 3 by the addition of *HCl*.
2. Fill to the third mark with *hydrogen peroxide*. Wait one minute.
3. Swirl to mix, and wait 20 seconds.
4. Measure the height of the bubble column (in millimeters), and record your results in Table 5.4.

Tube 2
1. Fill to the second mark with *water* adjusted to pH 7.
2. Wait one minute. Fill to the third mark with *hydrogen peroxide*.
3. Swirl to mix, and wait 20 seconds.
4. Measure the height of the bubble column (in millimeters), and record your results in Table 5.4.

Tube 3
1. Fill to the second mark with *water* adjusted to pH 11 by the addition of *NaOH*.
2. Fill to the third mark with *hydrogen peroxide*. Wait one minute.
3. Swirl to mix, and wait 20 seconds.
4. Measure the height of the bubble column (in millimeters), and record your results in Table 5.4.
5. Plot your results in Figure 5.3. Put pH on the X-axis and column height (mm) on the Y-axis.

Table 5.4		Effect of pH	
Tube	pH	Bubble Column Height (mm)	Explanation
1	3		
2	7		
3	11		

Figure 5.3 Effect of pH on enzyme activity.

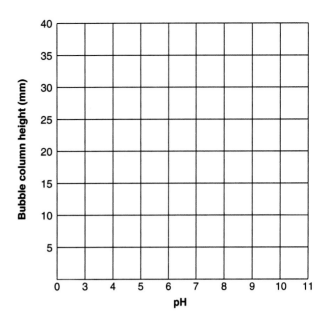

Conclusions

- The amount of bubbling corresponds to the degree of enzyme activity. Explain in Table 5.4 the degree of enzyme activity per tube.

- The results of which tube in Table 5.1 could be used as a control for Table 5.4? _____

  Why could this tube be considered a control? _____

Factors That Affect Enzyme Activity

In Table 5.5, summarize what you have learned about factors that affect the speed of an enzymatic reaction. For example, in general, what type of temperature promotes enzyme activity, and what type inhibits enzyme activity? Answer similarly for enzyme or substrate concentration and pH.

Table 5.5	Factors That Affect Enzyme Activity	
Factors	**Promote Enzyme Activity**	**Inhibit Enzyme Activity**
Temperature		
Enzyme or substrate concentration		
pH		

Conclusions

- Why does a warm temperature promote enzyme activity? _____

- Why does increasing enzyme concentration promote enzyme activity? _____

- Why does optimum pH promote enzyme activity? _____

1. What happens at the active site of an enzyme? _____

_____

2. On the basis of the active site, explain why the following conditions speed a chemical reaction:

   a. More enzyme _____

   b. More substrate _____

3. Name three other conditions (other than the ones mentioned in question 2) that maximize enzymatic reactions.

   a. _____

   b. _____

   c. _____

4. Explain the necessity for each of the three conditions you listed in question 3.

   a. _____

   b. _____

   c. _____

5. Lipase is a digestive enzyme that digests fat droplets in the basic conditions ($NaHCO_3$ is present) of the small intestine. Indicate which of the following test tubes would show digestion following incubation at 37°C, and explain why the others would not.

   Tube 1: Water, fat droplets _____

   Tube 2: Water, fat droplets, lipase _____

   Tube 3: Water, fat droplets, lipase, $NaHCO_3$ _____

   Tube 4: Water, lipase, $NaHCO_3$ _____

6. Fats are digested to fatty acids and glycerol. As the reaction described in question 5 proceeds, the solution will become what type pH? _____ Why? _____

_____

7. Given the following reaction:

$$2\ H_2O_2 \xrightarrow{\text{catalase}} 2\ H_2O + O_2$$

   hydrogen peroxide   water   oxygen

   a. Which substance is the substrate? _____

   b. Which substance is the enzyme? _____

   c. Which substances are the end products? _____

   d. Is this a synthetic or degradative reaction? _____

   How do you know? _____

6

Photosynthesis

Learning Outcomes

Introduction

The overall equation for **photosynthesis** is

$$CO_2 + H_2O \xrightarrow{\text{solar energy}} (CH_2O)_n + O_2$$

In this equation, (CH_2O) represents any general carbohydrate. Sometimes, this equation is multiplied by 6 so that glucose $(C_6H_{12}O_6)$ appears as an end product of photosynthesis. Photosynthesis takes place in chloroplasts (Fig. 6.1). Here membranous thylakoids are stacked in grana surrounded by the stroma. During the light reactions, pigments within the membranes, notably the chlorophylls, of thylakoids absorb solar energy, water (H_2O) is split, and oxygen (O_2) is released. The Calvin cycle reactions occur within the stroma. During these reactions, carbon dioxide (CO_2) is reduced and solar energy is now stored in a carbohydrate (CH_2O).

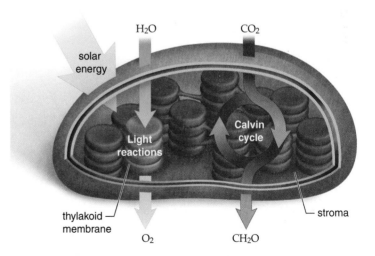

Figure 6.1 Overview of photosynthesis.
Photosynthesis includes the light reactions when energy is collected and O_2 is released, and the Calvin cycle reactions when carbohydrate (CH_2O) is formed.

6.1 Plant Pigments

The principal pigment in the thylakoids of plants is **chlorophyll *a*. Chlorophyll *b*, carotenes,** and **xanthophylls** play a secondary role by transferring the energy they absorb to chlorophyll *a* for use in photosynthesis.

    Chromatography is a technique that separates molecules from each other on the basis of their solubility in particular solvents. The solvents used in the following Experimental Procedure are petroleum ether and acetone, which have no charged groups and are therefore nonpolar. As a nonpolar solvent moves up the chromatography paper, the pigment moves along with it. The more nonpolar a pigment, the more soluble it is in a nonpolar solvent and the faster and farther it proceeds up the chromatography paper.

Experimental Procedure: Plant Pigments

> **Caution:** Ether (which is part of the chromatography solution) is toxic and extremely flammable. Do not breathe the fumes, and do not place the chromatography solution near any source of heat. A fume (ventilation) hood is recommended.

1. Assemble a chromatography apparatus (large, dry test tube and cork with a hook) and a strip of precut chromatography paper (handle by the top only) (Fig. 6.2). Attach the paper strip to the hook, and test for fit. The paper should hang straight and barely touch the bottom of the test tube; trim if necessary. Measure 2 cm from the bottom of the paper, and place a small dot with a pencil (not a pen). With a wax pencil, mark the test tube 1 cm below where the dot is with the stopper in place. Set the apparatus in a test tube rack.

Figure 6.2 Paper chromatography.

The paper must be cut to size and arranged to hang down without touching the sides of a dry tube. Then the pigment (chlorophyll) solution is applied to a designated spot. The chromatogram develops after the spotted paper is suspended in the chromatography solution.

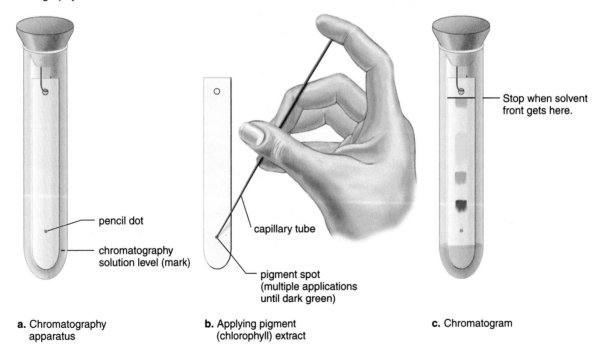

pencil dot

chromatography solution level (mark)

capillary tube

pigment spot (multiple applications until dark green)

Stop when solvent front gets here.

a. Chromatography apparatus

b. Applying pigment (chlorophyll) extract

c. Chromatogram

2. Prepare (or obtain) a plant *pigment extract,* as directed by your instructor.
3. Place the premarked chromatography paper strip onto a paper towel.
4. Fill a capillary tube by placing it into the extract. (It will fill by its own capillary action.)
5. Repeatedly apply the *pigment extract* to the pencil dot on the chromatographic strip. Let the spot dry between each application. Try to obtain a small dark green spot. (Placing your index finger over the wide end of the capillary tube will help keep the dot small.)
6. In a **fume hood,** add *chromatography solution* to the mark you made earlier. Do not submerge the pigment spot. Set the apparatus in a test-tube rack and close the chromatography apparatus tightly. Do not shake the test tube during the chromatography.
7. Allow approximately 10 minutes for your chromatogram to develop, but check it frequently so that the pigments do not reach the top of the paper.
8. When the solvent front has moved to within 1 cm of the upper edge of the paper (Fig. 6.2c), remove your chromatogram. Close the apparatus tightly. With a pencil, lightly mark the location of the solvent front, and allow the chromatogram to dry in the fume hood.
9. Identify the pigment bands. Beta-carotenes are represented by the bright orange-yellow band at the top. Xanthophylls are yellow and may be represented in multiple bands. The blue-green band is chlorophyll *a,* and the lowest, olive-green band is chlorophyll *b.* Which pigment is the most nonpolar (that is, has the greatest affinity for the nonpolar solvent)? _____
10. Calculate the R_f (ratio-factor) values for each pigment. For these calculations, mark the center of the initial pigment spot. This will be the starting point for all measurements. Also mark the midpoints of each pigment and the solvent front. Measure the distance between points for each pigment in millimeters (mm), and record these values in Table 6.1. Then use the following formula, and enter your R_f values in Table 6.1:

$$R_f = \frac{\text{distance moved by pigment}}{\text{distance moved by solvent}}$$

Table 6.1	Rf (Ratio-Factor) Values for Each Pigment	
Pigments	Distance Moved (mm)	R_f Values
Beta-carotenes		
Xanthophylls		
Chlorophyll *a*		
Chlorophyll *b*		
Solvent		

11. Do your results suggest that the chemical characteristics of these pigments might differ? _____

Why? _____

6.2 Solar Energy

During photosynthesis, solar energy is transformed into the chemical energy of a carbohydrate $(CH_2O)_n$. Without solar energy, photosynthesis would be impossible.

Release of oxygen from a plant indicates that the light reactions of photosynthesis are occurring. *Verify that photosynthesis releases oxygen by writing the overall equation for photosynthesis below.* The oxygen released from photosynthesis is taken up by a plant when cellular respiration occurs. This must be taken into account when the rate of photosynthesis is calculated.

Role of White Light

White (sun) light contains different colors of light, as is demonstrated when white light passes through a prism (Fig. 6.3). White light is the best for photosynthesis because it contains all the colors of light.

Figure 6.3 **White light.**
White light is made up of various colors, as can be seen when white light passes through a prism.

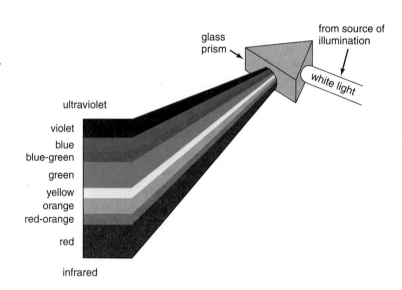

glass prism

from source of illumination

white light

ultraviolet
violet
blue
blue-green
green
yellow
orange
red-orange
red
infrared

Experimental Procedure: White Light

1. Place a generous quantity of *Elodea* with the cut end up (make sure the cuts are fresh) in a test tube with a rubber stopper containing a bent piece of glass tubing, as illustrated in Figure 6.4. When assembled, this is your volumeter for studying the need for light in photosynthesis. (Do not hold the volumeter in your hand, as body heat will also drive the reaction forward.) Your instructor will show you how to fix the volumeter in an upright position.

2. Before stoppering the test tube, add sufficient 3% *sodium bicarbonate* $(NaHCO_3)$ solution so that, when the rubber stopper is inserted into the tube, the solution comes to rest at about 1/4 the length of the bent glass tubing. Mark this location on the glass tubing with a wax pencil.

3. Place a beaker of plain water next to the *Elodea* tube to serve as a heat absorber. Place a lamp (150 watt) next to the beaker. The tube, beaker, and lamp should be as close to one another as possible.

4. Turn on the lamp. As soon as the edge of the solution in the tubing begins to move, time the reaction for 10 minutes. Be careful not to bump the tubing or to readjust the stopper, or your readings will be altered. After 10 minutes, mark the edge of the solution, and measure in

   millimeters the distance the edge moved: _____ mm/10 min. This is **net photosynthesis,** a measurement that does not take into account the oxygen that was used up for cellular

   respiration. Record your results in Table 6.2. Why did the edge move forward? _____

Figure 6.4 **Volumeter.**
A volumeter apparatus is used to study the role of light in photosynthesis.

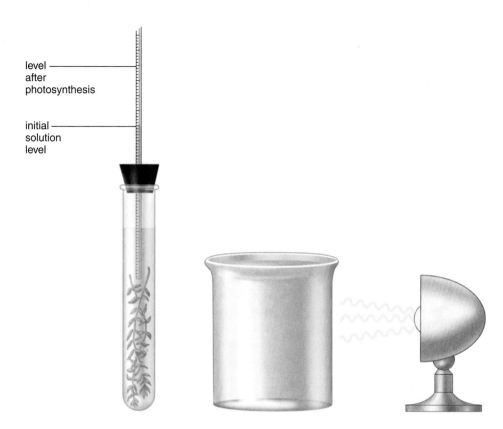

level after photosynthesis

initial solution level

5. Carefully wrap the tube containing *Elodea* in aluminum foil, and record here the length of time it takes for the edge of the solution in the tubing to recede 1 mm: _____. Convert your measurement to _____ mm/10 min, and record this value for **cellular respiration** in Table 6.2. (Do not use a minus sign, even though the edge receded.) Why does cellular respiration, which occurs in a plant all the time, cause the edge to recede? _____

_____

6. If the *Elodea* had not been respiring in step 4, how far would the edge have moved? _____ mm/10 min. This is **gross photosynthesis** (net photosynthesis + cellular respiration). Record this number in Table 6.2.

7. Calculate the **rate of photosynthesis** (mm/hr) by multiplying gross photosynthesis (mm/10 min) by 6 (that is, 10 min × 6 = 60 min = 1 hr): _____ mm/hr. Record this value in Table 6.2.

Table 6.2	Rate of Photosynthesis (White Light)
Data	
Net photosynthesis (white light)	
Cellular respiration (no light)	
Gross photosynthesis (net + cellular respiration)	(mm/10 min)
Rate of photosynthesis	(mm/hr)

Role of Green Light

Green light is only one part of white light (see Fig. 6.3). Plant pigments absorb certain colors of light better than other colors (Fig. 6.5). According to Figure 6.5, what color light do the chlorophylls absorb best? _____ Least? _____

What color light do the carotenoids (carotenes and xanthophylls) absorb best? _____

Least? _____

Does photosynthesis use green light? _____

The following Experimental Procedure will test your answer.

Figure 6.5 Action spectrum for photosynthesis.
The action spectrum for photosynthesis is the sum of the absorption spectrums for the pigments chlorophyll *a*, chlorophyll *b*, and carotenoids.

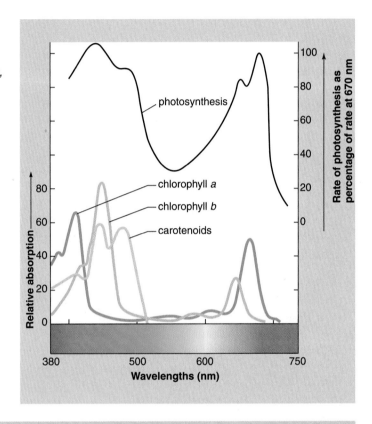

Experimental Procedure: Green Light

1. Add three drops of green dye (or use a green cellophane wrapper) to the beaker of water used in the previous Experimental Procedure until there is a distinctive green color. Remove all previous wax pencil marks from the glass tubing.
2. Record in Table 6.3 your data for gross photosynthesis (mm/10 min) and for rate of photosynthesis for white light (mm/hr) from Table 6.2.
3. Turn on the lamp. Mark the location of the edge of the solution on the glass tubing. As soon as the edge begins to move, time the reaction for 10 minutes. After 10 minutes, mark the edge of the solution, and measure in millimeters the distance the edge moved. Net photosynthesis

   for green light = _____ mm/10 min.
4. Carefully wrap the tube containing *Elodea* in aluminum foil, and record here the length of time

   it takes for the edge of the solution in the tubing to recede 1 mm: _____. Convert your

   measurement to _____ mm/10 min.

5. Calculate gross photosynthesis for green light (mm/10 min) as you did for white light, and record your data in Table 6.3.

6. Calculate rate of photosynthesis for green light (mm/hr) as you did for white light, and record your data in Table 6.3.

7. Average and record the Table 6.3 class data for both white light and green light, and record these averages in Table 6.3.

8. The following equation shows the rate of photosynthesis (green light) as a percentage of the rate of photosynthesis (white light):

$$\text{percentage} = \frac{\text{rate of photosynthesis (green light)}}{\text{rate of photosynthesis (white light)}} \times 100$$

Based on your data in Table 6.3, this percentage = _____. Based on

class data in Table 6.3, this percentage = _____. Record these values in Table 6.3.

Table 6.3 Rate of Photosynthesis (Green Light)	Your Data	Class Data
Gross photosynthesis (mm/10 min)		
White (from Table 6.2)		
Green		
Rate of photosynthesis (mm/hr)		
White (from Table 6.2)		
Green		

Conclusions

- Explain why the rate of photosynthesis with green light is only a portion of the rate of photosynthesis with white light. _____

_____

- How does the percentage based on your data differ from that based on class data?

_____

_____

6.3 Carbon Dioxide Uptake

During the Calvin cycle reactions of photosynthesis, the plant takes up carbon dioxide (CO_2) and reduces it to a carbohydrate, such as glucose ($C_6H_{12}O_6$). Therefore, the carbon dioxide in the solution surrounding *Elodea* should disappear as photosynthesis takes place.

Experimental Procedure: Carbon Dioxide Uptake

1. Temporarily remove the *Elodea* from the test tube. Empty the sodium bicarbonate ($NaHCO_3$) solution from the test tube, rinse the test tube thoroughly, and fill with a phenol red solution diluted to a faint pink. (Add more water if the solution is too dark.) Phenol red is a pH indicator that turns yellow in an acid and red in a base.

2. Blow *lightly* on the surface of the solution. Stop blowing as soon as the surface color changes to yellow. Then shake the test tube until the rest of the solution turns yellow.

   Blowing onto the solution adds what gas to the test tube? _____ When carbon dioxide combines with water, it forms carbonic acid. What causes the color change?

   _____

3. Thoroughly rinse the *Elodea* with distilled water, return it to the test tube with the phenol red solution, and assemble your volumeter as before.

4. If you used green dye, change the water in the beaker to remove the green solution.

5. Turn on the lamp, and wait until the edge of the solution just begins to move. Note the time. Observe until you note a change in color. Record your results in the appropriate column of Table 6.4.

6. Hypothesize why the solution in the test tube eventually turned red. _____

   _____

Use of a Control

Scientists are more confident of their results when an experimental procedure includes a control. Controls undergo all the steps in the experiment except the one being tested.

- Considering the test sample in Table 6.4, suggest a possible control sample for this experiment:

  _____

- Ask your instructor if you can actually perform this procedure. Both the control and test sample should be done at the same time.

- Record your results in Table 6.4. Why should all experiments have a control? _____

  _____

Table 6.4	Carbon Dioxide Uptake
Tube	**Time for Color Change**
Test sample: *Elodea* + phenol red solution + CO_2	
Control sample:	

6.4 Carbon Cycle

In this laboratory, you have demonstrated a relationship between cellular respiration and photosynthesis. Animals produce the carbon dioxide used by plants to carry out photosynthesis. Plants produce the food and oxygen that they and animals require to carry out cellular respiration. This relationship is illustrated in Figure 6.6 and can be represented by the following equation:

$$C_6H_{12}O_6 + 6\ O_2 \underset{\text{photosynthesis}}{\overset{\substack{\text{cellular}\\ \text{respiration}}}{\rightleftharpoons}} 6\ CO_2 + 6\ H_2O + \text{energy}$$

1. Which organelle in plants carries out the reaction in the previous equation in the reverse (right-to-left) direction? _____

2. Pertaining to photosynthesis, the energy in the equation is provided by _____.

3. Which organelle in plants and animals is involved in carrying out the reaction in this equation in the forward direction? _____

4. Pertaining to cellular respiration, the energy in the equation becomes chemical bond energy in what molecule? _____

5. Would it be correct to say that solar energy eventually becomes the chemical bond energy in ATP? _____ Why? _____

6. Considering that both plants and animals carry on cellular respiration, revise Figure 6.6 to improve its accuracy.

Figure 6.6 **Photosynthesis and cellular respiration.**
Animals are dependent on plants for a supply of oxygen, and plants are dependent on animals for a supply of carbon dioxide.

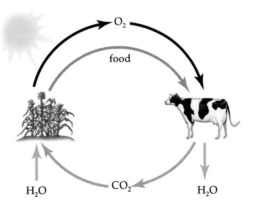

1. How are plant pigments involved in photosynthesis? _____
   _____

2. Why is it beneficial to have several different plant pigments involved in photosynthesis? _____
   _____

3. On what basis does chromatography separate substances? _____
   _____

4. Some types of red algae carry on photosynthesis several meters beneath the ocean surface. What color light do you predict does not penetrate to this depth? _____

5. Consider the following reaction:

$$CO_2 + H_2O \rightarrow \underset{\substack{\text{carbonic} \\ \text{acid}}}{H_2CO_3} \rightarrow H^+ + HCO_3^-$$

   a. Phenol red, a pH indicator, turns yellow (indicating acid) when you breathe into a solution. How does the reaction explain why the solution turned acidic? _____
   _____

   b. Phenol red turns back to red when a plant in light is added to the solution. In terms of the reaction, why does this occur? _____
   _____

6. Gas exchange occurs in both photosynthesis and cellular respiration. Contrast these two processes by completing the following table:

	Organelle	Gas Given Off	Gas Taken Up
Photosynthesis			
Cellular respiration			

7. What experimental conditions were used in this laboratory to test for cellular respiration in plant cells? _____
   _____

8. Suppose you replaced *Elodea* with animal cells in the experimental test tube. Would the results differ according to the use of a white light or no light? _____ Explain. _____
   _____

7

Cellular Respiration

Introduction

In this laboratory, you will study **fermentation** and **cellular respiration.** During fermentation, glucose is incompletely broken down, and much energy remains in the organic molecule that results. In cellular respiration, glucose is completely broken down to inorganic molecules. During fermentation, a small amount of chemical energy is converted to ATP molecules, and during cellular respiration much more chemical energy is converted to ATP molecules for use by the cell. ATP molecules are absolutely essential to cellular metabolism because they supply energy whenever a cell carries on activities, such as active transport, synthesis, muscle contraction, and nerve conduction.

In yeast and human beings, fermentation occurs in the following way:

$$C_6H_{12}O_6 \longrightarrow 2\ CO_2 + 2\ C_2H_5OH + 2\ ATP$$

$$\text{glucose} \qquad\qquad\qquad \text{carbon} \quad\ \text{ethanol}$$
$$\text{dioxide}$$

Yeast is used by the wine and beer industries to ferment the carbohydrates in fruits and grains to alcohol. In baking, the carbon dioxide given off from yeast fermentation causes bread to rise. Fermentation in animals and certain microbes produces lactic acid. This form of fermentation helps produce yogurt and many cheeses, as well as such products as sourdough breads, chocolate, and pickled foods.

In most organisms, cellular respiration occurs in the following way:

$$C_6H_{12}O_6 + 6\ O_2 \longrightarrow 6\ CO_2 + 6\ H_2O + 36\text{--}38\ ATP$$

$$\text{glucose} \quad\ \text{oxygen} \qquad\qquad \text{carbon} \quad\ \text{water}$$
$$\text{dioxide}$$

Notice that there is an uptake of oxygen during cellular respiration.

7.1 Fermentation

Yeasts (unicellular fungi) can use several types of sugars as an energy source. Glucose and fructose are monosaccharides; sucrose is a disaccharide that contains glucose and fructose. Fructose can be converted to glucose, the usual molecule acted on by yeast. In the following Experimental Procedure, you will test several of these food sources for their ability to ferment by measuring the amount of carbon dioxide given off in a respirometer. A **respirometer** is a device for measuring the amount of gas given off and/or consumed.

Experimental Procedure: Fermentation

Respirometer Practice

1. Completely fill a small tube (15 × 125 mm) with water (Fig. 7.1).
2. Invert a large tube (20 × 150 mm) over the small tube and, with your finger or a pencil, push the small tube up into the large tube until the upper lip of the small tube is in contact with the bottom of the large tube.
3. Quickly invert both tubes. Do not permit the small tube to slip away from the bottom of the large tube. A little water will leak out of the small tube and be replaced by an air bubble.
4. Practice this inversion until the bubble in the small tube is as small as you can make it.

Figure 7.1 **Respirometer for yeast experiment.**
Place a small tube inside a large tube. Hold the small tube in place as you rotate the entire apparatus; an air bubble will form in the small tube.

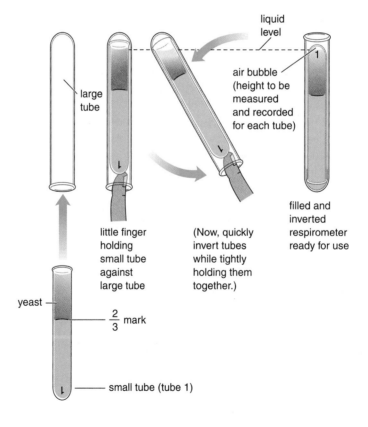

Testing Food Samples

Have ready four large test tubes. With a wax pencil, label and mark off a small test tube at the 2/3-full level. Use this tube to mark off three other small tubes at the same level.

1. Label and fill the small tubes as directed, and record the contents in Table 7.1.

   Tube 1   Fill to the mark with *glucose solution.*

   Tube 2   Fill to the mark with *fructose solution.*

   Tube 3   Fill to the mark with *sucrose solution.*

   Tube 4   Fill to the mark with *distilled water.*

2. Re-suspend a yeast solution each time, and fill all four tubes to the top with yeast suspension (Fig. 7.1).
3. Slide the large tubes over the small tubes, and invert them in the way you practiced. This will mix the yeast and sugar solutions.
4. Place the respirometers in a tube rack, and measure the initial height of the air space in the rounded bottom of the small tube. Record the height in Table 7.1.
5. Place the respirometers in an incubator or in a warm water bath maintained at 37°C. Note the time, and allow the respirometers to incubate about 20 minutes (incubator) or one hour (water bath). However, watch your respirometers and if they appear to be filling with gas quite rapidly, stop the incubation when appropriate. During the incubation period, begin the Experimental Procedure on cellular respiration (see page 70).
6. At the end of the incubation period, measure the final height of the gas bubble, and record it in Table 7.1. Calculate the net change, and record it in Table 7.1.

Table 7.1		Fermentation by Yeast			
Tube	Contents	Initial Gas Height	Final Gas Height	Net Change	Conclusion
1					
2					
3					
4					

Conclusions

- From your results, conclude how the sugars tested compare as a food source for yeast fermentation. Enter your conclusions in Table 7.1.

- How do you know that the yeast cells were respiring anaerobically (in the absence of oxygen) and not aerobically (in the presence of oxygen)? _____

  _____

- Why did the gas bubbles increase in size? _____

- Speculate on why sucrose is not as good a food source as fructose and glucose. _____

  _____

- Which respirometer was the control? _____

7.2 Cellular Respiration

As indicated in the cellular respiration equation given in the introduction to this laboratory, oxygen gas is consumed during cellular respiration, and carbon dioxide gas is given off. The uptake of oxygen is the evidence that an organism is carrying on cellular respiration. It is possible to use potassium hydroxide (KOH) to remove carbon dioxide as it is given off. The equation for this reaction is:

$$CO_2 + 2\ KOH \longrightarrow K_2CO_3 + H_2O$$

solid
potassium
carbonate

Experimental Procedure: Cellular Respiration

1. Obtain a volumeter, an apparatus that measures changes in gas volumes. Remove the three vials from the volumeter. Remove the stoppers from the vials and the vials from the volumeter. Label the vials 1, 2, and 3.

> **Caution:** Potassium hydroxide (KOH) is a strong, caustic base. Exercise care in using this chemical, and follow your instructor's directions for disposal of these tubes. If any potassium hydroxide should spill on your skin, rinse immediately with water.

2. Using the same amounts, place a small wad of absorbent cotton in the bottom of each vial. Without getting the sides of the vials wet, use a dropper to saturate the cotton with 15% potassium hydroxide (KOH). Place a small wad of dry cotton on top of the KOH-soaked absorbent cotton (Fig. 7.2).
3. Count 25 germinating soybean seeds and add to vial 1. Count 25 dry (nongerminating) soybean seeds and add to vial 2. Add glass beads to vial 2 so that the volume occupied in vial 2 is approximately the same as in vial 1. Add only glass beads to vial 3 to bring to approximately the same volume.
4. Each stopper should have a vent (rubber tube) with a clamp and a graduated side arm. Remove the clamps from the vents. Adjust the graduated side arm until only about 5 mm to 1 cm protrudes through the stopper (Fig. 7.3).
5. With a dropper, add a drop of *Brodie manometer fluid* (or water colored with vegetable dye and a small amount of detergent) to each side arm.

Figure 7.2 Vials.
In this experiment, three vials are filled as noted.

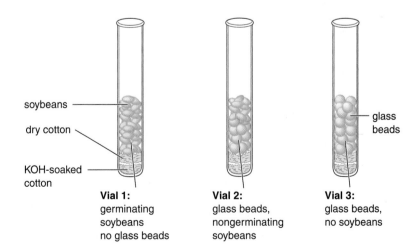

soybeans
dry cotton
KOH-soaked cotton

glass beads

Vial 1:
germinating
soybeans
no glass beads

Vial 2:
glass beads,
nongerminating
soybeans

Vial 3:
glass beads,
no soybeans

6. Firmly place the stoppers in the vials. Adjust the stoppers until the side arms are parallel to the lab bench.

7. Adjust the location of the marker drop in the side arms with the assistance of a dry dropper in the vent. The marker drop of vial 3 (thermobarometer) should be in the middle of the side arm. The marker drop of vials 1 and 2 should be between 0.80 and 0.90 ml. Close the vent with the clamp when the marker drop is at the correct location.

8. Allow the respirometers to equilibrate for 5 minutes, and then record in Table 7.2, to the nearest 0.01 ml, the initial position of the marker drop in each graduated side arm.

9. Wait 10 minutes, and then record in Table 7.2 any change in the position of the marker drop. Wait 10 more minutes, and then record in Table 7.2 any change in the position of the marker drop. Then record in Table 7.2 the net change for each vial—that is, the initial reading for each vial minus the vial's 20-minute reading.

Figure 7.3 **Volumeter containing three respirometers.** In this experiment, the respirometers are vials filled as per Fig. 7.2 with graduated side arms attached. Oxygen uptake is measured by movement of a marker drop in each side arm.

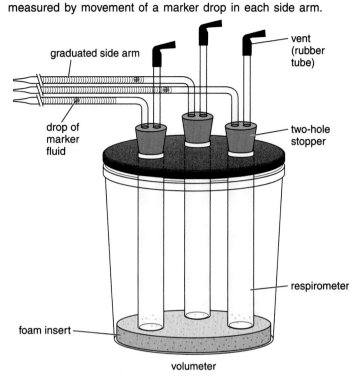

10. Did the marker drop change in vial 3 (glass beads)? _____ By how much? _____ Enter this number in the "Correction" column of Table 7.2, and use this number to correct the net change you observed in vials 1 and 2. (This is a correction for any change in volume due to atmospheric pressure changes or temperature changes.) This will complete Table 7.2.

Table 7.2	Cellular Respiration					
Vial Contents	Initial Reading	Reading After 10 Minutes	Reading After 20 Minutes	Net Change	Correction	(Corrected) Net Change
1 Germinating soybeans						
2 Dry (nongerminating) soybeans Glass beads						
3 Glass beads						

Conclusions

- In which vial did the water recede? _____ State the vial contents. _____

  Is this the vial that carried on cellular respiration? _____

- Why did the water recede? _____

- Why was it necessary to absorb the carbon dioxide? _____ 7—6
  _____

- In the soybean experiment, you were measuring the change in volume of what gas? _____

- Which respirometer in the soybean experiment was the control? _____

Laboratory Review 7

1. Both fermentation and cellular respiration ordinarily begin with what molecule? _____

2. How do the overall equations for these processes indicate that fermentation is anaerobic and that cellular respiration is aerobic? _____
   _____

3. Glucose breakdown results in the breaking of C—H bonds and stored energy is released. Contrast the end products of fermentation and cellular respiration in terms of their energy content. _____
   _____

4. Fermentation results in the net production of only 2 ATP, while cellular respiration results in the production of at least 36 ATP. Explain these results with reference to the end products of both of these processes. _____
   _____

5. In the Experimental Procedure: Fermentation, the gas bubble got larger. What gas was causing this increase in bubble size? _____

6. Why is it reasonable that, of the three sugars (glucose, fructose, and sucrose), glucose would result in the most activity during the fermentation experiment? _____
   _____

7. In the Experimental Procedure: Cellular Respiration, what gas was taken up by the soybeans? _____

8. Explain the role of each of the following components in the cellular respiration experiment:

   a. KOH _____

   b. Germinating soybeans _____

   c. Dry (nongerminating) soybeans _____

9. If you performed the cellular respiration experiment without soaking the cotton with KOH, what results would you predict? _____
   Why? _____

8

Mitosis and Meiosis

Learning Outcomes

8.1 The Cell Cycle
- Explain how mitosis is related to the cell cycle. 74
- Recognize individual stages of mitosis from microscope slides or images and discuss the events occurring in each. 75–80
- Contrast animal and plant cell mitosis and explain the distinct events of each. 75–80

8.2 Meiosis
- Compare the movement and resultant changes in chromosomes for cells as they undergo meiosis I and meiosis II. 81–85

8.3 Mitosis Versus Meiosis
- Evaluate why crossing-over can only occur in meiosis and not in mitosis, and explain what advanage this brings. 82
- Compare the results of mitosis with those for meiosis. 86–87

8.4 The Life Cycle of Human Beings
- Contrast the roles played by mitosis and meiosis in the human life cycle. 88–89

Introduction

Dividing cells experience nuclear division, cytoplasmic division, and a period of time between divisions called interphase. During **interphase,** the nucleus appears normal, and the cell is performing its usual cellular functions. Also, the cell is increasing all of its components, including such organelles as the mitochondria, ribosomes, and centrioles, if present. DNA replication (making an exact copy of the DNA) occurs toward the end of interphase. Thereafter, the chromosomes, which contain DNA, are duplicated and contain two chromatids held together at a **centromere.** These chromatids are called **sister chromatids.**

During nuclear division, called **mitosis,** the new nuclei receive the same number of chromosomes as the parental nucleus. When the cytoplasm divides, a process called **cytokinesis,** two daughter cells are produced. In multicellular organisms, mitosis permits growth and repair of tissues. In eukaryotic, unicellular organisms, mitosis is a form of asexual reproduction. Sexually reproducing organisms utilize another form of nuclear division, called **meiosis.** In animals, meiosis is a part of gametogenesis, the production of gametes (sex cells). The gametes are sperm in male animals and eggs in female animals. As a result of meiosis, the daughter cells have half the number of chromosomes as the parental cell. Because crossing-over of genetic material takes place and the chromosomes occur in various combinations in the daughter cells, meiosis contributes to recombination of genetic material and to variation among sexually reproducing organisms.

This laboratory examines both mitotic and meiotic cell division to show their similarities and differences. At the start of both types of divisions, the parental nucleus, surrounded by a double membrane (the nuclear envelope), contains one or more **nucleoli** (concentrated regions of RNA) and **chromatin** (threadlike strands of DNA) suspended in a transparent liquid called **nucleoplasm.** During division, chromatin condenses so that the chromosomes are visible, the nuclear envelope fragments, and a spindle appears. Spindle fibers assist the movement of chromosomes, which occurs at this time.

8.1 The Cell Cycle

As stated in the Introduction, the period of time between cell divisions is known as interphase. Because early investigators noted little visible activity between cell divisions, they dismissed this period of time as a resting state. But when they discovered that DNA replication and chromosome duplication occur during interphase, the **cell cycle** concept was proposed. Investigators have also discovered that cytoplasmic organelle duplication occurs during interphase, as does synthesis of the proteins involved in regulating cell division. Thus, the cell cycle can be broken down into four stages (Fig. 8.1). State the main event of each stage on the line provided:

G$_1$ ————————————————

S ————————————————

G$_2$ ————————————————

M ————————————————

The time required for the entire cell cycle varies according to the organism, but 18–24 hours is typical for animal cells. Mitosis (including cytokinesis, if it occurs) lasts less than an hour to slightly more than 2 hours; for the rest of the time, the cell is in interphase.

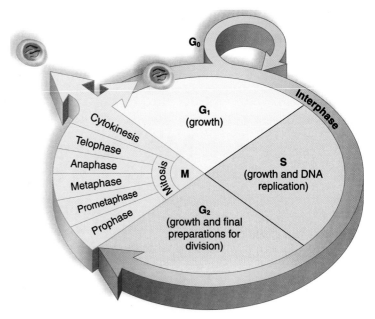

Figure 8.1 **The cell cycle.**
Immature cells go through a cycle that consists of four stages: G$_1$, S (for synthesis), G$_2$, and M (for mitosis). Eventually, some daughter cells "break out" of the cell cycle and become specialized cells.

Table 8.1	Structures Associated with Mitosis
Structure	**Description**
Nucleus	A large organelle containing the chromosomes and acting as a control center for the cells
Chromosome	Rod-shaped body in the nucleus that is seen during mitosis and meiosis and that contains DNA and, therefore the hereditary units, or genes
Nucleolus	An organelle found inside the nucleus; composed largely of RNA for ribosome formation
Spindle	Microtubule structure that brings about chromosome movement during cell division
Chromatids	The two identical parts of a chromosome following DNA replication
Kinetochore	Protein molecule that attaches a sister chromatid to a spindle fiber
Centromere	A constriction where duplicates (sister chromatids) of a chromosome are held together
Centrosome	The central microtubule-organizing center of cells; consists of granular material; in animal cells, contains two centrioles
Centrioles*	Short, cylindrical organelles in animal cells that contain microtubules and are associated with the formation of the spindle during cell division
Aster*	Short, radiating fibers produced by the centrioles

*Animal cells only

Mitosis

Mitosis is nuclear division that results in two new nuclei, each having the same number of chromosomes as the original nucleus. The **parent cell** is the cell that divides, and the resulting cells are called **daughter cells.**

When cell division is about to begin, chromatin starts to condense and compact to form visible, rodlike sister chromatids held together at the centromere (Fig. 8.2*a*). Label the sister chromatids, centromere, and kinetochore in the drawing of a duplicated chromosome in Figure 8.2*b*. This illustration represents a chromosome as it would appear just before nuclear division occurs.

Figure 8.2 Duplicated chromosomes.

DNA replication results in a duplicated chromosome that consists of two sister chromatids held together at a centromere. **a.** Scanning electron micrograph of a duplicated chromosome. **b.** Drawing of a duplicated chromosome.

1. _____

2. _____

3. _____

one chromatid

a.   9,850×   b.

Spindle

Table 8.1 lists the structures that play a role during mitosis. The spindle is a structure that appears and brings about an orderly distribution of chromosomes to the daughter cell nuclei. A spindle has fibers that stretch between two poles (ends). Spindle fibers are bundles of microtubules, protein cylinders found in the cytoplasm that can assemble and disassemble. The **centrosome,** the main microtubule-organizing center of the cell, divides before mitosis so that during mitosis, each pole of the spindle has a centrosome. Animal cells contain two barrel-shaped organelles called centrioles in each centrosome and asters, arrays of short microtubules radiating from the poles (see Fig. 8.3). The fact that plant cells lack centrioles suggests that centrioles are not required for spindle formation.

Figure 8.3 Phases of mitosis in animal and plant cells.
The colors signify which chromosomes were inherited from each parent.

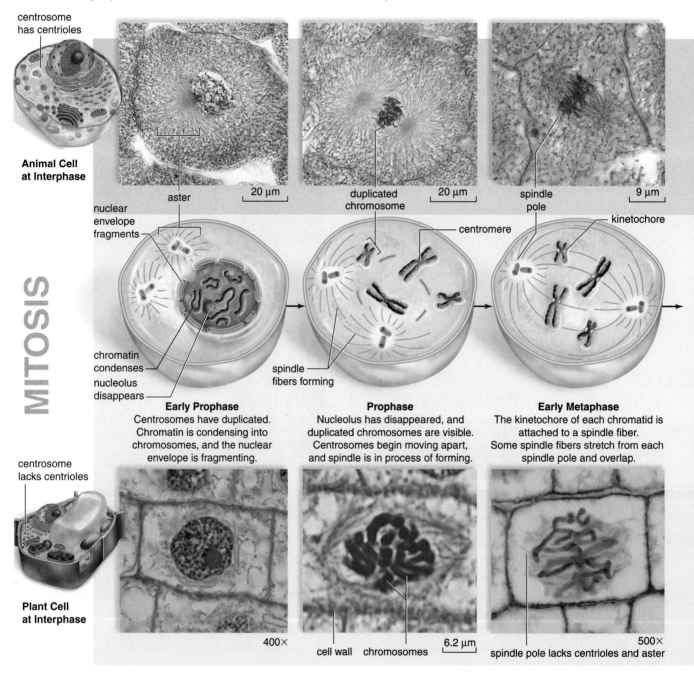

centrosome
has centrioles

**Animal Cell
at Interphase**

aster   20 μm

duplicated
chromosome   20 μm

spindle
pole   9 μm

nuclear
envelope
fragments

centromere

kinetochore

chromatin
condenses

nucleolus
disappears

spindle
fibers forming

Early Prophase
Centrosomes have duplicated.
Chromatin is condensing into
chromosomes, and the nuclear
envelope is fragmenting.

Prophase
Nucleolus has disappeared, and
duplicated chromosomes are visible.
Centrosomes begin moving apart,
and spindle is in process of forming.

Early Metaphase
The kinetochore of each chromatid is
attached to a spindle fiber.
Some spindle fibers stretch from each
spindle pole and overlap.

MITOSIS

centrosome
lacks centrioles

**Plant Cell
at Interphase**

400×

cell wall   chromosomes   6.2 μm

500×
spindle pole lacks centrioles and aster

The phases of mitosis are shown in Figure 8.3. Mitosis is the type of nuclear division that (1) occurs in the body (somatic) cells; (2) results in two daughter cells because there is only one round of division; and (3) keeps the chromosome number constant (same as the parent cell).

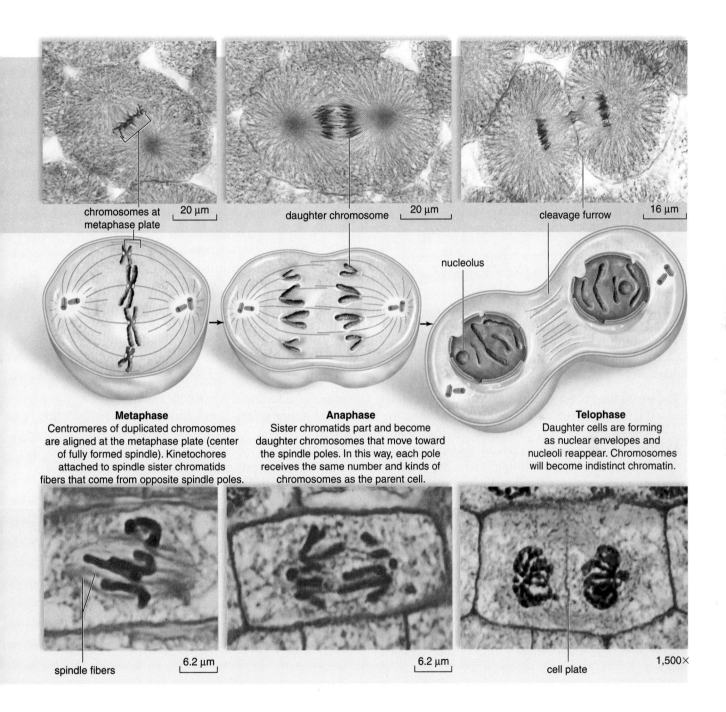

chromosomes at
metaphase plate   20 μm

daughter chromosome   20 μm

cleavage furrow   16 μm

nucleolus

Metaphase
Centromeres of duplicated chromosomes
are aligned at the metaphase plate (center
of fully formed spindle). Kinetochores
attached to spindle sister chromatids
fibers that come from opposite spindle poles.

Anaphase
Sister chromatids part and become
daughter chromosomes that move toward
the spindle poles. In this way, each pole
receives the same number and kinds of
chromosomes as the parent cell.

Telophase
Daughter cells are forming
as nuclear envelopes and
nucleoli reappear. Chromosomes
will become indistinct chromatin.

spindle fibers   6.2 μm

6.2 μm

cell plate   1,500×

Mitosis and Meiosis   Laboratory 8   **77**

Animal Mitosis Models

1. Using Figure 8.3 as a guide, identify the phases of animal cell mitosis in models of animal cell mitosis.

2. Each species has its own chromosome number. Counting the number of centromeres tells you the number of chromosomes in the models. What is the number of chromosomes in each

   of the cells in this model series? _____

Whitefish Blastula Slide

The blastula is an early embryonic stage in the development of animals. The **blastomeres** (blastula cells) shown are in different phases of mitosis (see Fig. 8.3).

1. Examine a prepared slide of whitefish blastula cells undergoing mitotic cell division.

2. Try to find a cell in each phase of mitosis. Have a partner or your instructor check your identification.

Observation: Plant Mitosis

Plant Mitosis Models

1. Identify the phases of plant cell mitosis using models of plant cell mitosis and Figure 8.3 as a guide.

2. Notice that plant cells do not have centrioles and asters. Plant cells do have centrosomes and this accounts for the formation of a spindle.

3. What is the number of chromosomes in each of the cells in this model series? _____

Onion Root Tip Slide

1. In plants, the root tip contains tissue that is continually dividing and producing new cells. Examine a prepared slide of onion root tip cells undergoing mitotic cell division. Try to find the phases that correspond to those shown in Figure 8.3.

2. Using high power, focus up and down on a cell in telophase. You may be able to just make out the cell plate, the region where a plasma membrane is forming between the two prospective daughter cells. Later, cell walls appear in this area.

Cytokinesis

Cytokinesis, division of the cytoplasm, usually accompanies mitosis. During cytokinesis, each daughter cell receives a share of the organelles that duplicated during interphase. Cytokinesis begins in anaphase, continues in telophase, and reaches completion by the start of the next interphase.

Cytokinesis in Animal Cells

In animal cells, a **cleavage furrow,** an indentation of the membrane between the daughter nuclei, begins as anaphase draws to a close (Fig. 8.4). The cleavage furrow deepens as a band of actin filaments called the contractile ring slowly constricts the cell, forming two daughter cells.

Were any of the cells of the whitefish blastula slide undergoing cytokinesis? _____

How do you know? _____

_____

_____

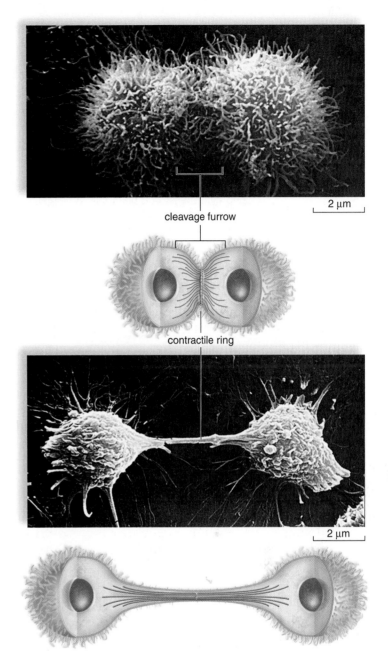

2 μm

cleavage furrow

contractile ring

2 μm

Figure 8.4 **Cytokinesis in animal cells.**
A single cell becomes two cells by a furrowing process. A contractile ring composed of actin filaments gradually gets smaller, and the cleavage furrow pinches the cell into two cells.
Copyright by R. G. Kessel and C. Y. Shih, *Scanning Electron Microscopy in Biology: A Students' Atlas on Biological Organization,* Springer-Verlag, 1974.

Cytokinesis in Plant Cells

After mitosis, the cytoplasm divides by cytokinesis. In plant cells, membrane vesicles derived from the Golgi apparatus migrate to the center of the cell and form a **cell plate** (Fig. 8.5), the location of a new plasma membrane for each daughter cell. Later, individual cell walls appear in this area.

Were any of the cells of the onion root tip slide undergoing cytokinesis? _____

How do you know? _____

Figure 8.5 Cytokinesis in plant cells.
During cytokinesis in a plant cell, a cell plate forms midway between two daughter nuclei and extends to the plasma membrane. Vesicles containing cell wall components fuse to form cell plate.

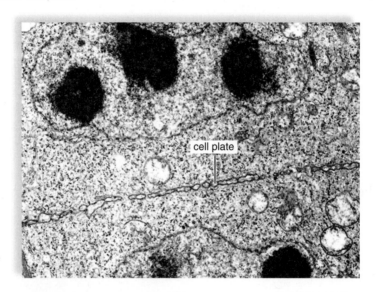

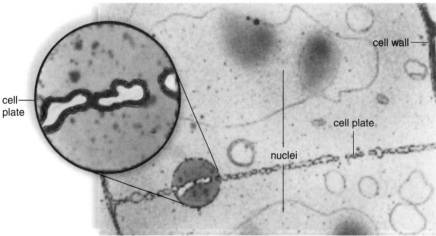

Summary of Mitotic Cell Division

1. The nuclei in the daughter cells have the _____ number of chromosomes as the parent cell had.

2. Mitosis is cell division in which the chromosome number _____.

8.2 Meiosis

Meiosis is a form of nuclear division in which the chromosome number is reduced by half. While the nucleus of the parent cell has the diploid number of chromosomes, the daughter nuclei, after meiosis is complete, have the haploid number of chromosomes. In sexually reproducing species, meiosis must occur or the chromosome number would double with each generation.

A diploid nucleus contains **homologues,** also called homologous chromosomes. Homologues look alike and carry the genes for the same traits. Before meiosis begins, the chromosomes are already double stranded—that is, they contain sister chromatids.

Experimental Procedure: Meiosis

In this exercise, you will use pop beads to construct homologues and move the chromosomes to simulate meiosis.

Building Chromosomes to Simulate Meiosis

1. Obtain the following materials: 48 pop beads of one color (e.g., red) and 48 pop beads of another color (e.g., blue) for a total of 96 beads; eight magnetic centromeres; and four centriole groups.
2. Build a homologous pair of duplicated chromosomes using Figure 8.6a as a guide. Each chromatid will have 16 beads. Be sure to bring the centromeres of two units of the same color together so that they attract and link to form one duplicated chromosome. (One member of the pair will be red, and the other will be blue.)
3. Build another homologous pair of duplicated chromosomes using Figure 8.6b as a guide. Each chromatid will have eight beads. Be sure to bring the centromeres of two units of the same color together so that they attract. (One member of the pair will be red, and the other will be blue.)
4. Note that your chromosomes are the same as those in Figure 8.6. The red chromosomes were inherited from one parent, and the blue chromosomes were inherited from the other parent.

Figure 8.6 Two pairs of homologues.
The chromosomes of these homologous pairs are duplicated.

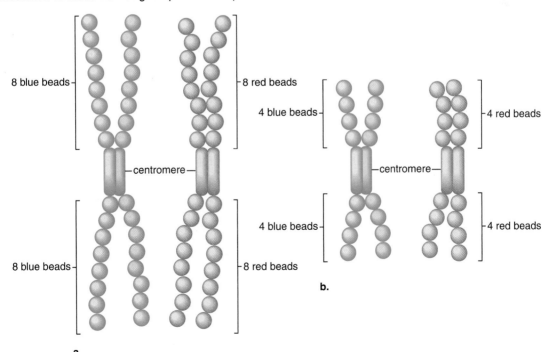

Meiosis I

Meiosis requires two nuclear divisions, and the first round during meiosis is called **meiosis I** (see Fig. 8.7).

During prophase of meiosis I, the spindle appears while the nuclear envelope and nucleolus disappear. Homologues line up next to one another during a process called synapsis. During **crossing-over,** the nonsister chromatids of a homologue pair exchange genetic material. At metaphase I, the homologue pairs line up at the metaphase plate of the spindle. During anaphase I, homologues separate and the chromosomes (still composed of two chromatids) move to each pole. In telophase I, the nuclear envelope and the nucleolus reappear as the spindle disappears. Each new nucleus contains one from each pair of chromosomes.

Prophase I

5. Using Figure 8.7 as a guide, put all four of the chromosomes you built in the center of your work area, which represents the nucleus. Place two pairs of centrioles outside the nucleus.
6. Separate the pairs of centrioles, and move one pair to opposite poles of the nucleus.
7. **Synapsis** is the pairing of homologues during prophase I. Simulate synapsis by bringing the homologues together.
8. **Crossing-over** is an exchange of genetic material between two homologues. It is a way to achieve genetic recombination during meiosis. Simulate crossing-over by exchanging the exact segments of two nonsister chromatids of a single homologous pair. Why use nonsister chromatids and not

    sister chromatids? _____

Metaphase I

Position the homologues at the metaphase plate in such a way that the homologues are prepared to move apart toward the centrioles.

Anaphase I

Separate the homologues, and move each one toward the opposite pole.

Telophase I

9. During telophase I, the chromosomes are at the poles. What combinations of chromosomes are at the poles? Fill in the following blanks with the words *red-long, red-short, blue-long,* and *blue-short:*

    Pole A: _____ and _____

    Pole B: _____ and _____
10. What other combinations would have been possible? (*Hint:* Alternate the colors at metaphase I.)

    Pole A: _____ and _____

    Pole B: _____ and _____

Conclusions

*   Do the chromosomes inherited from the mother or father have to remain together following

    meiosis I? _____
*   Name two ways that meiosis contributes to genetic variation:

    a. _____

    b. _____

Interkinesis

Interkinesis is the period between meiosis I and meiosis II. In some species, daughter cells do not form, and meiosis II follows right after meiosis I. *Does DNA replication occur during interkinesis?* _____ Explain. _____

Meiosis II

The second round of nuclear division during meiosis is called **meiosis II** (see Fig. 8.7).

During prophase of meiosis II, a spindle appears. Each chromosome attaches to the spindle individually. During metaphase II, the chromosomes are lined up at the metaphase plate. During anaphase II, the centromeres divide and the chromatids separate, becoming daughter chromosomes that move toward the poles. In telophase II, the spindle disappears as the nuclear envelope reappears. Notice that meiosis II is exactly like mitosis except that the nucleus of the parent cell and the daughter nuclei are haploid.

Prophase II

1. Using Figure 8.7 as a guide, choose the chromosomes from one pole (see #9, page 82) to represent those in the parent nucleus undergoing meiosis II.
2. Place two pairs of centrioles at opposite sides of these chromosomes to form the new spindle.

Metaphase II

Move the duplicated chromosomes to the metaphase II metaphase plate. *How many chromosomes are at the metaphase II metaphase plate?* _____

Anaphase II

Pull the two magnets of each duplicated chromosome apart. *What does this action represent?* _____

_____

Telophase II

Put the chromosomes—each having one chromatid—at the poles near the centrioles.

Conclusions

- You chose only one daughter nucleus from meiosis I to be the nucleus that divides. In reality both daughter nuclei go on to divide again. Therefore, how many nuclei are usually present when meiosis II is complete? _____

- In this exercise, how many chromosomes were in the parent cell nucleus undergoing meiosis II?

  _____

- How many chromosomes are in the daughter nuclei? _____ Explain. _____

  _____

Summary of Meiotic Cell Division

1. The parent cell has the diploid (2n) number of chromosomes, and the daughter cells have the _____ (n) number of chromosomes.
2. Meiosis is cell division in which the chromosome number _____ .
   Whereas meiosis reduces the chromosome number, fertilization restores the chromosome number.
3. A zygote contains the same number of chromosomes as the parent, but are these exactly the same chromosomes? _____
4. What is another way that sexual reproduction results in genetic variation? _____

   _____

Meiosis Phases

Figure 8.7 Meiosis I and II in plant cell micrographs and animal cell drawings. Crossing-over occurred during meiosis I.

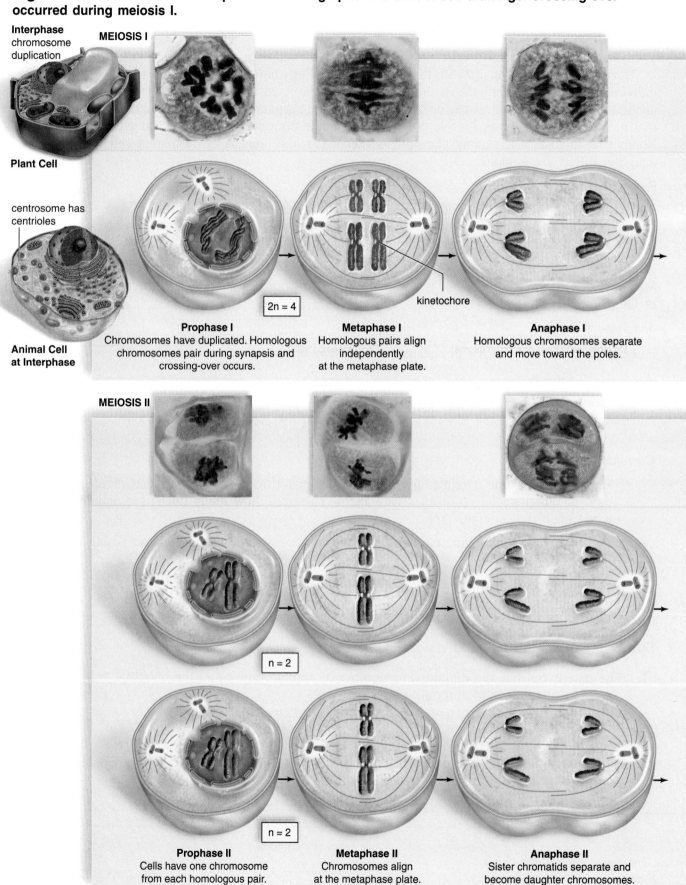

Interphase chromosome duplication

Plant Cell

centrosome has centrioles

Animal Cell at Interphase

MEIOSIS I

2n = 4

kinetochore

Prophase I
Chromosomes have duplicated. Homologous chromosomes pair during synapsis and crossing-over occurs.

Metaphase I
Homologous pairs align independently at the metaphase plate.

Anaphase I
Homologous chromosomes separate and move toward the poles.

MEIOSIS II

n = 2

n = 2

Prophase II
Cells have one chromosome from each homologous pair.

Metaphase II
Chromosomes align at the metaphase plate.

Anaphase II
Sister chromatids separate and become daughter chromosomes.

The phases of meiosis are shown in Figure 8.7. Meiosis is the type of nuclear division that (1) occurs in the sex organs (testes and ovaries); (2) results in four cells because there are two rounds of cell division; and (3) reduces the chromosome number to half that of the parent cell.

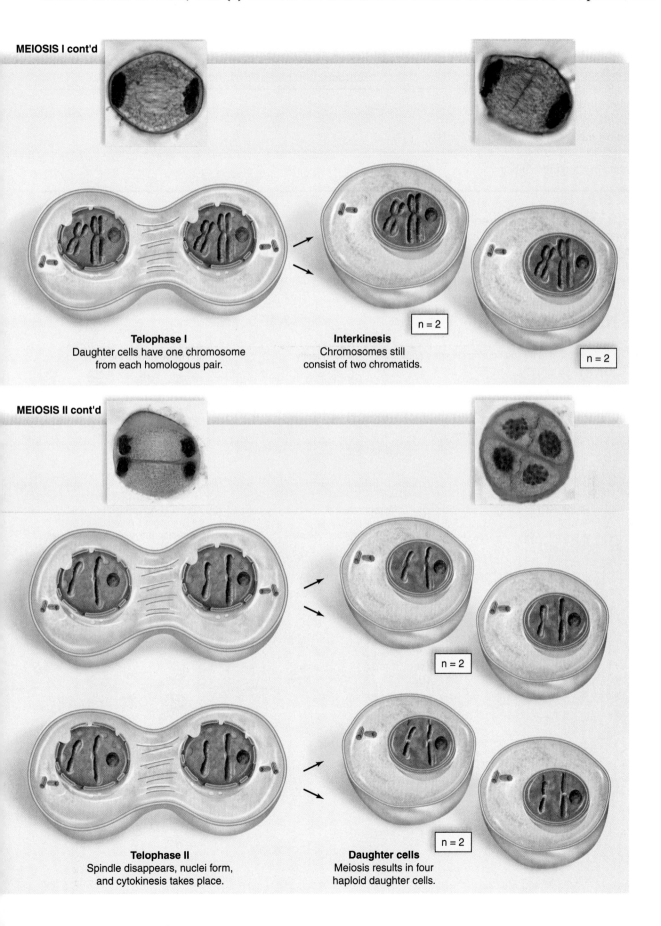

MEIOSIS I cont'd

Telophase I
Daughter cells have one chromosome
from each homologous pair.

Interkinesis
Chromosomes still
consist of two chromatids.

n = 2

n = 2

MEIOSIS II cont'd

n = 2

Telophase II
Spindle disappears, nuclei form,
and cytokinesis takes place.

Daughter cells
Meiosis results in four
haploid daughter cells.

n = 2

8.3 Mitosis Versus Meiosis

In comparing mitosis to meiosis it is important to note that meiosis requires two nuclear divisions but mitosis requires only one nuclear division. Therefore, mitosis produces two daughter cells and

Table 8.2	Differences Between Mitosis and Meiosis		
		Mitosis	Meiosis
1.	Number of divisions		
2.	Chromosome number in daughter cells		
3.	Number of daughter cells		

Figure 8.8 Meiosis I compared to mitosis.

Compare metaphase I of meiosis to metaphase of mitosis. Only in metaphase I are the homologous chromosomes paired at the metaphase plate. Members of homologous chromosome pairs separate during anaphase I, and therefore the daughter cells are haploid. The blue chromosomes were inherited from one parent, and the red chromosomes were inherited from the other parent. The exchange of color between nonsister chromatids represents the crossing-over that occurred during meiosis I.

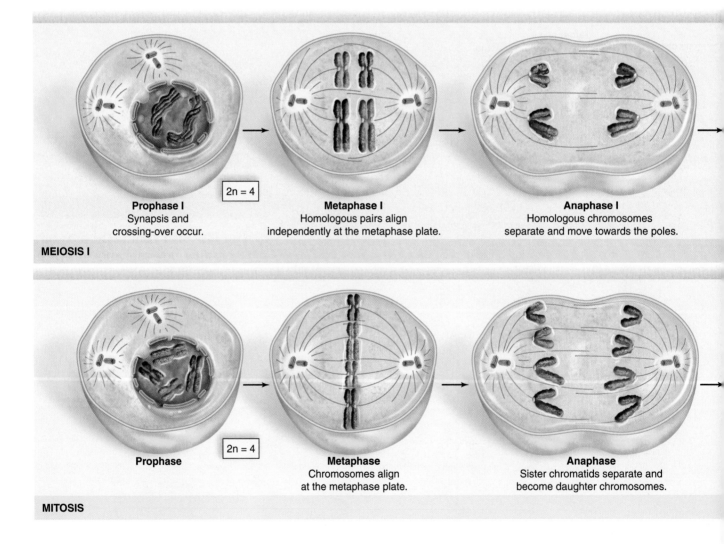

MEIOSIS I

| **Prophase I** | **Metaphase I** | **Anaphase I** |
| Synapsis and crossing-over occur. | Homologous pairs align independently at the metaphase plate. | Homologous chromosomes separate and move towards the poles. |

2n = 4

MITOSIS

| **Prophase** | **Metaphase** | **Anaphase** |
| | Chromosomes align at the metaphase plate. | Sister chromatids separate and become daughter chromosomes. |

2n = 4

meiosis produces four daughter cells. Following mitosis, the daughter cells are still diploid, but following meiosis, the daughter cells are haploid. Figure 8.8 explains why. Fill in Table 8.2 to indicate general differences between mitosis and meiosis and complete Table 8.3 to indicate specific differences between mitosis and meiosis I. Mitosis need only be compared with meiosis I because the exact events occur during both mitosis and meiosis II except that the cells are diploid during mitosis and haploid during meiosis II.

Table 8.3	Mitosis Compared with Meiosis I
Mitosis	**Meiosis I**
Prophase: No pairing of chromosomes	Prophase I: _____
Metaphase: Duplicated chromosomes at metaphase plate	Metaphase I: _____
Anaphase: Sister chromatids separate	Anaphase I: _____
Telophase: Chromosomes have one chromatid	Telophase I: _____

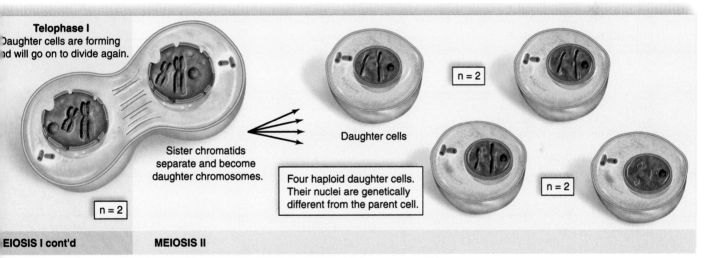

Telophase I
Daughter cells are forming and will go on to divide again.

n = 2

Sister chromatids separate and become daughter chromosomes.

Daughter cells

n = 2

n = 2

Four haploid daughter cells. Their nuclei are genetically different from the parent cell.

MEIOSIS I cont'd   MEIOSIS II

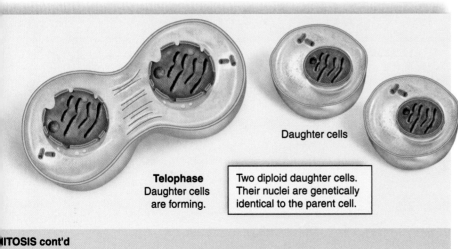

Daughter cells

Telophase
Daughter cells are forming.

Two diploid daughter cells. Their nuclei are genetically identical to the parent cell.

MITOSIS cont'd

8.4 The Life Cycle of Human Beings

The term **life cycle** in sexually reproducing organisms refers to all the reproductive events that occur from one generation to the next. The human life cycle involves both mitosis and meiosis **(Fig. 8.9).**

During development and after birth, mitosis is involved in the continued growth of the child and the repair of tissues at any time. As a result of mitosis, each somatic (body) cell has the diploid number of chromosomes (2n).

During gamete formation, meiosis reduces the chromosome number from the diploid to the haploid number (n) in such a way that the gametes (sperm and egg) have one chromosome derived from each pair of homologues. In males, meiosis is a part of **spermatogenesis,** which occurs in the testes and produces sperm. In females, meiosis is a part of **oogenesis,** which occurs in the ovaries and produces eggs. After the sperm and egg join during fertilization, the resulting cell, called the **zygote,** has homologous pairs of chromosomes. The zygote then undergoes mitosis with differentiation of cells to become a fetus, and eventually a new human being.

Meiosis keeps the number of chromosomes constant between the generations, and it also, as we have seen, causes the gametes to be different from one another. Therefore, due to sexual reproduction, there are more variations among individuals.

Complete Figure 8.9 by placing either the term *mitosis* or *meiosis* in the boxes.

Summary of Human Life Cycle

1. Complete Figure 8.9 by placing either the term *mitosis* or *meiosis* in the boxes.
2. Fill in this flow chart based on Figure 8.9 to ensure your understanding of the role of meiosis and mitosis in humans.

	Male	Female
Name of organ that produces gametes	——————	——————
Name of process that produces gametes	——————	——————
Type of cell division involved in process	——————	——————
Name of gamete	——————	——————
Number of chromosomes in gamete	——————	——————

Fertilization

Results of fertilization	——————
Number of chromosomes	——————
Type of cell division needed for growth	——————
Result of growth	——————

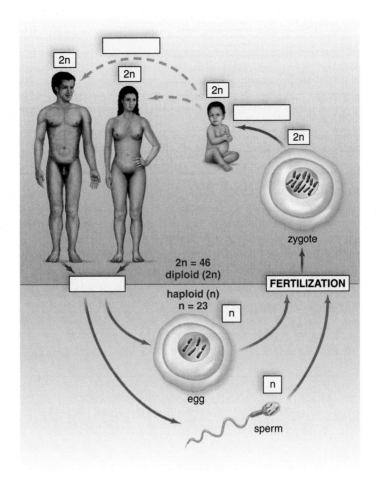

Figure 8.9 Life cycle of humans.

3. Match these definitions with the terms in Fig. 8.9.

   a. Female gamete with the n number of chromosomes _____

   b. Type of cell division that keeps the chromosome number the same and occurs during growth and repair _____

   c. First cell that results after fertilization _____

   d. Type of cell division that reduces the chromosome number and occurs during gamete formation _____

   e. Half the diploid number of chromosomes _____

   f. Male gamete with the n number of chromosomes _____

1. Do the chromosomes have one or two chromatids as they move toward the poles

   a. during anaphase of mitosis? _____

   b. during anaphase of meiosis I? _____

   c. during anaphase of meiosis II? _____

2. Asexual reproduction of a haploid protozoan can be described as n → n. Explain. _____
   _____

3. Explain why furrowing is a suitable mechanism for cytokinesis of animal cells but not plant cells.
   _____

4. A student is simulating meiosis I with chromosomes that are red-long and yellow-long; red-short and yellow-short. Why would you *not* expect to find both red-long and yellow-long in one resulting daughter cell? _____
   _____

5. Assume that you have built a homologous pair of chromosomes, each having two chromatids. One homologue contains red beads, while the other contains yellow beads. Describe the appearance of two nonsister chromatids following crossing-over. _____
   _____

6. What is the function of mitosis and meiosis in the human life cycle? _____
   _____

9

Mendelian Genetics

Introduction

Gregor Mendel, sometimes called the "father of genetics," formulated the basic laws of genetics examined in this laboratory. He determined that individuals have two alternate forms of a gene (two **alleles,** in modern terminology) for each trait in their body cells. Today, we know that alleles are on the chromosomes (Fig. 9.1). An individual can be homozygous dominant (two dominant alleles, *GG*), homozygous recessive (two recessive alleles, *gg*), or heterozygous (one dominant and one recessive allele, *Gg*). **Genotype** refers to an individual's genes, while **phenotype** refers to an individual's appearance. Homozygous dominant and heterozygous individuals show the dominant phenotype; homozygous recessive individuals show the recessive phenotype.

Assume that the chromosomes shown in Figure 9.1 are those of a fruit fly, and *G* = gray body and *g* = ebony (black) body. What is the genotype of the fly?

_____

What is the phenotype of this fly? _____

Explain. _____

_____

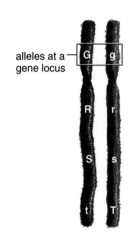

Figure 9.1
Gene locus.
Each allelic pair, such as *Gg*, is located on a pair of chromosomes at a particular gene locus.

9.1 One-Trait Crosses

A single pair of alleles is involved in one-trait crosses. Mendel found that reproduction between two heterozygous individuals *(Aa),* called a **monohybrid cross,** results in both dominant and recessive phenotypes among the offspring. The expected phenotypic ratio among the offspring was 3:1. Three offspring had the dominant phenotype for every one that had the recessive phenotype.

Mendel realized that these results were obtainable only if the alleles of each parent segregated (separated from each other) during meiosis (otherwise, all offspring would inherit a dominant allele, and no offspring would be homozygous recessive). Therefore, Mendel formulated his first law of inheritance:

Law of Segregation
Each organism contains two alleles for each trait, and the alleles segregate during the formation of gametes. Each gamete then contains only one allele for each trait. When fertilization occurs, the new organism has two alleles for each trait, one from each parent.

Inheritance is a game of chance. Just as there is a 50% probability of heads or tails when tossing a coin, there is a 50% probability that a sperm or egg will have an *A* or an *a* when the parent is *Aa.* The chance of an equal number of heads or tails improves as the number of tosses increases. In the same way, the chance of an equal number of gametes with *A* and *a* improves as the number of gametes increases. Therefore, the 3:1 ratio among offspring is more likely when a large number of sperm fertilize a large number of eggs.

Color of Tobacco Seedlings

In tobacco plants, a dominant allele *(C)* for chlorophyll gives the plants a green color, and a recessive allele *(c)* for chlorophyll causes a plant to appear white. If a tobacco plant is homozygous for the recessive allele *(c),* it cannot manufacture chlorophyll and thus appears white (Fig. 9.2).

Figure 9.2 **Monohybrid cross.**
These tobacco seedlings are growing on an agar plate. The white plants cannot manufacture chlorophyll.

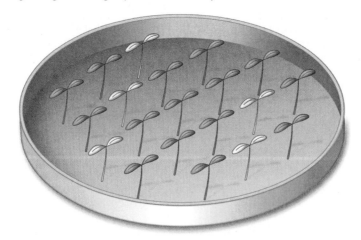

Experimental Procedure: Color of Tobacco Seedlings

1. Obtain a numbered agar plate on which tobacco seedlings are growing. They are the offspring of the cross *Cc* × *Cc.* Complete the Punnett square.

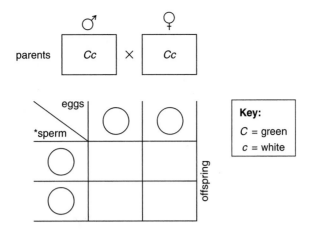

parents: Cc × Cc

eggs / *sperm → offspring

Key:
C = green
c = white

What is the expected phenotypic ratio? _____

2. Using a binocular dissecting microscope, view the seedlings, and count the number that are green and the number that are white. Record the plate number and your results in Table 9.1.

3. Repeat steps 1 and 2 for two additional plates. Total the number that are green and the number that are white.

4. Complete Table 9.1 by recording the class data.

Table 9.1	Color of Tobacco Seedlings	
	Number of Offspring	
	Green Color	White Color
Plate # _____		
Plate # _____		
Plate # _____		
Totals		
Class data		

Conclusions

- Calculate the actual phenotypic ratio you observed. _____ Do your results differ from the expected ratio? _____ Explain. _____

_____

- Repeat these steps using the class data. Do your class data give a ratio that is closer to the expected ratio? _____ Explain. _____

_____

*In flowering plants, the sperm are not flagellated.

Drosophila melanogaster Characteristics

Both the adults and the larvae of *Drosophila melanogaster* (the fruit fly) feed on plant sugars and on the wild yeasts that grow on rotting fruit. The female flies first lay **eggs** on these same materials. After a day or two, the eggs hatch into small **larvae** that feed and grow for about eight days, depending on the temperature. During this period, they **molt** twice. Therefore, three larval periods of growth, called **instars,** occur between molts. When fully grown, the third-stage instar larvae cease feeding and **pupate.** During pupation, which lasts about four days, the larval tissues are reorganized to form those of the adult. The life cycle is summarized in Figure 9.3.

Figure 9.3 **Life cycle of *Drosophila*.**
The adult female lays eggs that go through three larval stages before pupating. Metamorphosis, which occurs during pupation, produces the adult fly.

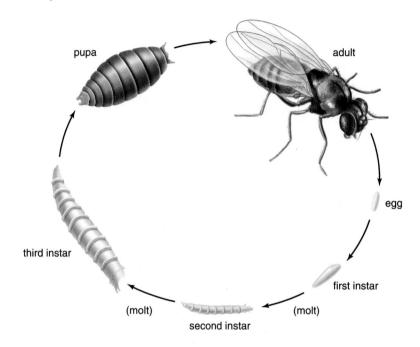

Observation: Drosophila melanogaster

Culture
You will be provided with a stock culture of *Drosophila melanogaster* to examine. Answer the following questions:

1. Where in the culture vial are the adult flies? _____

2. The eggs? _____

3. The larvae? _____

4. The pupae? _____

Flies
You will be provided with either slides, frozen flies, or live flies to examine. If the flies are alive, follow the directions on page 96 for using FlyNap™ to anesthetize them.

1. Put frozen or anesthetized flies on a white card, and use a camel-hair brush to move them around. Use a binocular dissecting microscope or a hand lens to see the flies clearly.
2. If you are looking at slides, you may use the scanning lens of your compound light microscope to view the flies.

3. Examine wild-type flies. Complete Table 9.2 for wild-type flies. Long wings extend beyond the body, and vestigial (short) wings do not extend beyond the body.

4. Examine mutant flies. Complete Table 9.2 for all the mutant flies you examined.

Table 9.2	Characteristics of Wild-Type and Mutant Flies				
	Wild-Type	**Ebony Body**	**Vestigial-Wing**	**Sepia-Eye**	**White-Eye**
Wing length					
Color of eyes					
Color of body					

5. Use the following characteristics to distinguish male flies from female flies (Fig. 9.4):

- The male is generally smaller.
- The male has a more rounded abdomen than the female. The female has a pointed abdomen.
- The male has sex combs on the forelegs.
- Dorsally, the male is seen to have a black-tipped abdomen, whereas the female appears to have dark lines only at the tip.
- Ventrally, the abdomen of the male has a dark region at the tip due to the presence of claspers; this dark region is lacking in the female.

Figure 9.4 *Drosophila* male versus *Drosophila* female.

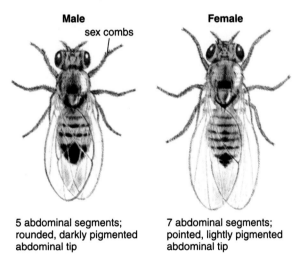

Male
sex combs

Female

5 abdominal segments;
rounded, darkly pigmented
abdominal tip

7 abdominal segments;
pointed, lightly pigmented
abdominal tip

Ventral Views of Abdomens

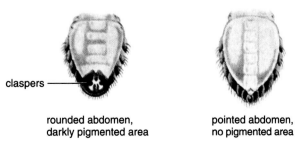

claspers

rounded abdomen,
darkly pigmented area

pointed abdomen,
no pigmented area

Anesthetizing Flies

The use of FlyNap™ allows flies to be anesthetized for at least 50 minutes without killing them.

1. Dip the absorbent end of a swab into the FlyNap™ bottle, as shown in Figure 9.5.
2. Tap the bottom of the culture vial on the tabletop to knock the flies to the bottom of the vial. (If the medium is not firm, you may wish to transfer the flies to an empty bottle before anesthetizing them. Ask your instructor for assistance, if this is the case.)
3. With one finger, push the plug slightly to one side. Remove the swab from the FlyNap™, and quickly stick the anesthetic end into the culture vial beside the plug so that the anesthetic tip is below the plug. Keep the culture vial upright with the swab in one place.
4. Remove the plug and the swab immediately after the flies are anesthetized (approximately 2 minutes in an empty vial, 4 minutes in a vial with medium), and spill the flies out onto a white file card. The length of time the flies remain anesthetized depends on the amount of FlyNap™ on the swab and on the number and age of the flies in the culture vial.
5. Transfer the anesthetized flies from the white file card onto the glass plate of a binocular dissecting microscope for examination, or use a hand lens.

Figure 9.5 **FlyNap™.**
Flies can be anesthetized for at least 50 minutes without being killed in the FlyNap™.

a. Moisten swab in FlyNap™.

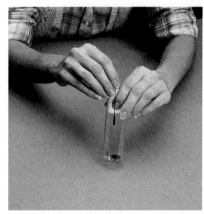
b. Insert swab into culture bottle.

c. Remove flies as shown.

Wing Length in *Drosophila*

In fruit flies, the allele for long wings *(L)* is dominant over the allele for vestigial (short) wings *(l)*. You will be examining the results of the cross *Ll* × *Ll*. Complete this Punnett square:

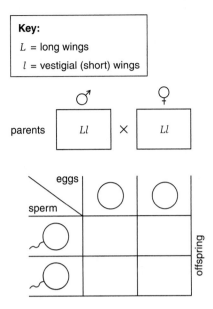

Key:

L = long wings

l = vestigial (short) wings

parents | ♂ Ll × ♀ Ll

eggs / sperm

offspring

What is the expected phenotypic ratio among the offspring? _____

Experimental Procedure: Wing Length in Drosophila

The cross described here will take three weeks. The experiment can be done in one week, however, if your instructor provides you with a vial that already contains the results of the cross *Ll* × *Ll*. In this case, proceed directly to step 3.

1. **Week 1:** Place heterozygous flies in a prepared culture vial. Your instructor will show you how to use instant medium and dry yeast to prepare the vial. Label your culture. What is the

   phenotype of heterozygous flies? _____

   What is the genotype of heterozygous flies? _____

2. **Week 2:** Remove the heterozygous flies from the vial before their offspring pupate. Why is it

   necessary to remove these flies before you observe your results? _____

   _____

3. **Week 3:** Observe the results of the cross by counting the offspring. Follow the directions on page 96 for anesthetizing and removing flies. When counting, use the binocular dissecting microscope or a hand lens. Divide your flies into those with long wings and those with vestigial (short) wings. Record your results and the class results in Table 9.3.

Table 9.3	Wing Length in *Drosophila*	
	Number of Offspring	
	Long Wings	Vestigial Wings
Your data		
Class data		

Conclusions

- Calculate the actual phenotypic ratio you observed. _____ Do your results differ from the expected ratio? _____ Explain. _____

- Repeat these steps using the class data. Do your class data give a ratio that is closer to the expected ratio? _____ Explain. _____

9.2 Two-Trait Crosses

Two-trait crosses involve two pairs of alleles. Mendel found that during a **dihybrid cross,** when two dihybrid individuals *(AaBb)* reproduce, the phenotypic ratio among the offspring is 9:3:3:1, representing four possible phenotypes. He realized that these results could only be obtained if the alleles of the parents segregated independently of one another when the gametes were formed. From this, Mendel formulated his second law of inheritance:

> **Law of Independent Assortment**
>
> Members of an allelic pair segregate (assort) independently of members of another allelic pair. Therefore, all possible combinations of alleles can occur in the gametes.

Color and Texture of Corn

In corn plants, the allele for purple kernel *(P)* is dominant over the allele for yellow kernel *(p)*, and the allele for smooth kernel *(S)* is dominant over the allele for rough kernel *(s)* (Fig. 9.6).

Figure 9.6 Dihybrid cross.
Four types of kernels are seen on an ear of corn following a dihybrid cross: purple smooth, purple rough, yellow smooth, and yellow rough.

20 mm

1. Obtain an ear of corn from the supply table. You will be examining the results of the cross *PpSs* × *PpSs*. Complete this Punnett square:

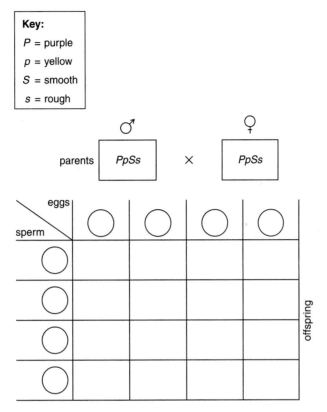

Key:
P = purple
p = yellow
S = smooth
s = rough

♂ parents PpSs × ♀ PpSs

eggs / sperm / offspring

What is the expected phenotypic ratio among the offspring? _____

2. Count the number of kernels of each possible phenotype listed in Table 9.4. Record the sample number and your results in Table 9.4. Use three samples, and total your results for all samples. Also record the class data.

Table 9.4	Color and Texture of Corn			
	Number of Kernels			
	Purple Smooth	**Purple Rough**	**Yellow Smooth**	**Yellow Rough**
Sample # _____				
Sample # _____				
Sample # _____				
Totals				
Class data				

Conclusions

- From your data, which two traits seem dominant? _____ and _____

  Which two traits seem recessive? _____ and _____

- Calculate the actual phenotypic ratio you observed. _____ Do your results differ from the

  expected ratio? _____ Explain. _____

- Repeat these steps using the class data. Do your class data give a ratio that is closer to the

  expected ratio? _____ Explain. _____

  _____

Wing Length and Body Color in *Drosophila*

In *Drosophila*, long wings *(L)* are dominant over vestigial (short) wings *(l)*, and gray body *(G)* is dominant over ebony (black) body *(g)*. You will be examining the results of the cross *LlGg* × *LlGg*. Complete this Punnett square:

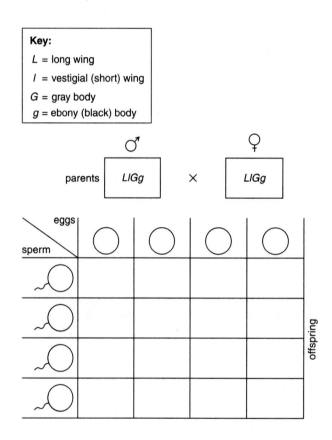

What are the expected phenotypic results of this cross? _____

The cross described here will take three weeks. The experiment can be done in one week if your instructor provides you with a vial that already contains the results of the cross *LlGg* × *LlGg*. In this case, proceed directly to step 3.

1. **Week 1:** Place heterozygous flies in a prepared culture vial. Your instructor will show you how to use instant medium and dry yeast to prepare the vial. Label your culture.

   What is the phenotype of heterozygous flies? _____

   What is the genotype of heterozygous flies? _____

2. **Week 2:** Remove the heterozygous flies from the vial before their offspring pupate. Why is it

   necessary to remove these flies before you observe your results? _____

3. **Week 3:** Observe the result of the cross by counting the offspring. Follow the standard directions (see page 96) for anesthetizing and removing flies. Find one fly of each phenotype, and check with the instructor that you have identified them correctly before proceeding. When counting, use the binocular dissecting microscope or a hand lens. Divide your flies into the groups indicated in Table 9.5, and record your results. Also record the class data.

Table 9.5	Wing Length and Body Color in *Drosophila*			
	Phenotypes			
	Long Wings Gray Body	Long Wings Ebony Body	Vestigial Wings Gray Body	Vestigial Wings Ebony Body
Number of offspring				
Class data				

Conclusions

- Calculate the actual phenotypic ratio you observed. _____ Do your results differ from the expected ratio? _____ Explain. _____

  _____

- Repeat these steps using the class data. Do your class data give a ratio that is closer to the expected ratio? _____ Explain. _____

  _____

9.3 X-Linked Crosses

In most animals, including fruit flies and humans, males usually have an X and Y chromosome while females have two X chromosomes. Some alleles, called X-linked alleles, occur only on the X chromosome. Males with a normal chromosome inheritance are never heterozygous for X-linked alleles and if they inherit a recessive X-linked allele, it will be expressed.

Red/White Eye Color in *Drosophila*

In fruit flies, red eyes (X^R) are dominant over white eyes (X^r). You will be examining the results of the cross $X^R Y \times X^R X^r$. Complete this Punnett square:

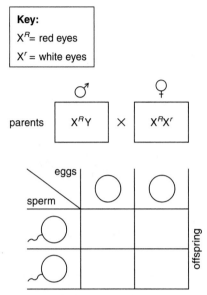

Key:
X^R = red eyes
X^r = white eyes

parents: $X^R Y$ ♂ × $X^R X^r$ ♀

What are the expected phenotypic results of this cross?

Females: _____ Males: _____

Experimental Procedure: Red/White Eye Color in Drosophila

The cross described here will take three weeks. The experiment can be done in one week if your instructor provides you with a vial that already contains the results of the cross $X^R Y \times X^R X^r$. In this case, proceed directly to step 3.

1. **Week 1:** Place the parental flies ($X^R Y \times X^R X^r$) in a prepared culture vial. Your instructor will show you how to use instant medium and dry yeast to prepare the vial. Label your culture.

   What is the phenotype of the female and male flies you are using? _____

   What is the genotype of the female flies? _____

   What is the genotype of the male flies? _____

2. **Week 2:** Remove the parental flies from the vial before their offspring pupate.

   Why is it necessary to remove these flies before you observe your results? _____

   _____

3. **Week 3:** Observe the results of the cross by counting the offspring. Follow the standard directions (see page 96) for anesthetizing and removing flies. When counting, use the binocular dissecting microscope or a hand lens. Divide your flies into the following groups: (1) red-eyed males, (2) red-eyed females, (3) white-eyed males, and (4) white-eyed females. Record your results in Table 9.6. Also record the class data.

Table 9.6	Red/White Eye Color in *Drosophila*	
	Number of Offspring	
Your data	**Red Eyes**	**White Eyes**
Males		
Females		
Class data		
Males		
Females		

Conclusions

- Calculate the actual phenotypic ratio you observed for males and females separately.

  Males: _____

  Females: _____

- Do your results differ from the expected results? _____

- Repeat these steps using the class data. Do your class data give a ratio that is closer to the expected ratio? _____ Explain. _____

  _____

1. If offspring exhibit a 3:1 phenotypic ratio, what are the genotypes of the parents? _____

2. In fruit flies, which of the characteristics that you studied was X-linked? _____

3. If offspring exhibit a 9:3:3:1 phenotypic ratio, what are the genotypes of the parental generation?

   _____

   _____

4. If the F_2 generation consists of 90 long-winged flies to 30 vestigial-winged flies, what was the phenotype of the F_1 flies? _____

   _____

5. Briefly describe the life cycle of *Drosophila*. _____

   _____

6. When doing a genetic cross, why is it necessary to remove parent flies before the pupae have hatched?

   _____

7. What is the genotype of a white-eyed male fruit fly? _____

8. Suppose you count 40 green tobacco seedlings and 2 white tobacco seedlings in one agar plate. Do your results show that both parent plants were heterozygous for the color allele? _____
   Explain. _____

9. Suppose you count tobacco seedlings in six agar plates, and your data are as follows: 125 green plants and 39 white plants. What is the phenotypic ratio? _____

   _____

10. Suppose that students in the laboratory periods before you removed some of the purple and yellow corn kernels on the ears of corn as they were performing the Experimental Procedure. What specific effects would this have on your results?

    _____

    _____

10

Human Genetics

Introduction

As shown in Figure 9.1, genes are arranged in a linear fashion along the chromosomes. A zygote receives 23 pairs of chromosomes when the gametes unite during fertilization. One of each pair is inherited from the male parent and the other from the female parent. Thereafter, due to the process of cell division, each body cell contains copies of these chromosomes in the nucleus. Twenty-two pairs of the chromosomes are **autosomes,** and one pair is the **sex chromosomes.** Males normally receive an X and Y chromosome, while females receive two X chromosomes. The sex chromosomes carry genes, just as the autosomes do. Some of these genes determine the sex of the individual (that is, whether the individual has testes or ovaries), but most of the genes on the sex chromosomes control traits unrelated to gender. They are called sex-linked genes because they are on the sex chromosomes. Most of the sex-linked genes are on the X chromosomes, called X-linked genes.

In this laboratory, we will study chromosomal inheritance and genetic inheritance in humans. We will see that the same rules of inheritance apply to all organisms, whether they are plants, fruit flies, or humans. Chromosomal inheritance has a marked effect on the general anatomy and physiology of the individual. If by chance a human has one less chromosome than usual, called a monosomy, or one more chromosome than usual, called a trisomy, a syndrome results. A **syndrome** is a group of symptoms that appear together and tend to indicate the presence of a particular disorder. Monosomies and trisomies arise due to nondisjunction (nonseparation) of chromosomes during gametogenesis. Gametogenesis in females is called *oogenesis* (egg production) and in males is called *spermatogenesis* (sperm production).

10.1 Chromosomal Inheritance

To view an individual's chromosomal inheritance, cells can be treated and photographed just prior to division. Before birth, cells can be obtained by amniocentesis, a procedure in which a physician uses a long needle to withdraw a portion of the amniotic fluid containing fetal cells. In chorionic villi sampling, cells are removed from the chorion. In any case, the fetal cells are cultured, and then a karyotype of the chromosomes is prepared (Fig. 10.1). Karyotypes can also be done using white blood cells from an adult. A karyotype displays and numbers the homologous chromosomes plus the sex chromosomes. In the male karyotype, it is possible to see that the **X chromosome** is the larger and the **Y chromosome** is the smaller of the sex chromosomes.

Figure 10.1 **Human karyotype preparation.**

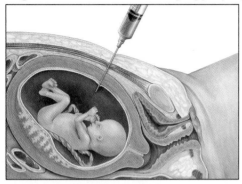

a. During amniocentesis, a long needle is used to withdraw amniotic fluid containing fetal cells.

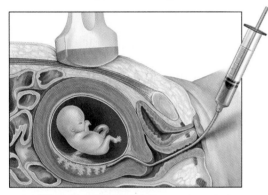

b. During chorionic villi sampling, a suction tube is used to remove cells from the chorion, where the placenta will develop.

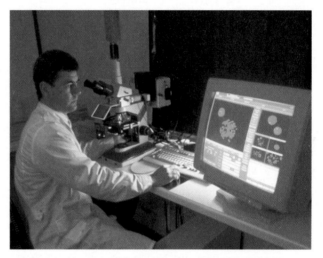

c. Cells are microscopically examined and photographed. A computer is used to arrange chromosomes by pairs.

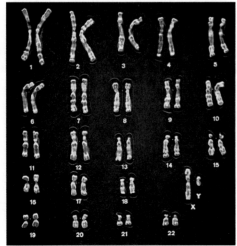

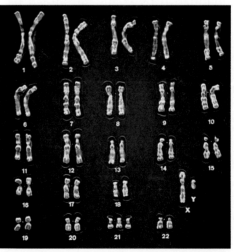

d. Normal male karyotype with 46 chromosomes.

Down syndrome karyotype with an extra chromosome 21.

Numerical Chromosome Abnormalities

Gamete formation in humans involves meiosis, the type of cell division that reduces the chromosome number by one-half because the homologues separate during meiosis I. When homologues fail to separate during meiosis I, called **nondisjunction,** gametes with too few (n − 1) or too many (n + 1) chromosomes result. Nondisjunction can also occur during meiosis II if the chromatids fail to separate and the daughter chromosomes go into the same daughter cell.

Down syndrome (Trisomy 21)

Down syndrome, the most common autosomal trisomy in humans, is due to the inheritance of three number 21 chromosomes (Fig. 10.1*d*). Therefore, the syndrome is also called trisomy 21. Usually, nondisjunction occurs during oogenesis, and the egg contains two instead of one chromosome 21.

A person with Down syndrome is short; has an eyelid fold; a flat face; stubby fingers; a large, fissured tongue; and, unfortunately, mental retardation, which can sometimes be severe.

Sex Chromosome Abnormalities

A female with **Turner syndrome** (XO) has only one sex chromosome, an X chromosome; the O signifies the absence of the second sex chromosome. Because the ovaries never become functional, these females do not undergo puberty or menstruation, and their breasts do not develop. Generally, females with Turner syndrome have a short build, folds of skin on the back of the neck, difficulty recognizing various spatial patterns, and normal intelligence. With hormone supplements, they can lead fairly normal lives.

When an egg having two X chromosomes is fertilized by an X-bearing sperm, an individual with **poly-X syndrome** results. The body cells have three X chromosomes and therefore 47 chromosomes. Although they tend to have learning disabilities, poly-X females have no apparent physical abnormalities, and many are fertile and have children with a normal chromosome count.

When an egg having two X chromosomes is fertilized by a Y-bearing sperm, a male with **Klinefelter syndrome** results. This individual is male in general appearance, but the testes are underdeveloped, and the breasts may be enlarged. The limbs of XXY males tend to be longer than average, muscular development is poor, body hair is sparse, and many XXY males have learning disabilities.

Jacob syndrome can be due to nondisjunction during meiosis II of spermatogenesis. These males are usually taller than average, suffer from persistent acne, and tend to have speech and reading problems. At one time, it was suggested that XYY males were likely to be criminally aggressive, but the incidence of such behavior has been shown to be no greater than that among normal XY males.

Complete Table 10.1 to show how a physician would recognize each of these syndromes from a karyotype.

Table 10.1	Numerical Sex Chromosome Abnormalities	
Syndrome	**Comparison With Normal Number**	
Turner:	Normal male:	Normal female:
Poly-X:	Normal male:	Normal female:
Klinefelter:	Normal male:	Normal female:
Jacob:	Normal male:	Normal female:

Building the Chromosomes

1. Obtain the following materials: 36 red pop beads, 26 blue (or green) pop beads, and 8 magnetic centromeres.
2. Build four duplicated sex chromosomes (3 Xs and 1 Y) as follows:

   Two red X chromosomes: Each chromatid will have nine red pop beads. Place the centromeres so that two beads are above each centromere and seven beads are below each centromere. Bring the centromeres together.

   One blue X chromosome: Each chromatid will have nine blue pop beads. Place the centromere so that two beads are above each centromere and seven beads are below each centromere. Bring the centromeres together.

   Y chromosome: Each chromatid will have four blue pop beads. Place the centromeres so that two beads are above each centromere and two beads are below each centromere. Bring the centromeres together.

Simulating Meiosis During Normal Oogenesis

Place one blue X and one red X chromosome together in the middle of your work area. (The blue X chromosome came from the father, and the red X chromosome came from the mother.) Have the chromosomes go through meiosis I and meiosis II. What is the chromosome constitution of each of the four meiotic products? Each egg has _____ (number) _____ (type) chromosome(s).

Simulating Meiosis During Normal Spermatogenesis

Place a red X and a blue Y chromosome together in the middle of your work area. (The red X chromosome came from the mother, and the blue Y chromosome came from the father.) Have the chromosomes go through meiosis I and meiosis II. What is the chromosome constitution of each of the four meiotic products? Two sperm have one _____ chromosome, and two sperm have one

_____ chromosome.

Simulating Fertilization

In the Punnett square provided here, fill in the products of fertilization using the *type* of gamete that resulted from normal oogenesis and the *types* of gametes that resulted from normal spermatogenesis. (Disregard the color of the chromosomes.)

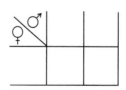

---

*Exercise courtesy of Victoria Finnerty, Sue Jinks-Robertson, and Gregg Orloff of Emory University, Atlanta, GA.

Figure 10.2 Nondisjunction of sex chromosomes.

a. Nondisjunction during oogenesis produces the types of eggs shown. Fertilization with normal sperm results in the syndromes noted. (Nonviable means existence is not possible.) **b.** Nondisjunction during meiosis I of spermatogenesis results in the types of sperm shown. Fusion with normal eggs results in the syndromes noted. Nondisjunction during meiosis II of spermatogenesis results in the types of sperm shown. Fusion with normal eggs results in the syndromes shown.

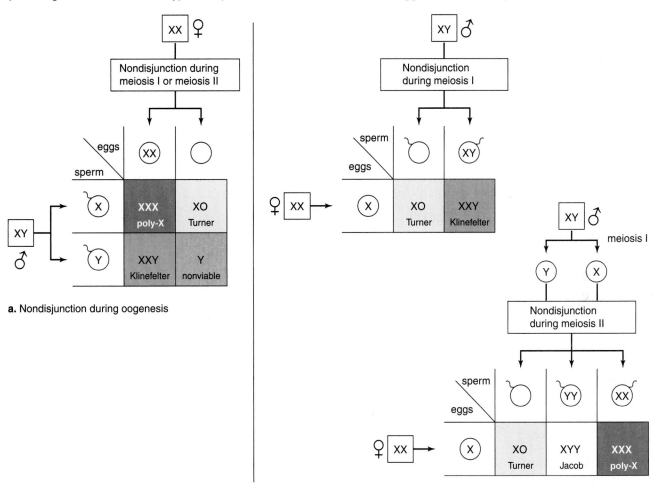

Simulating Nondisjunction During Meiosis I

1. Simulate meiosis during normal oogenesis and spermatogenesis as before except as follows. Assume that nondisjunction occurs during meiosis I, but the chromatids separate at the centromere during meiosis II (Fig. 10.2a). What is the chromosome constitution of each of the four meiotic products:

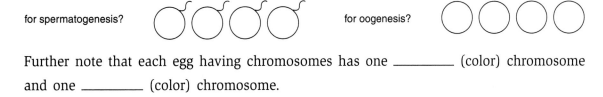

Further note that each egg having chromosomes has one _____ (color) chromosome and one _____ (color) chromosome.

2. In the space provided here, draw a Punnett square, and fill in the products of fertilization using (a) normal sperm × *types* of abnormal eggs as in Figure 10.2a, and (b) normal egg × *types* of abnormal sperm, as in Figure 10.2b for nondisjunction during meiosis I. (Disregard the color of the chromosomes.)

a.   b.

Conclusions

- What syndromes are the result of (a)? _____
- Are all offspring viable (capable of living)? _____ Explain. _____
  _____
- What syndromes are the result of (b)? _____
- Are all offspring viable? _____ Explain. _____
  _____

Simulating Nondisjunction During Meiosis II

1. Simulate meiosis during normal oogenesis and spermatogenesis as before except as follows. Assume that meiosis I is normal, but the chromatids of the chromosomes fail to separate during meiosis II. What is the chromosome constitution of each of the four meiotic products:

for spermatogenesis?   for oogenesis?

2. In the space provided here, draw a Punnett square, and fill in the products of fertilization using (a) normal sperm × *types* of abnormal eggs, as in Figure 10.2a, and (b) normal egg × *types* of abnormal sperm, as in Figure 10.2b for nondisjunction during meiosis II.

a.   b.

Conclusions

- What syndromes are the result of (a)? _____
- Are all offspring viable? _____ Explain. _____
  _____
- What syndromes are the result of (b)? _____
- Are all offspring viable? _____ Explain. _____
  _____

10.2 Genetic Inheritance

Just as we inherit pairs of chromosomes, we inherit pairs of alleles, alternate forms of a gene. As before, a **dominant allele** is assigned a capital letter, while a **recessive allele** is given the same letter lowercased.

The **genotype** tells the alleles of the individual, while the **phenotype** describes the appearance of the individual.

Autosomal Dominant and Recessive Traits

The alleles for autosomal traits are carried on the nonsex chromosomes. If individuals are homozygous dominant (*AA*) or heterozygous (*Aa*), their phenotype is the dominant trait. If individuals are homozygous recessive (*aa*), their phenotype is the recessive trait.

Experimental Procedure: Autosomal Traits

1. For this Experimental Procedure, you will need a lab partner to help you determine your phenotype for the traits listed in the first column of Table 10.2. Figure 10.3 illustrates some of these traits. Record your phenotypes by circling them in the first column of the table.

Figure 10.3 Examples of human phenotypes.

Widow's peak

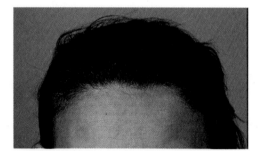

Straight hairline

Bent little finger

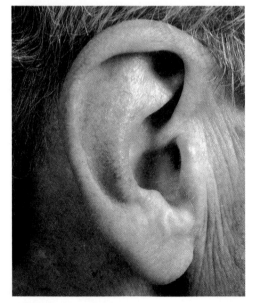

Unattached earlobes

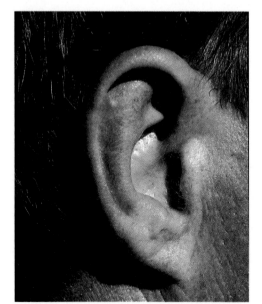

Attached earlobes

Straight little finger

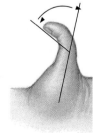

Hitchhiker's thumb

2. Determine your probable genotype. If you have the recessive phenotype, you know your genotype. If you have the dominant phenotype, you may be able to decide whether you are homozygous dominant or heterozygous by recalling the phenotype of your parents, siblings, or children. Circle your probable genotype in the second column of Table 10.2.

3. Your instructor will tally the class's phenotypes for each trait so that you can complete the third column of Table 10.2.

4. Complete Table 10.2 by calculating the percentage of the class with each trait. Are dominant phenotypes always the most common in a population? _____ Explain. _____

_____

Table 10.2	Autosomal Human Traits		
Trait: d = Dominant r = Recessive	**Possible Genotypes**	**Number in Class**	**Percentage of Class with Trait**

Hairline:
    Widow's peak (d)
    Straight hairline (r)
	WW or *Ww*
	ww
		_____	

Earlobes:
    Unattached (d)
    Attached (r)
	EE or *Ee*
	ee
		_____	

Skin pigmentation:
    Freckles (d)
    No freckles (r)
	FF or *Ff*
	ff
		_____	

Hair on back of hand:
    Present (d)
    Absent (r)
	HH or *Hh*
	hh
		_____	

Thumb hyperextension—"hitchhiker's thumb":
    Last segment cannot be bent
    backward (d)
    Last segment can be bent
    back to 60° (r)
	TT or *Tt*
	
	tt
		_____	

Bent little finger:
    Little finger bends toward
    ring finger (d)
    Straight little finger (r)
	LL or *Ll*
	
	ll
		_____	

Interlacing of fingers:
    Left thumb over right (d)
    Right thumb over left (r)
	II or *Ii*
	ii
		_____	

Genetics Problems

1. Nancy and the members of her immediate family have attached earlobes. Her maternal grandfather has unattached earlobes. What is the genotype of her maternal grandfather? _____ Nancy's maternal grandmother is no longer living. What could have been the genotype of her maternal grandmother? _____

2. Joe does not have a bent little finger, but his parents do. What is the expected phenotypic ratio among the parents' children? _____

3. Henry is adopted. He has hair on the back of his hand. Could both of his parents have had hair on the back of the hand? _____ Could both of his parents have had no hair on the back of the hand? _____ Explain. _____

_____

_____

Sex Linkage

The sex chromosomes carry genes that affect traits other than the individual's sex. Genes on the sex chromosomes are called **sex-linked genes.** The vast majority of sex-linked genes have alleles on the X chromosome and are called **X-linked alleles.** Most often, the abnormal condition is recessive.

Color blindness is an X-linked, recessive trait. The possible genotypes and phenotypes are as follows:

Females

$X^B X^B$ = normal vision

$X^B X^b$ = normal vision (carrier)

$X^b X^b$ = color blindness

Males

$X^B Y$ = normal vision

$X^b Y$ = color blindness

Experimental Procedure: X-Linked Traits

1. Your instructor will provide you with a color blindness chart. Have your lab partner present the chart to you. Write down the words or symbols you see, but do not allow your partner to see what you write, and do not discuss what you see. This is important because color-blind people see something different than do people who are not color blind. _____

2. Now test your lab partner as he or she has tested you. _____

3. Are you color blind? _____ If so, what is your genotype? _____

4. If you are a female and are not color blind, you can judge whether you are homozygous or heterozygous by knowing if any member of your family is color blind. If your father is color blind, what is your genotype? _____ If your mother is color blind, what is your genotype? _____ If you know of no one in your family who is color blind, what is your probable genotype? _____

Genetics Problems

1. The only color-blind member of Arlene's family is her brother. What is her brother's genotype? _____ What is her father's genotype? _____ What is her mother's genotype? _____ What is Arlene's genotype if she later has a color-blind son? _____

2. A person with Klinefelter syndrome is color blind. Both his father and mother have normal vision, but his maternal uncle is color blind. In which parent and at what meiotic division did sex chromosome nondisjunction occur? _____
(*Hint:* Use Figure 10.2 to help solve this problem.)

3. A person with Turner syndrome has hemophilia. Her mother does not have hemophilia, but her father does. In which parent did nondisjunction occur, considering that the single X came from the father? _____ Is it possible to tell if nondisjunction occurred during meiosis I or meiosis II? _____ Explain. _____

_____

(*Hint:* Use Figure 10.2 to help solve this problem.)

Pedigrees

A **pedigree** shows the inheritance of a genetic disorder within a family and can help determine whether any particular individual has an allele for that disorder. Then a Punnett square can be done to determine the chances of a couple producing an affected child.

In a pedigree, Roman numerals indicate the generation, and Arabic numerals indicate particular individuals in that generation. The symbols used to indicate normal and affected males and females, reproductive partners, and siblings are shown in Figure 10.4.

Figure 10.4 Pedigree symbols.

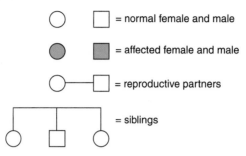

○ □ = normal female and male

● ■ = affected female and male

○—□ = reproductive partners

= siblings

Pedigree Analyses

For each of the following pedigrees, determine how a genetic disorder is passed. Use Table 10.3 to help with this determination. Is the inheritance pattern autosomal dominant, autosomal recessive, or X-linked recessive? Also, decide the genotype of particular individuals in the pedigree. Remember that the *genotype* indicates the dominant and recessive alleles present and the *phenotype* is the actual physical appearance of the trait in the individual. A pedigree indicates the phenotype, and you can reason out the genotype.

Table 10.3	Pedigree Solution Chart		
Inheritance Pattern	**Notes**	**Clues**	**Possible Genotypes**
Autosomal dominant (any chromosome except X or Y)	If at least one chromosome has the allele, the individual will be affected.	One or both parents are affected. Many of the children are affected.	AA or Aa = affected aa = normal
Autosomal recessive (any chromosome except X or Y)	Both chromosomes must have the recessive allele for the individual to be affected.	Neither parent is affected. Few of the children are affected.	AA or Aa = normal Aa = carrier* aa = affected
X-linked recessive (only the X chromosome)	The trait is only carried on the X chromosome. There must be a recessive allele on the X chromosome for the trait to be expressed.	Trait is primarily found in males. It is often passed from grandfather to grandson.	$X^A X^A$ and $X^A X^a$ = normal female; $X^A X^a$ = carrier female* $X^A Y$ = normal male $X^a X^a$ = affected female $X^a Y$ = affected male

*A carrier is one who does not show the trait but has the ability to pass it on to his or her offspring.

1. Study the following pedigree:

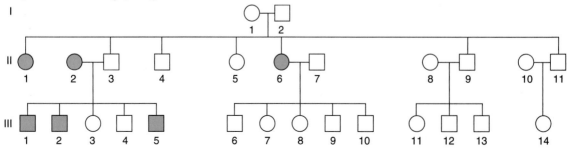

a. What is the inheritance pattern for this genetic disorder? _____

b. What is the genotype of the following individuals? Use *A* for the dominant allele and *a* for the recessive allele.

Generation I, individual 1: _____

Generation II, individual 1: _____

Generation III, individual 8: _____

2. Study the following pedigree:

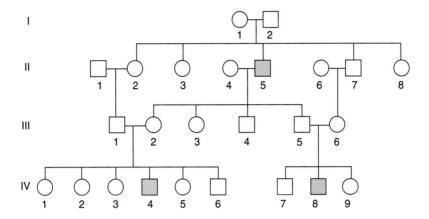

a. What is the inheritance pattern for this genetic disorder? _____

b. What is the genotype of the following individuals?

Generation I, individual 1: _____

Generation II, individual 8: _____

Generation III, individual 1: _____

3. Study the following pedigree:

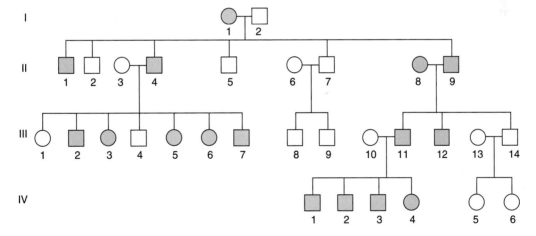

a. What is the inheritance pattern for this genetic disorder? _____

b. What is the genotype of the following individuals?

Generation I, individual 1: _____

Generation II, individual 7: _____

Generation III, individual 4: _____

Generation III, individual 11: _____

1. Name one pair of chromosomes not homologous in a normal male karyotype. _____

2. Which one could produce an egg in which two X chromosomes carry the same alleles: nondisjunction during meiosis I, or nondisjunction during meiosis II? Explain. _____
   _____

3. Which one could produce a sperm with two X chromosomes: nondisjunction during meiosis I, or nondisjunction during meiosis II? Explain. _____

4. If an individual exhibits the dominant trait, do you know the genotype? _____
   Why or why not? _____
   _____

5. A son is color blind, but both parents are normal. Give the genotype of the mother _____ and the father _____. Explain the pattern of inheritance. _____
   _____

6. What pattern of inheritance in a pedigree would allow you to decide that a trait is X-linked?
   _____

7. What pattern of inheritance in a pedigree would allow you to decide that a trait is autosomal recessive?
   _____

11

DNA Biology and Technology

Learning Outcomes

11.1 DNA Structure and Replication
- Explain how the structure of DNA facilitates replication. 118–120
- Explain why DNA replication is semiconservative and discuss the significance of this feature. 120

11.2 RNA Structure
- Discuss the structure of RNA and compare it to the structure of DNA. 121–122

11.3 DNA and Protein Synthesis
- Describe how DNA stores information. 122
- Compare the events of transcription with those of translation during protein synthesis. 122–125

11.4 Isolation of DNA
- Describe how DNA can be isolated before it is sequenced. 126

11.5 Genetic Disorders
- Describe the relationship between abnormal DNA base sequence and a genetic disorder. 127–128
- Demonstrate an understanding of the process of gel electrophoresis. 128–129

Introduction

This laboratory pertains to molecular genetics and biotechnology. Molecular genetics is the study of the structure and function of **DNA (deoxyribonucleic acid),** the genetic material. Biotechnology is the manipulation of DNA for the benefit of human beings and other organisms.

First we will study the structure of DNA and see how that structure facilitates DNA replication in the nucleus of cells. DNA replicates prior to cell division; following cell division, each daughter cell has a complete copy of the genetic material. DNA replication is also needed to pass genetic material from one generation to the next. You may have an opportunity to use models to see how replication occurs.

Then we will study the structure of **RNA (ribonucleic acid)** and how it differs from that of DNA, before examining how DNA, with the help of RNA, specifies protein synthesis. The linear construction of DNA, in which nucleotide follows nucleotide, is paralleled by the linear construction of the primary structure of protein, in which amino acid follows amino acid. Essentially, we will see that the sequence of nucleotides in DNA codes for the sequence of amino acids in a protein. We will also review the role of three types of RNA in protein synthesis. DNA's code is passed to messenger RNA (mRNA), which moves to the ribosomes containing ribosomal RNA (rRNA). Transfer RNA (tRNA) brings the amino acids to the ribosomes, and they become sequenced in the order directed by mRNA.

We now understand that a mutated gene has an altered DNA base sequence, which can lead to a genetic disorder. You will have an opportunity to carry out a laboratory procedure that detects whether an individual is normal, has sickle cell disease, or is a carrier.

11.1 DNA Structure and Replication

The structure of DNA lends itself to **replication,** the process that makes a copy of a DNA molecule. DNA replication is a necessary part of chromosome duplication, which precedes cell division. It also makes possible the passage of DNA from one generation to the next.

DNA Structure

DNA is a polymer of nucleotide monomers (Fig. 11.1). Each nucleotide is composed of three molecules: deoxyribose (a 5-carbon sugar), a phosphate, and a nitrogen-containing base.

Figure 11.1 Overview of DNA structure.

Diagram of DNA double helix shows that the molecule resembles a twisted ladder. Sugar-phosphate backbones make up the sides of the ladder, and hydrogen-bonded bases make up the rungs of the ladder. Complementary base pairing dictates that A is bonded to T and G is bonded to C and vice versa. *Label the boxed nucleotide pair as directed in the next Observation.*

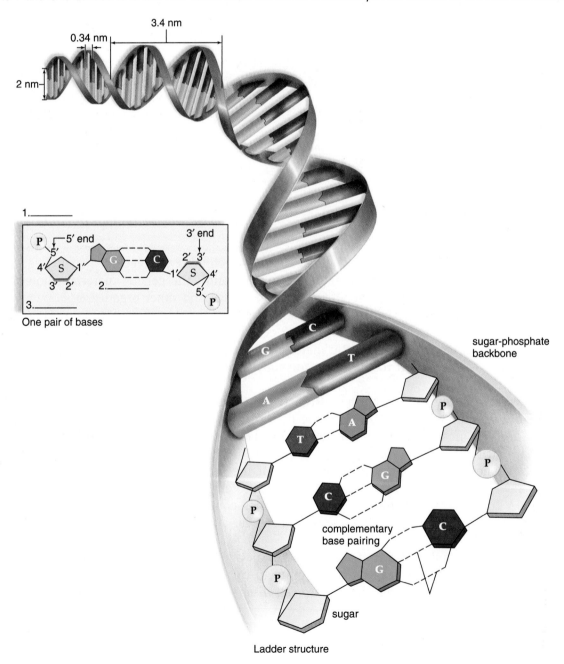

One pair of bases

Ladder structure

1. A boxed nucleotide pair is shown in Figure 11.1. If you are working with a kit, draw a representation of one of your nucleotides here. *Label phosphate, base pair, and deoxyribose in your drawing and in Figure 11.1, 1–3.*

2. Notice the four types of bases: cytosine (C), thymine (T), adenine (A), and guanine (G). What is the color of each of the four types of bases in Figure 11.1? In your kit? Complete Table 11.1 by writing in the colors of the bases.

Table 11.1	Base Colors	
	In Figure 11.1	In Your Kit
Cytosine		
Thymine		
Adenine		
Guanine		

3. Using Figure 11.1 as a guide, join several nucleotides together. Observe the entire DNA molecule. What type of molecules make up the backbone (uprights of ladder) of DNA (Fig. 11.1)?

   _____ and _____ In the backbone, the phosphate of one nucleotide is bonded to a sugar of the next nucleotide.

4. Using Figure 11.1 as a guide, join the bases together with hydrogen bonds. Label a hydrogen bond in Figure 11.1. Dashes are used to represent hydrogen bonds in Figure 11.1 because hydrogen bonds are (strong or weak)? _____

5. Notice in Figure 11.1 and in your model, that the base A is always paired with the base

   _____, and the base C is always paired with the base _____. This is called complementary base pairing.

6. In Figure 11.1, what molecules make up the rungs of the ladder? _____

7. Each half of the DNA molecule is a DNA strand. Why is DNA also called a double helix

   (Fig. 11.1)? _____

DNA Replication

During replication, the DNA molecule is duplicated so that there are two DNA molecules. We will see that complementary base pairing makes replication possible.

Observation: DNA Replication

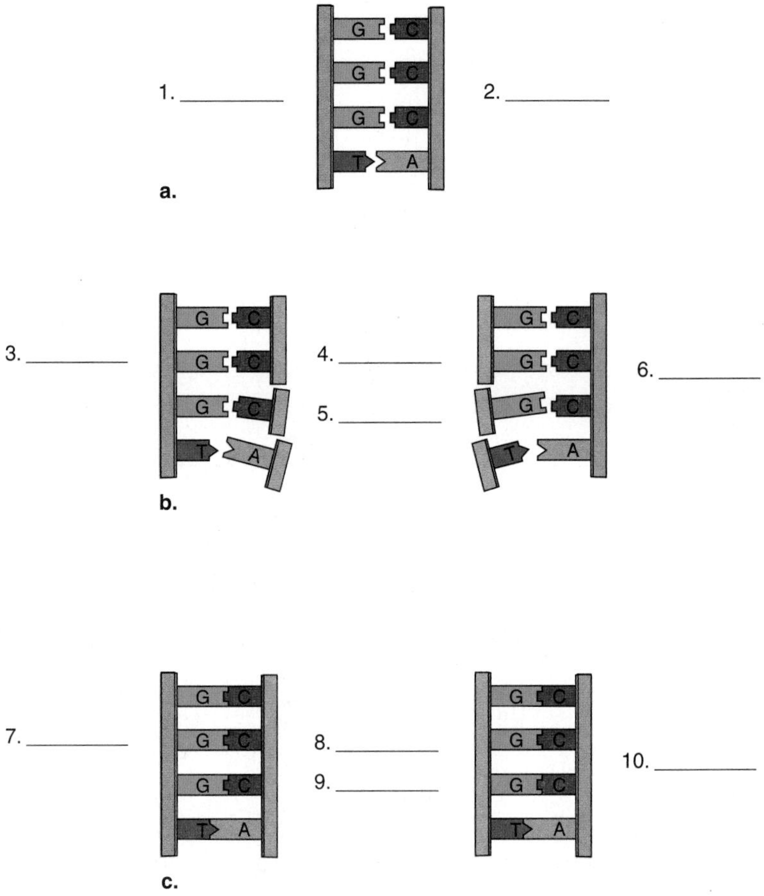

1. Before replication begins, DNA is unzipped. Using Figure 11.2a as a guide, break apart your two DNA strands. What bonds are broken in order to unzip the DNA strands? _____

2. Using Figure 11.2b as a guide, attach new complementary nucleotides to each strand using complementary base pairing.

3. Show that you understand complementary base pairing by completing Table 11.2. You now have two DNA molecules (Fig. 11.2c). Are your molecules identical? _____

4. Because of complementary base pairing, each new double helix is composed of an _____ strand and a _____ strand. *Write old or new beside each strand in Figures 11.2a, b, and c, 1–10. Conservative* means to save something from the past. Why is DNA replication called semiconservative?

_____

_____

_____

_____

Figure 11.2 DNA replication.

Use of the ladder configuration better illustrates how replication takes place. **a.** The parental DNA molecule. **b.** The "old" strands of the parental DNA molecule have separated. New complementary nucleotides available in the cell are pairing with those of each old strand. **c.** Replication is complete.

5. Genetic material has to be inherited from cell to cell and organism to organism. Consider that because of DNA replication a chromosome is composed of two chromatids and each chromatid is a complete DNA molecule. The chromatids separate during cell division so that each daughter cell receives a copy of each chromosome. Does replication provide a means for passing DNA from cell to cell and organism to organism?

_____

Explain. _____

_____

Table 11.2	DNA Replication																									
Old strand	G	G	G	T	T	C	C	A	T	T	A	A	A	T	T	C	C	A	G	A	A	A	T	C	A	T	A
New strand																											

11.2 RNA Structure

Like DNA, RNA is a polymer of nucleotides (Fig. 11.3). In an RNA nucleotide, the sugar ribose is attached to a phosphate molecule and to a nitrogen-containing base, C, U, A, or G. In RNA, the base uracil replaces thymine as one of the pyrimidine bases. RNA is single stranded, whereas DNA is double stranded.

Figure 11.3 Overview of RNA structure.
RNA is a single strand of nucleotides. *Label the boxed nucleotide as directed in the next Observation.*

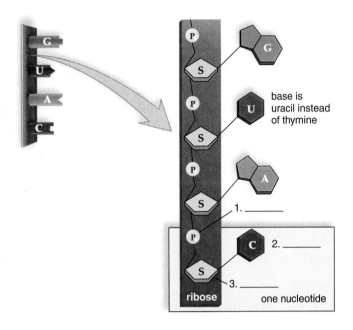

1. Describe the backbone of an RNA molecule. _____

2. Where are the bases located in an RNA molecule? _____

3. Complete Table 11.3 to show the complementary DNA bases for the RNA bases.

Table 11.3	DNA and RNA Bases			
RNA Bases	C	U	A	G
DNA Bases				

Observation: RNA Structure

1. If you are using a kit, draw a nucleotide for the construction of mRNA. *Label the ribose (the sugar in RNA), the phosphate, and the base in your drawing and in Figure 11.3, 1–3.*

2. Complete Table 11.4 by writing in the colors of the bases in Figure 11.3 and in your kit.

Table 11.4	Base Colors	
	In Figure 11.3	In Your Kit
Cytosine		
Uracil		
Adenine		
Guanine		

3. The base uracil substitutes for the base thymine in RNA. Complete Table 11.5 to show the several other ways RNA differs from DNA.

Table 11.5	DNA Structure Compared with RNA Structure	
	DNA	RNA
Sugar	Deoxyribose	
Bases	Adenine, guanine, thymine, cytosine	
Strands	Double stranded with base pairing	
Helix	Yes	

11.3 DNA and Protein Synthesis

Protein synthesis requires the processes of transcription and translation. During **transcription,** which takes place in the nucleus, an RNA molecule called **messenger RNA (mRNA)** is made complementary to one of the DNA strands. This mRNA leaves the nucleus and goes to the ribosomes in the cytoplasm. Ribosomes are composed of **ribosomal RNA (rRNA)** and proteins in two subunits.

During **translation,** RNA molecules called **transfer RNA (tRNA)** bring amino acids to the ribosome, and they join in the order prescribed by mRNA.

The final sequence of amino acids in a protein is specified by DNA. This is the information that DNA, the genetic material, stores.

Transcription

During transcription, complementary RNA is made from a DNA template (Fig. 11.4). A portion of DNA unwinds and unzips at the point of attachment of the enzyme RNA polymerase. A strand of mRNA is produced when complementary nucleotides join in the order dictated by the sequence of bases in DNA. Transcription occurs in the nucleus, and the mRNA passes out of the nucleus to enter the cytoplasm.

Label Figure 11.4. For number 1, note the name of the enzyme that carries out mRNA synthesis. For number 2, note the name of this molecule. For number 3, note where this molecule will be active.

Observation: Transcription

1. If you are using a kit, unzip your DNA model so that only one strand remains. This strand is the **template strand,** the strand that is transcribed.

2. Using Figure 11.4 as a guide, construct a messenger RNA (mRNA) molecule by first lining up RNA nucleotides complementary to the template strand of your DNA molecule. Join the nucleotides together to form mRNA.

3. A portion of DNA has the sequence of bases shown in Table 11.6. *Complete Table 11.6 to show the sequence of bases in mRNA.*

4. If you are using a kit, unzip mRNA transcript from the DNA. Locate the end of the strand that will move into the cytoplasm.

Figure 11.4 **Messenger RNA (mRNA).**
Messenger RNA complementary to a section of DNA forms during transcription.

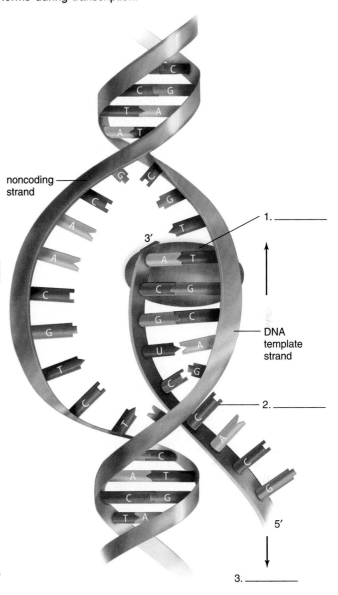

noncoding strand

3′

1.———

DNA template strand

2.———

5′

3.———

Table 11.6	Transcription
DNA	T A C A C G A G C A A C T A A C A T
mRNA	

Translation

DNA specifies the sequence of amino acids in a polypeptide because every three bases stand for an amino acid. Therefore, DNA is said to have a **triplet code.** The bases in mRNA are complementary to the bases in DNA. Every three bases in mRNA are called a **codon.** One codon of mRNA represents one amino acid. Thus, the sequence of DNA bases serves as the blueprint for the sequence of amino acids assembled to make a protein. The correct sequence of amino acids in a polypeptide is the message that mRNA carries.

Messenger RNA leaves the nucleus and proceeds to the ribosomes, where protein synthesis occurs. Transfer RNA (tRNA) molecules are so named because they transfer amino acids to the ribosomes. Each tRNA has a specific amino acid at one end and a unique **anticodon** at the other end (Fig. 11.5). *Label Figure 11.5,* where the amino acid is represented as a colored ball, the tRNA is green, and the anticodon is the sequence of three bases. (The anticodon is complementary to the mRNA codon.)

Figure 11.5 **Transfer RNA (tRNA).**
Transfer RNA carries amino acids to the ribosomes.

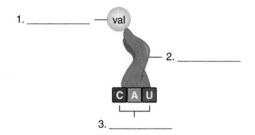

Observation: Translation

1. Figure 11.6 shows seven tRNA–amino acid complexes. Every amino acid has a name; in the figure, only the first three letters of the name are inside the ball. Using the mRNA sequence given in Table 11.7, number the tRNA–amino acid complexes in the order they will come to the ribosome.

2. If you are using a kit, arrange your tRNA–amino acid complexes in the proper order. Complete

    Table 11.7. Why are the codons and anticodons in groups of three? _____

    _____

Figure 11.6 **Transfer RNA diversity.**
Each type of tRNA carries only one particular amino acid, designated here by the first three letters of its name.

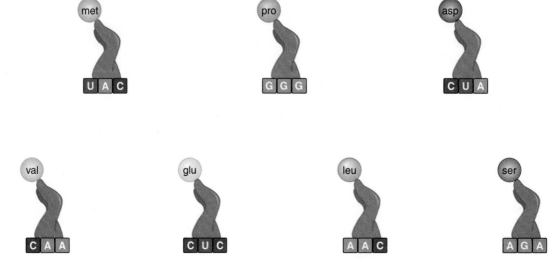

Table 11.7	Translation						
mRNA codons	AUG	CCC	GAG	GUU	GAU	UUG	UCU
tRNA anticodons							
Amino acid*							

*Use three letters only. See Table 11.8.

Table 11.8	Names of Amino Acids
Abbreviation	**Name**
met	methionine
pro	proline
asp	aspartate
val	valine
glu	glutamate
leu	leucine
ser	serine

3. Figure 11.7 shows the manner in which the polypeptide grows. A ribosome has three binding sites for tRNAs. They are the A (amino acid) site, the P (peptide) site, and the E (exit) site. A tRNA leaves from the E site after it has passed its amino acid or peptide to the second tRNA–amino acid complex. Then the ribosome moves forward, making room for the next tRNA–amino acid. This sequence of events occurs over and over until the entire polypeptide is borne by the last tRNA to come to the ribosome. Then a release factor releases the polypeptide chain from the ribosome. *In Figure 11.7, label the ribosome, the mRNA, and the peptide. Also, indicate the A, P, and E sites.*

Figure 11.7 Protein synthesis.

1. A ribosome has room for two tRNA–amino acid complexes. 2. Before a tRNA leaves, an RNA passes its attached peptide to its neighboring tRNA–amino acid complex. 3. The ribosome moves forward, and the next tRNA–amino acid complex arrives.

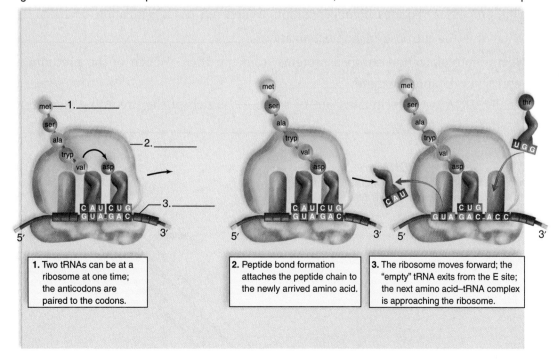

1. met — 1. _____
2. _____
3. _____

1. Two tRNAs can be at a ribosome at one time; the anticodons are paired to the codons.

2. Peptide bond formation attaches the peptide chain to the newly arrived amino acid.

3. The ribosome moves forward; the "empty" tRNA exits from the E site; the next amino acid–tRNA complex is approaching the ribosome.

11.4 Isolation of DNA

In the following Experimental Procedure, you will isolate DNA from the cells of an organism using a modified procedure like that used worldwide in biotechnology laboratories. You will extract DNA from a vegetable or fruit filtrate that contains DNA in solution. To prepare the filtrate, your instructor homogenized the vegetable or fruit with a detergent. The detergent emulsifies and forms complexes with the lipids and proteins of the plasma membrane, causing them to precipitate out of solution. Cell contents, including DNA, become suspended in solution. The cellular mixture is then filtered to produce the filtrate that contains DNA and its adhering proteins.

The DNA molecule is easily degraded (broken down), so it is important to closely follow all instructions. Handle glassware carefully to prevent nucleases in your skin from contaminating the glassware.

Experimental Procedure: Isolating DNA

1. Obtain a pair of gloves and wear them when doing this procedure.
2. Obtain a large, clean test tube, and place it in an ice bath. Let stand for a few minutes to make sure the test tube is cold. Everything must be kept very cold.
3. Obtain approximately 4 ml of the *filtrate,* and add it to your test tube while keeping the tube in the ice bath.
4. Obtain and add 2 ml of cold *meat tenderizer solution* to the solution in the test tube, and mix the contents slightly with a stirring rod or Pasteur pipette. Let stand for 10 minutes so the enzyme has time to strip the DNA of protein.
5. Use a graduated cylinder or pipette to slowly add an equal volume (approximately 6 ml) of ice-cold *95% ethanol* along the inside of the test tube. Keep the tube in the ice bath, and tilt it to a 45° angle. You should see a distinct layer of ethanol over the white precipitate, the DNA. Let the tube sit for 2 to 3 minutes.
6. Insert a glass rod or a Pasteur pipette into the tube until it reaches the bottom of the tube. *Gently* swirl the glass rod or pipette, always in the same direction. (You are not trying to mix the two layers; you are trying to wind the DNA onto the glass rod like cotton candy.) This process is called "spooling" the DNA. The stringy, slightly gelatinous material that attaches to the pipette is DNA (Fig. 11.8). If the DNA has been damaged, it will still precipitate, but as white flakes that cannot be collected on the glass rod.
7. Answer the following questions:

   a. This procedure requires homogenization. When did homogenization occur? _____

   What was the purpose of homogenization? _____

   b. Next, deproteinization stripped proteins from the DNA. Which of the preceding steps represents deproteinization? _____

   c. Finally, DNA was precipitated out of solution. Which of the preceding steps represents the precipitation of DNA? _____

Figure 11.8 Isolation of DNA.
The addition of ethanol causes DNA to come out of solution so that it can be spooled onto a glass rod.

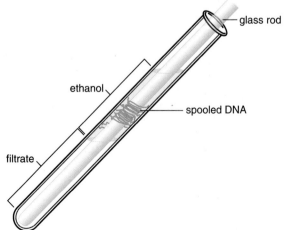

11.5 Genetic Disorders

The base sequence of DNA in all the chromosomes is an organism's genome. Now that the Human Genome Project is finished, we know the normal order of all the 3.6 billion nucleotide bases in the human genome. Someday it will be possible to sequence anyone's genome within a relatively short time, and thereby determine what particular base sequence alterations signify that he or she has a disorder or will have one in the future. In this laboratory, you will study the alteration in base sequence that causes a person to have sickle cell disease.

In persons with sickle cell disease, the red blood cells aren't biconcave disks like normal red blood cells—they are sickle-shaped. Sickle-shaped cells can't pass along narrow capillary passageways. They clog the vessels and break down, causing the person to suffer from poor circulation, anemia, and poor resistance to infection. Internal hemorrhaging leads to further complications, such as jaundice, episodic pain in the abdomen and joints, and damage to internal organs.

Sickle-shaped red blood cells are caused by an abnormal hemoglobin (Hb^S). Individuals with the $Hb^A Hb^A$ genotype are normal; those with the $Hb^S Hb^S$ genotype have sickle cell disease, and those with the $Hb^A Hb^S$ have sickle cell trait. Persons with sickle cell trait do not usually have sickle-shaped cells unless they experience dehydration or mild oxygen deprivation.

Genomic Sequence for Sickle Cell Disease

Examine Figures 11.9a and b, which show the DNA base sequence, the mRNA codons, and the amino acid sequence for a portion of Hb^A and the same portion for Hb^S.

Figure 11.9 Sickle cell disease.
a. When red blood cells are normal, the base sequence (in one location) for Hb^A alleles is CTC. **b.** In sickle cell disease, the sequence at these locations is CAC.

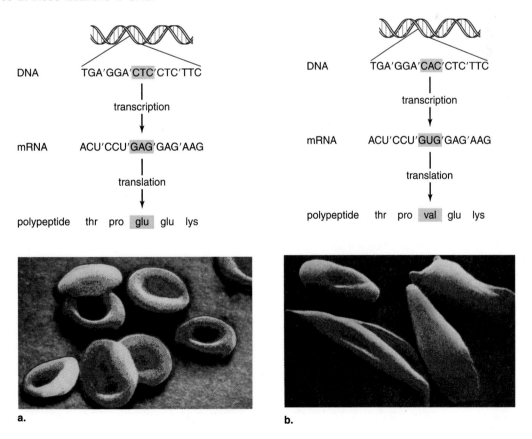

DNA Biology and Technology Laboratory 11

1. In what three-DNA-base sequence does Hb^A differ from Hb^S? Hb^A _____ Hb^S _____

2. What are the codons for these three bases? Hb^A _____ Hb^S _____

3. What is the amino acid difference? Hb^A _____ Hb^S _____

This one amino acid difference causes the polypeptide chain in sickle cell hemoglobin to pile up as firm rods that push against the plasma membrane and deform the red blood cell into a sickle shape:

$$- CH_2 - CH_2 - C \begin{matrix} {}^{\diagup\diagup O} \\ {}_{\diagdown O^-} \end{matrix}$$

glutamate
(polar *R* group)

$$- CH_2 \begin{matrix} {}^{\diagup CH_3} \\ {}_{\diagdown CH_3} \end{matrix}$$

valine
(nonpolar *R* group)

Gel Electrophoresis

The two most widely used techniques for separating molecules in biotechnology are chromatography and gel electrophoresis. Chromatography separates molecules on the basis of their solubility and size. **Gel electrophoresis** separates molecules on the basis of their charge and size (Fig. 11.10).

During gel electrophoresis, charged molecules migrate across a span of gel (gelatinous slab) because they are placed in a powerful electrical field. In the present experiment, the fragment mixture for each DNA sample is placed in a small depression in the gel called a well. The gel is placed in a powerful electrical field. The electricity causes equal length DNA fragments, which are negatively charged, to move through the gel to the positive pole at a faster rate than those that have no charge.

Almost all gel electrophoresis is carried out using horizontal gel slabs. First, the gel is poured onto a glass plate, and the wells are formed. After the samples are added to the wells, the gel and the glass plate are put into an electrophoresis chamber, and buffer is added. The fragments begin to migrate after the electrical current is turned on. With staining, the fragments appear as a series of bands spread from one end of the gel to the other.

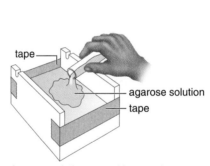

a. Agarose solution poured into casting tray

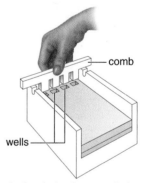

b. Comb that forms wells for samples

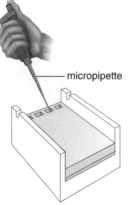

c. Wells that can be loaded with samples

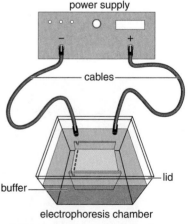

d. Electrophoresis chamber and power supply

Figure 11.10 Equipment and procedure for gel electrophoresis.

In this procedure, you will perform gel electrophoresis, if instructed to do so, and analyze your data to come to a conclusion.

Performing Gel Electrophoresis

> **Caution:** **Gel electrophoresis** Students should wear personal protective equipment: safety goggles and smocks or aprons while loading gels and during electrophoresis and protective gloves while staining.

1. Obtain three samples of hemoglobin provided by your kit. They are labeled samples A, B, and C.
2. If so directed by your instructor, carry out gel electrophoresis of these samples (see Fig. 11.10).

Analyzing the Electrophoresed Gel

1. Sickle cell hemoglobin (Hb^S) migrates slower toward the positive pole than normal hemoglobin (Hb^A) because the amino acid valine has no polar R groups, whereas the amino acid glutamate does have a polar R group.
2. In Figure 11.11, which lane contains only Hb^S, signifying that the individual is Hb^SHb^S? _____
3. Which lane contains only Hb^A, signifying that the individual is Hb^AHb^A? _____
4. Which lane contains both Hb^S and Hb^A, signifying that the individual is Hb^AHb^S? _____

Figure 11.11 **Gel electrophoresis of hemoglobins.**

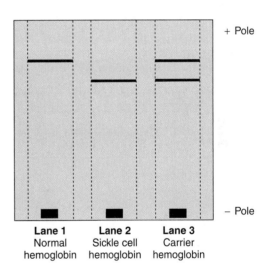

Lane 1
Normal
hemoglobin

Lane 2
Sickle cell
hemoglobin

Lane 3
Carrier
hemoglobin

Conclusion

- You are a genetic counselor. A young couple seeks your advice because sickle cell disease occurs among the family members of each. You order DNA base sequencing to be done. The results come back that at one of the loci for normal hemoglobin, each has the abnormal sequence CAC instead of CTC. The other locus is normal. What are the chances that this couple will have a child with sickle cell disease? _____

1. Explain why DNA is said to have a structure that resembles a ladder.

   _____

2. How is complementary base pairing different when pairing DNA to DNA than when pairing DNA to mRNA?

   _____

   _____

3. Explain why the genetic code is called a triplet code. _____

4. What role does each of the following molecules play in protein synthesis?

   a. DNA _____

   b. mRNA _____

   c. tRNA _____

   d. Amino acids _____

5. Which of the molecules listed in number 4 are involved in transcription? _____

6. Which of the molecules listed in number 4 are involved in translation? _____

7. During the isolation of DNA, what role was played by these substances?

   a. Detergent _____

   b. Meat tenderizer _____

   c. Ethanol _____

8. What is the purpose of gel electrophoresis? _____

   _____

9. Why does sickle cell hemoglobin (Hb^S) migrate slower than normal hemoglobin (Hb^A) during gel electrophoresis? _____

   _____

10. Why are red blood cells sickle-shaped in a person with sickle cell disease? _____

    _____

12

Evidences of Evolution

Learning Outcomes

12.1 Fossil Record
- Describe fossils and how they increase our understanding of evolution in the past. 133–136
- Explain how scientists use fossils to establish relationships between different forms of life. 136

12.2 Comparative Anatomy
- Explain how comparative anatomy gives evidence of common descent. 137–142
- Compare the human skeleton with the chimpanzee skeleton, and illustrate how the differences between the two reflect adaptation to different ways of life. 139–142

12.3 Molecular Evidence
- Explain how molecular evidence provides additional support to the concept of common descent. 143
- Explain how biochemistry aids the study of evolutionary relationships among organisms. 143–145

Introduction

Evolution is the process by which life has changed through time. A **species** is a group of similarly constructed organisms that share common genes, and a **population** is all the members of a species living in a particular area. When new variations arise that allow certain members of a population to capture more resources, these individuals tend to survive and to have more offspring than the other, unchanged members. Therefore, each successive generation will include more members with the new variation. Eventually, most members of a population and then the species will have the same **adaptations,** structures, physiology, and behavior that make an organism suited to its environment.

Adaptations to various ways of life explain why life is so diverse. However, evolution, which has been ongoing since the origin of life, is also an explanation for the unity of life. All organisms share the same characteristics of life because they can trace their ancestry to the first cell or cells. Many different lines of evidence support this hypothesis of common descent, and the more varied the evidence supporting the hypothesis, the more certain the hypothesis becomes.

In this laboratory, you will study three types of data that support the hypothesis of common descent: (1) the fossil record, (2) comparative anatomy (embryological and adult), and (3) biochemical comparison. **Fossils** are the remains or evidence of some organism that lived long ago. Fossils can be used to trace the history of life on Earth. A comparative study of the anatomy of modern groups of organisms has shown that each group has structures of similar construction called **homologous structures.** For example, all vertebrate animals have essentially the same type of skeleton. Homologous structures signify relatedness through evolution. Living organisms use the same basic molecules including ATP, the carrier of energy in cells; DNA, which makes up genes; and proteins, such as enzymes and antibodies. In this laboratory, you will analyze the activity of enzymes to determine evolutionary relationships among organisms.

Table 12.1				The Geological Timescale: Major Divisions of Geological Time and Some of the Major Evolutionary Events That Occurred	
Era	**Period**	**Epoch**	**Millions of Years Ago**	**Plant Life**	**Animal Life**
Cenozoic*	Quaternary	Holocene	0.01–Present	Human influence on plant life.	Age of *Homo sapiens*.
				Significant Mammalian Extinction	
		Pleistocene	1.8–0.01	Herbaceous plants spread and diversify.	Presence of ice age mammals. Modern humans appear.
	Tertiary	Pliocene	5.3–1.8	Herbaceous angiosperms flourish.	First hominids appear.
		Miocene	23.03–5.3	Grasslands spread as forests contract.	Apelike mammals and grazing mammals flourish; insects flourish.
		Oligocene	33.9–23.03	Many modern families of flowering plants evolve.	Browsing mammals and monkeylike primates appear.
		Eocene	55.8–33.9	Subtropical forests with heavy rainfall thrive.	All modern orders of mammals are represented.
		Paleocene	65.5–55.8	Flowering plants continue to diversify.	Primitive primates, herbivores, carnivores, and insectivores appear.
Mesozoic				*Mass Extinction: Dinosaurs and Most Reptiles*	
	Cretaceous		145.5–65.5	Flowering plants spread; conifers persist.	Placental mammals appear; modern insect groups appear.
	Jurassic		199.6–145.5	Flowering plants appear.	Dinosaurs flourish; birds appear.
				Mass Extinction	
	Triassic		251–199.6	Forests of conifers and cycads dominate.	First mammals appear; first dinosaurs appear; corals and molluscs dominate seas.
Paleozoic				*Mass Extinction*	
	Permian		299–251	Gymnosperms diversify.	Reptiles diversify; amphibians decline.
	Carboniferous		359.2–299	Age of great coal-forming forests: Ferns, club mosses, and horsetails flourish.	Amphibians diversify; first reptiles appear; first great radiation of insects.
				Mass Extinction	
	Devonian		416–359.2	First seed plants appear. Seedless vascular plants diversify.	First insects and first amphibians appear.
	Silurian		443.7–416	Seedless vascular plants appear.	Jawed fishes diversify and dominate the seas.
				Mass Extinction	
	Ordovician		488.3–443.7	Nonvascular plants appear on land.	First jawless and then jawed fishes appear.
	Cambrian		542–488.3	Marine algae flourish.	All invertebrate phyla present; first chordates appear.
Precambrian Time			630	Oldest soft-bodied invertebrate fossils.	
			1,000	Protists evolve and diversify.	
			2,200	Oldest eukaryotic fossils.	
			2,700	O_2 accumulates in atmosphere.	
			3,000	Oldest known fossils (prokaryotes).	
			4,570	Earth forms.	

*Many authorities divide the Cenozoic era into the Paleogene period (contains Paleocene, Eocene, Oligocene epochs) and the Neogene period (contains the Miocene, Pliocene, Pleistocene and Holocene).

12.1 Fossil Record

The **geological timescale** (Table 12.1) pertains to the history of Earth from its formation 4 to 4.5 billion years ago to the present. The ages of rocks can be measured in years by analyzing naturally occurring radioactive elements found in minute quantities in certain rocks and minerals. Before this method was discovered, geologists depended on a relative dating system; that is, they reasoned that any given stratum (layer of sediment) is younger than the stratum of Earth's crust just beneath it.

Because certain fossils are associated with particular strata, geologists are able to relate various strata around the world. *Fossils* are the remains and traces of past life or any other direct evidences of past life. By the 1860s, fossil-containing rocks in western Europe had been divided into three great eras: Paleozoic (ancient life), Mesozoic (middle life), and Cenozoic (recent life). Each era was divided into periods, and periods, in turn, were divided into epochs. The first period of the Paleozoic is called the Cambrian, and the time before the Cambrian period is called the Precambrian. The oldest fossils that have been found date from the Precambrian. For reasons that are still being explored, the fossil record improves dramatically starting with the Cambrian period.

Notice in Table 12.1 that divisions of the geological timescale generally tend to become increasingly shorter: For example, the Cenozoic era is much shorter than the Paleozoic era, and the Quaternary period is much shorter than the Tertiary period. Notice also that time is measured as "millions of years ago," and therefore, larger numbers represent an earlier time than smaller numbers.

Use Table 12.1 to answer the following questions:

1. During the _____ era and the _____ period, the first vascular plants appeared. How many million years ago was this? _____

2. During the _____ era and the _____ period, angiosperm (flowering plant) diversity occurred. How many million years ago was this? _____

   Use Figure 12.1 to answer the following questions:

3. During what period did the trilobites appear? _____
   When did they become extinct? _____

4. During what period did the insects first appear? _____
   How many million years ago was this? _____

   Use Figure 12.2 to answer the following questions:

5. During what period were the cycads most abundant? _____
   What types of animals were also prevalent at this time? _____

6. During what period did the angiosperms evolve? _____
   How do you know they are the most prominent plants today? _____
   _____

Figure 12.1 Geological history of selected animals.

The relative abundance of each group during any particular time period is indicated by the width of the band.

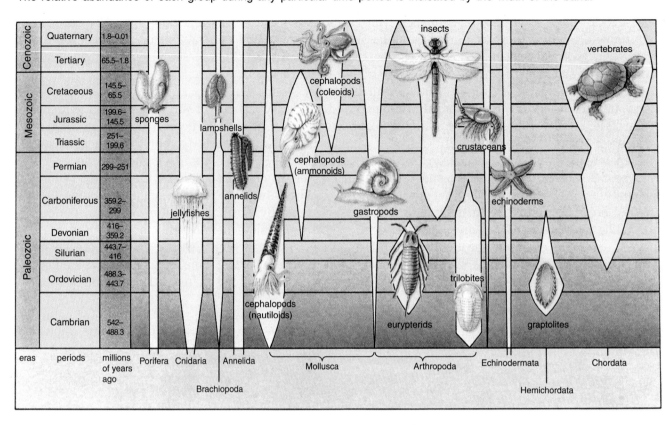

Figure 12.2 Geological history of selected algae, fungi, and plants.

The relative abundance of each group during any particular time period is indicated by the width of the band.

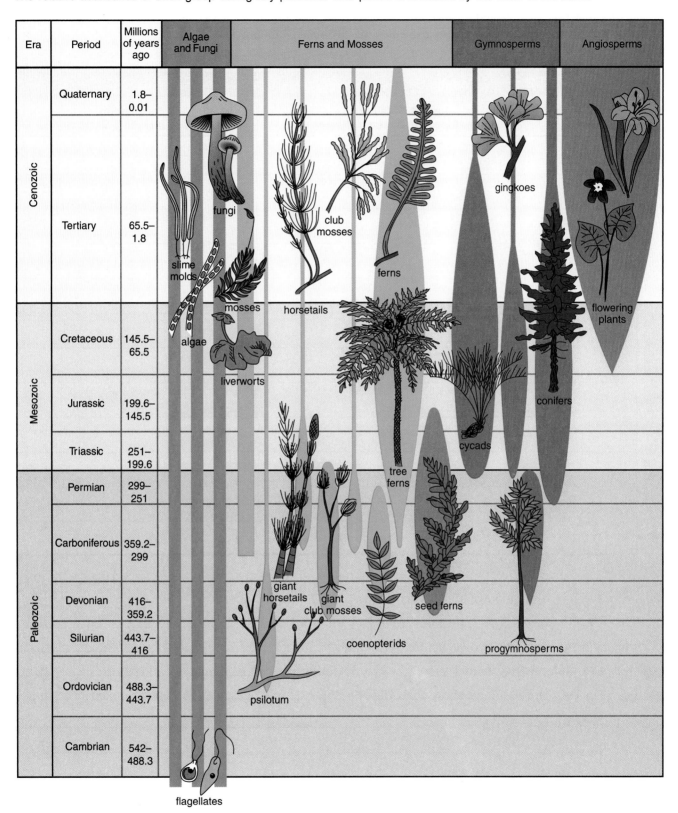

1. Obtain a box of selected fossils from each geological era (Cenozoic, Mesozoic, Paleozoic). Fill in Table 12.2 from the identification key or the fossil labels.

Era	Period	Type of Fossil (Phylum, Class, or Common Name)	Description (May Include a Sketch)
Table 12.2		**Fossils**	
Cenozoic			
Mesozoic			
Paleozoic			

2. From your observation of the fossils, answer the following questions:

   a. Did organisms first appear in the sea or on land? _____

   b. Did invertebrates evolve before vertebrates? _____

   c. Did cone-bearing plants evolve before flowering plants? _____

   d. Do your observations show that, despite observed changes, fossils can be linked over time because of a similarity in form? _____

   _____

12.2 Comparative Anatomy

In the study of evolutionary relationships, organisms or parts of organisms are said to be **homologous** if they exhibit similar basic structures and embryonic origins. If these organisms or parts of organisms are similar in function only, they are said to be **analogous.** Only homologous structures indicate an evolutionary relationship and are used to classify organisms.

Comparison of Adult Vertebrate Forelimbs

The limbs of vertebrates are homologous structures. Homologous structures share a basic pattern, although there may be specific differences. The similarity of homologous structures is explainable by descent from a common ancestor.

Observation: Vertebrate Forelimbs

1. The central diagram in Figure 12.3 represents the forelimb bones of the ancestral vertebrate. The basic components are the humerus (h), ulna (u), radius (r), carpals (c), metacarpals (m), and phalanges (p) in the five digits.
2. Carefully compare and label in Figure 12.3 the corresponding forelimb bones of the frog, the lizard, the bird, the bat, the cat, and the human. In particular, note the specific modifications that have occurred in some of the bones to meet the demands of a particular way of life.
3. Fill in Table 12.3 to indicate which bones in each specimen appear to most resemble the ancestral condition and which most differ from the ancestral condition.

Table 12.3	Comparison of Vertebrate Forelimbs	
Animal	**Bones That Resemble Common Ancestor**	**Bones That Differ from Common Ancestor**
Frog		
Lizard		
Bird		
Bat		
Cat		
Human		

4. Relate the change in bone structure to mode of locomotion in two examples.
   Example 1: _____
   Example 2: _____

Figure 12.3 Vertebrate forelimbs.

Because all vertebrates evolved from a common ancestor, their forelimbs share homologous structures.

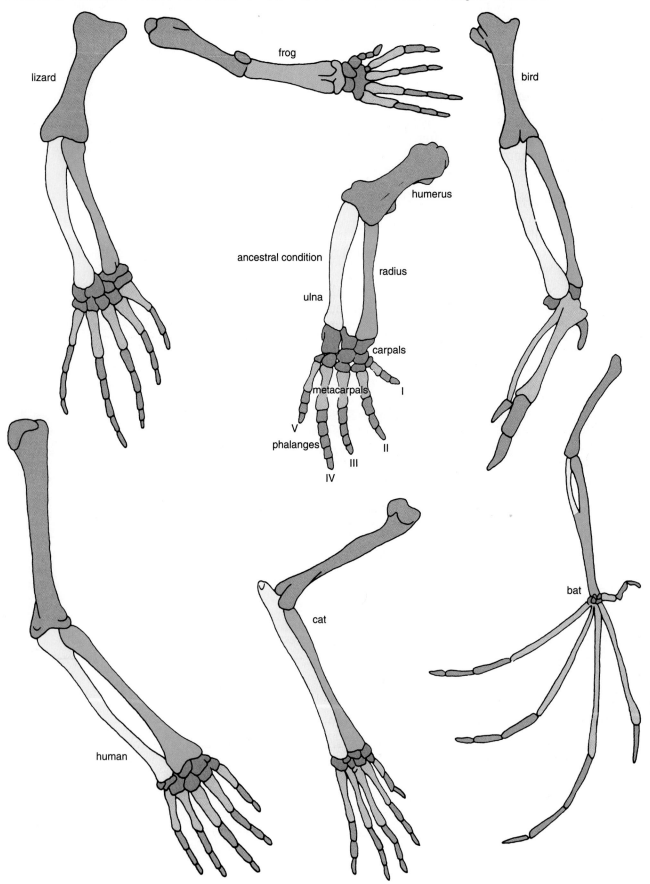

Comparison of Chimpanzee and Human Skeletons

Chimpanzees and humans are closely related, as is apparent from the comparison of the classifications of chimpanzees and humans in Table 12.4. Are chimpanzees and humans both primates?

_____ At what category does the classification of chimpanzees and the classification of humans first differ? _____

Table 12.4	Comparison of Chimpanzee and Human Classifications	
Classification	Chimpanzees	Humans
Domain	Eukarya	Eukarya
Kingdom	Animalia	Animalia
Phylum	Chordata	Chordata
Class	Mammalia	Mammalia
Order	Primate	Primate
Family	Pongidae	Hominidae
Genus	*Pan*	*Homo*
Species	*troglodytes*	*sapiens*

Observation: Chimpanzee and Human Skeletons

Comparison of Skeletons

Chimpanzees are arboreal and climb in trees. While on the ground, they tend to knuckle-walk, with their hands bent. Humans are terrestrial and walk erect.

Examine chimpanzee and human skeletons (Fig. 12.4), and answer the following questions:

1. **Head and torso:** Where are the head and trunk with relation to the hips and legs—thrust forward over the hips and legs or balanced over the hips and legs? *Record your answer in Table 12.5.*
2. **Spine:** Which animal has an S-shaped spine? Which has only a slight curve? *Record your answer in Table 12.5.*

   How does this contribute to an erect posture in humans? _____

   _____

3. **Pelvis:** Apes sway when they walk because lifting one leg throws them off balance. Which animal has a narrow and long pelvis, and which has a broad and short pelvis? *Record your answer in Table 12.5.*
4. **Femur:** In humans, the femur better supports the trunk. In which animal is the femur angled inward between articulations with the pelvic girdle and the knee? In which animal is the femur angled out a bit? *Record your answer in Table 12.5.*
5. **Knee joint:** In humans, the knee joint is modified to support the body's weight. In which animal is the femur larger at the bottom and the tibia larger at the top? *Record your answer in Table 12.5.*
6. **Foot:** In humans, the foot is adapted for walking long distances and running with less chance of injury. Which foot has an arch? *Record your answer in Table 12.5.*

Figure 12.4 **Adult skeletons.**
Adaptations for standing erect.

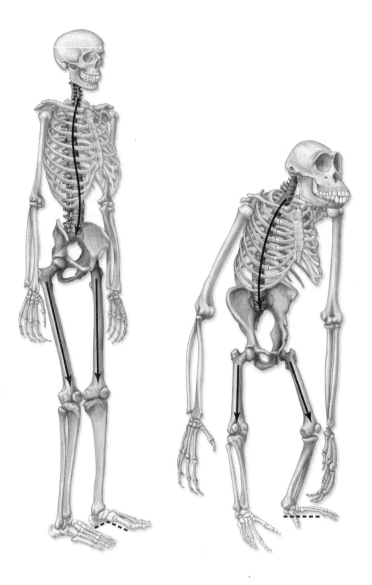

Human spine exits from the center; ape spine exits from rear of skull.

Human spine is S-shaped; ape spine has a slight curve.

Human pelvis is bowl-shaped; ape pelvis is longer and more narrow.

Human femurs angle inward to the knees; ape femurs angle out a bit.

Human knee can support more weight than ape knee.

Human foot has an arch; ape foot has no arch.

Table 12.5	Comparison of Chimpanzee and Human Skeletons	
Skeletal Part	**Chimpanzee**	**Human**
Head and torso		
Spine		
Pelvis		
Femur		
Knee joint		
Foot:		
Opposable toe		
Arch		

Figure 12.5 Chimpanzee skull compared with a human skull.

Although the general shapes of these skulls are different, they have similarities.

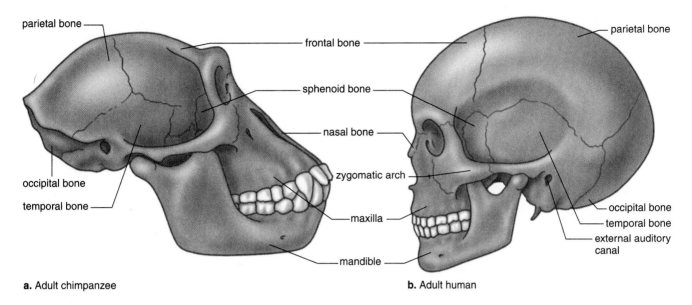

a. Adult chimpanzee   b. Adult human

7. Examine the position and shape of the parietal bones in both the chimpanzee and human skulls (Fig. 12.5). How does the chimpanzee skull differ from the human skull in this respect? _____

8. Compare the shape and position of the occipital bones in the chimpanzee and human skulls.

    _____

9. How does the difference in the position of the foramen magnum, a large opening in the base of the skull for the spinal cord, correlate with the posture and stance of the two organisms (see Fig. 12.4)?

    _____

10. Compare the slope of the frontal bones of the chimpanzee and human skulls. How are they different? _____

11. For which skull is the supraorbital ridge (the region of frontal bone just above the eye socket) thicker? _____

12. What is the position of the mouth and chin in relation to the profile for each skull? _____

    _____

    What effect has the evolutionary change in the positions of these bones had on the shape of the face? _____

13. Examine the teeth in the adult chimpanzee and adult human skulls. Are the shapes and types of teeth similar in both? _____ Diet can account for many of the observed differences. Humans are omnivorous. A diet rich in meat does not require strong grinding teeth or well-developed facial muscles. Chimpanzees are vegetarians, and a vegetarian diet requires strong facial muscles that attach to bony projections.

Comparison of Vertebrate Embryos

The anatomy shared by vertebrates extends to their embryological development. For example, as embryos, they all have a postanal tail and paired pharyngeal pouches. In aquatic animals, these pouches become functional gills (Fig. 12.6). In humans, the first pair of pouches becomes the cavity of the middle ear and auditory tube, the second pair becomes the tonsils, and the third and fourth pairs become the thymus and parathyroid glands.

Figure 12.6 Vertebrate embryos.
During early developmental stages, vertebrate embryos have certain characteristics in common.

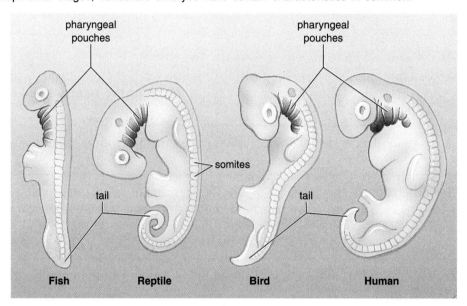

Observation: Chick and Pig Embryos

1. Obtain prepared slides of vertebrate embryos at comparable stages of development. Observe each of the embryos using a binocular dissecting microscope.
2. List five similarities of vertebrate embryos:

   a. _____

   b. _____

   c. _____

d. _____ _____

e. _____

3. Why do the embryos resemble one another so closely? _____

_____

12.3 Molecular Evidence

Almost all living organisms use the same basic cellular molecules, including DNA, ATP, and many identical and nearly identical enzymes. In addition, living organisms utilize the same DNA triplet code and the same 20 amino acids in their proteins. There is no obvious functional reason these elements need to be so similar. Therefore, their similarity is best explained by descent from a common ancestor.

Protein Differences

According to the **molecular clock hypothesis,** the number of amino acid changes between organisms is proportional to the length of time since two organisms began evolving separately from a

common ancestor. Why should that be? _____

The sequence of amino acids in **cytochrome c,** a carrier of electrons in the electron transport chain found in mitochondria and chloroplasts, has been determined in a variety of organisms. Figure 12.7 lists the number of differences between amino acid sequences for several of these.

Figure 12.7 Significance of molecular differences.

The branch points in this diagram indicate the number of amino acids that differ between human cytochrome c and the organisms depicted. These molecular data are consistent with those provided by a study of the fossil record and comparative anatomy.

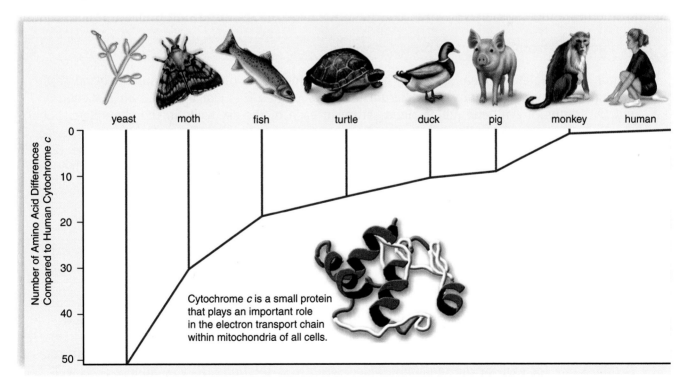

Cytochrome c is a small protein that plays an important role in the electron transport chain within mitochondria of all cells.

Protein Similarities

The immune system makes **antibodies** (proteins) that react with foreign proteins, termed **antigens.** Antigen-antibody reactions are specific. An antibody will react only with its particular antigen. In today's laboratory, this reaction can be observed when a precipitate, a substance separated from the solution, appears.

Biochemists have used the antibody-antigen reaction to determine the degree of relatedness between animals. In one technique, human serum (containing human proteins) is injected into the bloodstream of a rabbit, and the rabbit makes antibodies against the human serum. Some of the rabbit's blood is then drawn off, and the sensitized serum that contains the antibodies is separated from it. This sensitized rabbit serum will react strongly (determined by the amount of precipitate) against a new sample of human blood serum. The rabbit serum also will react against serum from other animals. **The more closely related these animals are to humans, the more the precipitate forms** (Fig 12.8).

Figure 12.8 **Antigen-antibody reaction.**
When antigens react with antibodies, a complex forms that appears as a precipitate.

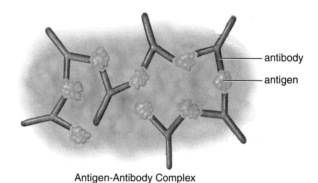

Antigen-Antibody Complex

Experimental Procedure: Protein Similarities

1. Obtain a chemplate (a clear glass tray with wells), one bottle of synthetic *human blood serum,* one bottle of synthetic *rabbit blood serum,* and five bottles (I–V) of *blood serum test solution.*
2. Put two drops of synthetic rabbit blood serum in each of the six wells in the chemplate. Label the wells 1–6. See the yellow circles in Figure 12.9.
3. Add 2 drops of synthetic human blood serum to each well. See the red circles in Figure 12.9. Stir with the plastic stirring rod that was attached to the chemplate. The rabbit serum has now been "sensitized" to human serum. (This simulates the production of antibodies in the rabbit's bloodstream in response to the human blood proteins.)
4. Rinse the stirrer. (The large cavity of the chemplate may be filled with water to facilitate rinsing.)
5. Add 4 drops of *blood serum test solution III* (tests for human blood proteins) to well 6.

   Describe what you see. _____

   _____

   This well will serve as the basis by which to compare all the other samples of test blood serum.
6. Now add 4 drops of *blood serum test solution I* to well 1. Stir and observe. Rinse the stirrer. Do the same for each of the remaining *blood serum test solutions (II–V)*—adding II to well 2, III to well 3, and so on. Be sure to rinse the stirrer after each use.

7. At the end of 10 and 20 minutes, record the amount of precipitate in each of the six wells in Figure 12.9. Well 6 is recorded as having ++++ amount of precipitate after both 10 and 20 minutes. Compare the other wells with this well (+ = trace amount; 0 = none). Holding the plate slightly above your head at arm's length and looking at the underside toward an overhead light source will allow you to more clearly determine the amount of precipitate.

Figure 12.9 Molecular evidence of evolution.
The greater the amount of precipitate, the more closely related an animal is to humans.

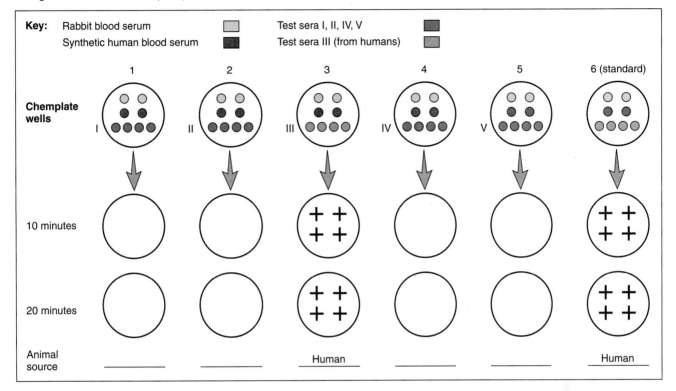

Conclusions

- The last row in Figure 12.9 tells you that the test serum in well 3 is from a human. How do your test results confirm this? _____
- Aside from humans, the test sera (supposedly) came from a pig, a monkey, an orangutan, and a chimpanzee. Which is most closely related to humans—the pig or the chimpanzee?

  _____

- Judging by the amount of precipitate, complete the last row in Figure 12.9 by indicating which serum you believe came from which animal. On what do you base your conclusions?

  _____

  _____

1. List three types of evidence suggesting that various types of organisms are related through common descent. _____

2. Why would you *not* expect a fossil buried millions of years ago to look exactly like a modern-day organism? _____

3. A horseshoe crab has changed little in approximately 200 million years of existence. Would you expect to find that the environment of the horseshoe crab has changed minimally? Explain. _____
_____

4. If a characteristic is found in bacteria, fungi, pine trees, snakes, and humans, when did it most likely evolve? _____ Why? _____
_____

5. What are homologous structures, and what do they show about relatedness? _____
_____

6. Why do humans and chicks develop similarly to reptiles? _____
_____
_____

7. What do DNA mutations have to do with amino acid changes in a protein? _____
_____

8. How can the antigen-antibody reaction help determine the degree of relatedness between species? _____
_____
_____

9. Using plus (+) symbols, show the amount of reaction you would expect when an antibody against human serum is tested against sera from a pig, monkey, and chimpanzee. _____
_____

10. Define the following terms:

    fossil _____

    common descent _____

    comparative anatomy _____

    adaptation _____

    biochemistry _____

13

Diversity: Bacteria, Protists, and Fungi

Learning Outcomes

Introduction

In today's laboratory, we are studying groups of organisms—bacteria, protists, and fungi—in which at least some, if not all or many, members are microscopic. You may recall that all living things are classified as either prokaryotes or eukaryotes on the basis of whether they lack or have a nucleus. Bacteria are prokaryotes in the domain Bacteria. Protists, fungi, plants, and animals are eukaryotes in the domain Eukarya. This means that bacteria, placed in a different domain from the other two, are distantly related to them. In this laboratory, we will study bacteria chiefly as pathogenic organisms. The protists are a diverse group because they include the photosynthetic algae, the often motile protozoans, and the somewhat funguslike slime molds. Unlike the other two groups, which are unicellular, fungi are usually multicellular. Fungi reproduce by producing windblown spores, and along with bacteria are most often saprotrophic. That means that they release digestive enzymes into the environment and absorb the products of digestion across the plasma membrane.

	Domain	Cell Structure	Nutrition
Bacteria	Bacteria	Unicellular	Most heterotrophic
Protists			
Protozoans	Eukarya	Unicellular	Heterotrophic
Algae	Eukarya	Unicellular, colonial, filamentous, or multicellular	Photosynthetic
Slime molds	Eukarya	Unicellular stage and multinucleated plasmodium	Heterotrophic
Fungi	Eukarya	Multicellular	Heterotrophic

13.1 Bacteria

In this laboratory you will first relate the general structure of a bacterium to its ability to cause disease. The specific shape, growth habit, and staining characteristics of bacteria are often used to identify them. Therefore, you will observe a variety of bacteria using the microscope. You will also perform one of the most important and discriminating tests, the Gram-staining protocol. Aside from their medical importance, bacteria are essential in ecosystems because, along with fungi, they are decomposers that break down dead organic remains, and thereby return inorganic nutrients to plants.

Pathogenic Bacteria

Pathogenic bacteria are infectious agents that cause disease. Infectious bacteria are able to invade and multiply within a host. Some also produce a toxin. Antibiotic therapy is often an effective treatment against a bacterial infection.

We will explore how it is possible to relate the structure of a bacterium to its ability to be invasive and avoid destruction by the immune system. We will also consider what morphophysiological attributes allow bacteria to be resistant to antibiotics and to pass the necessary genes on to other bacteria.

Observation: Structure of a Bacterium

1. Study the structure of a generalized bacterium in Figure 13.1 and, if available, examine a model or view a CD-ROM of a bacterium.
2. Identify:

    capsule, a gel-like coating outside the cell wall. Capsules often allow bacteria to stick to surfaces such as teeth. They also prevent phagocytic white blood cells from taking them up and destroying them.

    fimbriae, hairlike bristles that allow adhesion to surfaces. This can be how a bacterium clings to and gains access to the body prior to an infection.

    sex pilus, elongated, hollow appendage used to transfer DNA to other cells. Genes that allow bacteria to be resistant to antibiotics can be passed in this manner.

    flagellum, a rotating filament that pushes the cell forward.

    cell wall, a structure that provides support and shapes the cell. Antibiotics that prevent the formation of a cell wall are most effective against Gram-positive rather than Gram-negative bacteria.

    plasma membrane, a sheet that surrounds the cytoplasm and regulates entrance and exit of molecules. Resistance to antibiotics can be due to plasma membrane alterations that do not allow the drug to bind to the membrane or cross the membrane, or to a plasma membrane that increases the elimination of the drug from the bacteria.

    ribosomes, site of protein synthesis. Some bacteria possess antibiotic-inactivating enzymes that make them resistant to antibiotics.

    nucleoid, the location of the bacterial chromosome.

Figure 13.1 Generalized structure of a bacterium.

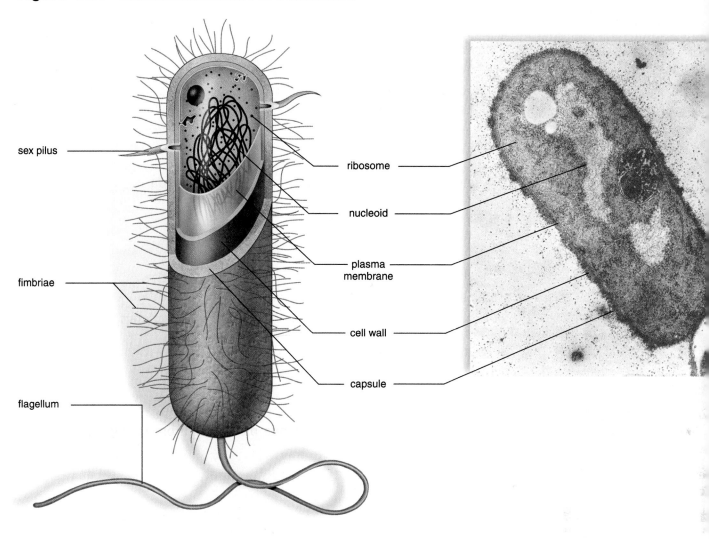

sex pilus

fimbriae

flagellum

ribosome

nucleoid

plasma membrane

cell wall

capsule

Conclusions

- Which portions of a bacterial cell aid the ability of a bacterium to cause infections?

  _____

- Which portions of a bacterial cell aid the ability of a bacterium to be resistant to
  antibiotics? _____

Also, some bacteria contain **plasmids,** small rings of DNA that replicate independently of the chromosomes and can be passed to other bacteria. Genes that allow bacteria to be resistant to antibiotics are often located in a plasmid.

Identification of Bacteria by Morphology

Shape and arrangement help identify bacteria. Most bacteria have one of three shapes. Bacteria that are spherical are called cocci (sing., **coccus**)(Fig. 13.2a). Bacteria that are rod-shaped are called bacilli (sing., **bacillus**) (Fig. 13.2b). Spiral bacteria are called spirilla, (sing., **spirillum**)(Fig. 13.2c). The bacteria featured in Figure 13.2 are all single cells; therefore, they are unicellular.

Bacteria have a variety of shapes and arrangements in addition to the three mentioned. The cocci and bacilli can appear in a variety of sizes: large or small spheres, long or short rods. There are some intermediate forms called coccobacilli. Curved rods are called vibrios. Some bacteria exist as chains, packets, or clusters of cells. Diplococcus means that a coccus forms a chain consisting of only two bacteria. A coccus that forms a longer chain is called a *streptococcus*. A chain of bacilli is called a *streptobacillus*. Packets of four cells are called a tetrad, while packets of eight cells are called a sarcina. More common are bacteria that form clusters. When the cluster is made of cocci, the organism is a *staphylococcus*.

Observation: Cell Morphologies

1. View the microscope slides of bacteria that are on display. What magnification is required to view bacteria? _____
2. Identify the three different shapes of bacteria, using Figure 13.2 as a guide.

Figure 13.2 Shapes of bacteria.
a. Streptococci, which exist as chains of cocci, cause a number of illnesses, including strep throat. **b.** *Escherichia coli,* which lives in your intestine, is a bacillus with flagella. **c.** *Treponema pallidum,* the cause of syphillis, is a spirillum.

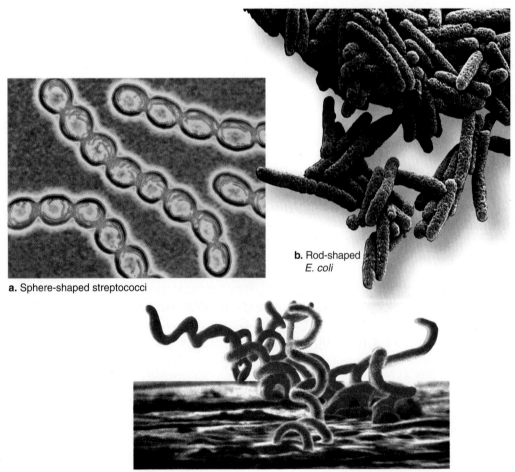

a. Sphere-shaped streptococci

b. Rod-shaped E. coli

c. Spirillum, T. pallidum

Agar Plates

Agar is a semisolid medium used to grow bacteria.

1. View agar plates that have been inoculated with bacteria and then incubated. Notice the "colonies" of bacteria growing on the plates. Each colony contains cells that are all descended from one original cell.
2. Compare the colonies' color, surface, and margin, and note your observations in Table 13.1.

Table 13.1	Agar Staining
Plate Number	**Description of Colonies**

Identification of Bacteria by Gram Stain

Most bacterial cells are protected by a cell wall that contains a unique molecule called peptidoglycan. Bacteria are commonly differentiated by using the Gram-stain procedure, which distinguishes bacteria that have a thick layer of peptidoglycan (Gram-positive) from those that have a thin layer of peptidoglycan (Gram-negative). Gram-positive bacteria retain a crystal violet-iodine complex and stain blue-purple, whereas Gram-negative bacteria decolorize and counterstain red-pink with safranin.

Experimental Procedure: Gram Stain

1. Use one designated square of a slide that has six squares.
2. With a sterile cotton swab, obtain a sample from around your teeth or inside your nose.
3. Carefully roll the swab across your allotted square. Body samples must be spread out thinly and evenly on the slide.
4. Allow the smear to air-dry.
5. Fix the smear by flooding the slide with *absolute methanol* for 1 minute. Allow the smear to dry before staining.
6. Flood the smear with *Gram Crystal Violet,* and wait for 1 to 2 minutes.
7. Gently rinse off the crystal violet with cold tap water.
8. Flood the smear with *Gram Iodine,* and allow it to react for 1 minute.
9. Gently rinse off the iodine with cold tap water.
10. Gently rinse the smear with *Gram Decolorizer* until the solution rinses colorlessly from the slide (approximately 20 to 30 seconds).
11. Immediately rinse the smear with cold tap water.
12. Flood the smear with *Gram Safranin,* and allow it to stain for 15 to 30 seconds.
13. Gently rinse off the safranin with cold tap water.
14. Blot off excess water with a paper towel, and allow the smear to air-dry.
15. Examine microscopically. (This will require the use of the oil immersion lens.)

Conclusion

The Gram stain is one way to distinguish bacteria from each other. Are these bacteria Gram-positive or Gram-negative? _____

Cyanobacteria

Some bacteria are photosynthesizers that use solar energy to produce their own food. Cyanobacteria are believed to have arisen some 3.7 billion years ago and are thought to have been the first organisms to release oxygen into the atmosphere. Their importance as a source of oxygen, even today, should not be underemphasized. At one time, cyanobacteria were identified as blue-green algae, but now we know they are a type of bacterium. Keep in mind that they have isolated thylakoids and not chloroplasts.

Observation: Cyanobacteria

Oscillatoria

1. Prepare a wet mount of an *Oscillatoria* culture, if available, or examine a prepared slide, using high power (45×) or oil immersion (if available). This is a filamentous cyanobacterium with individual cells that resemble a stack of pennies (Fig. 13.3a).
2. *Oscillatoria* takes its name from the characteristic oscillations that you may be able to see if your sample is alive. If you have a living culture, are oscillations visible? _____

Anabaena

1. Prepare a wet mount of an *Anabaena* culture, if available, or examine a prepared slide, using high power (45×) or oil immersion (if available). This is also a filamentous cyanobacterium, although its individual cells are barrel-shaped (Fig. 13.3b,c).
2. Note the thin nature of this strand. If you have a living culture, what is its color? _____ (Prepared slides are artificially stained and may not resemble the color of the living cyanobacterium.)

Figure 13.3 *Oscillatoria* **and** *Anabaena.*
a. *Oscillatoria* is a filamentous cyanobacterium (magnification 120×). b., c. *Anabaena* is a filamentous cyanobacterium with heterocysts where atmospheric nitrogen (N_2) is converted to ammonia.

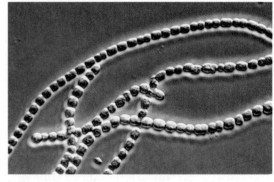

b. Photomicrograph

heterocyst

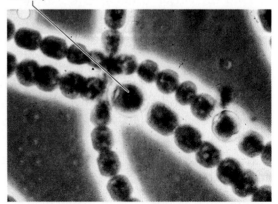

c. Photomicrograph

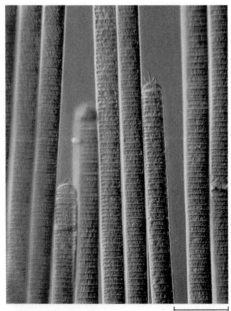

a. Photomicrograph

500 μm

13.2 Protists

Protists (kingdom Protista) are all eukaryotes, even though they may be unicellular. The protists include the protozoans, which are heterotrophic by ingestion; the slime molds, which creep along the forest floor as a plasmodium; and the algae, which photosynthesize in the same manner as plants.

Protozoans

The term protozoan is often restricted to unicellular, heterotrophic eukaryotes that ingest food by forming **food vacuoles.** Other vacuoles, such as **contractile vacuoles** that rid the cell of excess water, are also typical. Usually protozoans have some form of locomotion; some use **pseudopodia,** some move by **cilia,** and some use **flagella** (Fig. 13.4). Sporozoans—such as *Plasmodium vivax,* which causes a common form of malaria—do not locomote at all.

Figure 13.4 Protozoan diversity.

The protozoans are unicellular and heterotrophic by ingestion. Some protozoans move by pseudopods, cilia, or flagella. **a.** *Amoeba;* **b.** *Paramecium;* **c.** *Trypanosoma.*

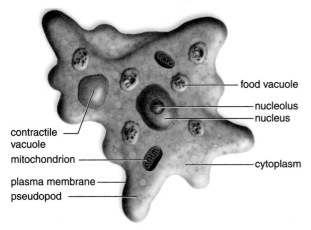

a. *Amoeba* moves by pseudopods

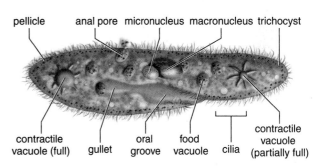

b. *Paramecium* moves by cilia

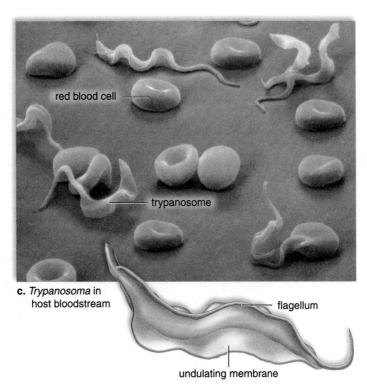

c. *Trypanosoma* in host bloodstream

d. *Trypanosoma* moves by flagella

1. You may already have had the opportunity to observe *Euglena* in Laboratory 2. However, your instructor may want you to observe these organisms again.
2. If a video, CD-ROM, or film is available, watch it, and note the various forms of protozoans.
3. Prepare wet mounts or examine prepared slides of protozoans as directed by your instructor.
4. Note the means of locomotion for each protozoan shown in Figure 13.4. _____

Pond Water

Pond water typically contains various examples of protozoans and algae, the protists studied in this laboratory.

Figure 13.5 Microorganisms found in pond water.

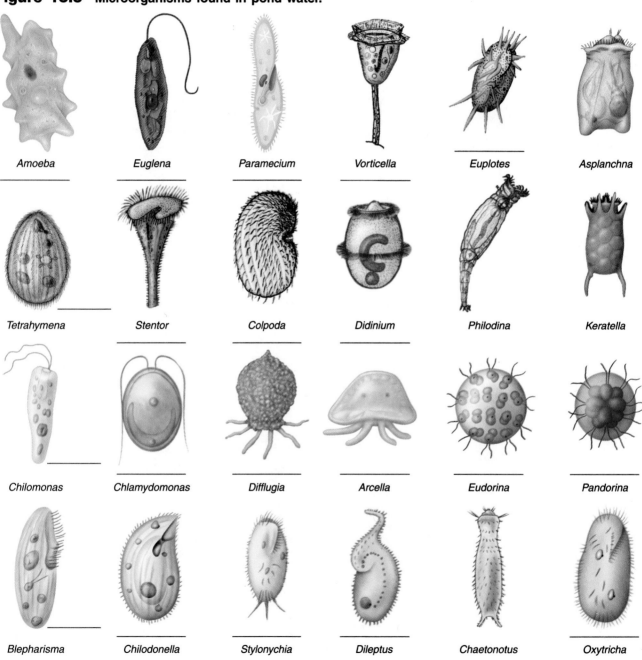

| Amoeba | Euglena | Paramecium | Vorticella | Euplotes | Asplanchna |

| Tetrahymena | Stentor | Colpoda | Didinium | Philodina | Keratella |

| Chilomonas | Chlamydomonas | Difflugia | Arcella | Eudorina | Pandorina |

| Blepharisma | Chilodonella | Stylonychia | Dileptus | Chaetonotus | Oxytricha |

1. Prepare a wet mount of a sample of pond water. Be sure to select some of the sediment on the bottom and a few strands of filamentous algae.
2. Identify the organisms you see by consulting Figure 13.5. *On the lines provided, write in the means of locomotion for each protist.* Those with chloroplasts are algae, and those without chloroplasts are protozoans. *Difflugia* and *Arcella* are amoebas. No line is provided for *Asplanchna, Philodina, Keratella,* and *Chaetonotus* because these are actually very small animals.

Slime Molds

Slime molds are called the funguslike protists. Like the fungi, they are saprotrophic, obtaining their energy from dead and decaying plant and animal material, and they form spores during some part of their life cycle.

There are two types of slime molds: plasmodial slime molds and cellular slime molds. **Plasmodial slime molds** usually exist as a **plasmodium**, a fan-shaped, multinucleated mass of cytoplasm. The plasmodium of a plasmodial slime mold creeps along, phagocytizing decaying plant material in a forest or an agricultural field. During times unfavorable for growth, such as a drought, the plasmodium develops many sporangia. A **sporangium** is a reproductive structure that produces spores by meiosis. In some plasmodial slime molds, the spores become flagellated cells, and in others, they are amoeboid. In any case, they fuse to form a zygote that develops into a plasmodium (Fig. 13.6). **Cellular slime molds** usually exist as individual amoeboid cells, which aggregate on occasion to form a pseudoplasmodium.

1. Obtain a plate of *Physarum* growing on agar. Examine the plate carefully under the dissecting microscope.

2. Describe what you see. _____

_____

Figure 13.6 **Plasmodial slime molds.**
a. The plasmodium, multinucleated cytoplasm, creeps along forest floor. **b.** During sexual reproduction when conditions are unfavorable for growth, the diploid adult forms sporangia. Haploid spores germinate, releasing haploid amoeboid or flagellated cells that fuse to form a plasmodium.

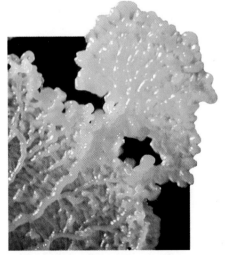

a. Plasmodium phagocytizes food

b. Sporangia produce spores

1 mm

Algae

The algae, whether green algae, red algae, brown algae, or golden-brown algae, all photosynthesize, as do plants. Why aren't they considered plants? Because they never protect the zygote and other reproductive structures the way plants do. Aside from releasing oxygen into the environment, algae play an important role in aquatic ecosystems—both freshwater and marine—because they are producers. Producers produce food for themselves and all members of an ecosystem.

Observation: A Sampling of Algae

To exemplify algae, you will examine a filamentous form (*Spirogyra*), a colonial form (*Volvox*), and a unicellular form (diatoms). The seaweeds seen along the coasts are also algae.

1. Obtain and examine a slide of *Spirogyra* (Fig. 13.7). The most prominent feature of the cells is the spiral, ribbonlike chloroplast. The nucleus is in the center of the cell, anchored by cytoplasmic strands. Your slide may show **conjugation,** a sexual means of reproduction illustrated in Figure 13.7b. If it does not, obtain a slide that does show this process. Conjugation tubes form between two adjacent filaments, and the contents of one set of cells enter the other set. As the nuclei fuse, a zygote is formed. The zygote overwinters, and in the spring, meiosis and subsequently germination occur.

Figure 13.7 *Spirogyra.*

a. *Spirogyra,* a filamentous green alga, in which each cell has a ribbonlike chloroplast. **b.** During conjugation, the cell contents of one filament enter the cells of another filament. Zygote formation follows.

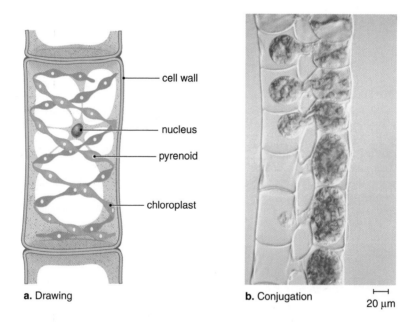

a. Drawing

b. Conjugation

20 μm

cell wall

nucleus

pyrenoid

chloroplast

2. Obtain and examine a slide of *Volvox* (Fig. 13.8). *Volvox* is a green algal colony. It is motile (capable of locomotion) because the thousands of cells that make up the colony have flagella. These cells are connected by delicate cytoplasmic extensions.

   Volvox is capable of both asexual and sexual reproduction. Certain cells of the adult colony can divide to produce **daughter colonies** (Fig. 13.8c) that reside for a time within the parental colony. A daughter colony escapes the parental colony by releasing an enzyme that dissolves away a portion of the matrix of the parental colony. During sexual reproduction, some colonies of *Volvox* have cells that produce sperm, and others have cells that produce eggs.

Figure 13.8 *Volvox.*
(a) The adult colony contains **(b)** many individual cells and produces **(c)** daughter colonies.

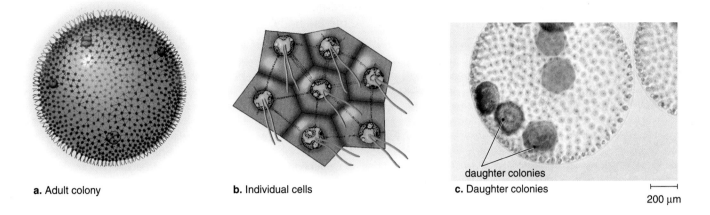

a. Adult colony   **b.** Individual cells   **c.** Daughter colonies

daughter colonies

200 µm

3. Obtain and examine a slide of diatoms (Fig. 13.9). Diatoms possess a yellow-brown pigment in addition to chlorophyll (see Fig. 13.9). The diatom cell wall is in two sections, with the larger one fitting over the smaller as a lid fits over a box. Since the cell wall is impregnated with silica, diatoms are said to "live in glass houses." The glass cell walls of diatoms do not decompose, so they accumulate in thick layers subsequently mined as diatomaceous earth and used in filters and as a natural insecticide. Diatoms, being photosynthetic and extremely abundant, are important food sources for the small heterotrophs (organisms that must acquire food from external sources) in both marine and freshwater environments.

Figure 13.9 Diatom shells.
The overlapping shells of a diatom are impregnated with silica. Scientists use their delicate markings to identify the particular species.

13.3 Fungi

Fungi (kingdom Fungi) (Fig. 13.10) are **saprotrophic** in the same manner as bacteria. Both fungi and bacteria are often referred to as "organisms of decay" because they break down dead organic matter and release inorganic nutrients for plants. A fungal body, called a **mycelium,** is composed of many strands, called **hyphae** (Fig. 13.11). Sometimes, the nuclei within a hypha are separated by walls called septa.

Fungi produce windblown **spores** (small, haploid bodies with a protective covering) when they reproduce sexually or asexually.

Figure 13.10 Diversity of fungi.
a. Scarlet hood, an inedible mushroom. **b.** Spores exploding from a puffball. **c.** Common bread mold. **d.** A morel, an edible fungus.

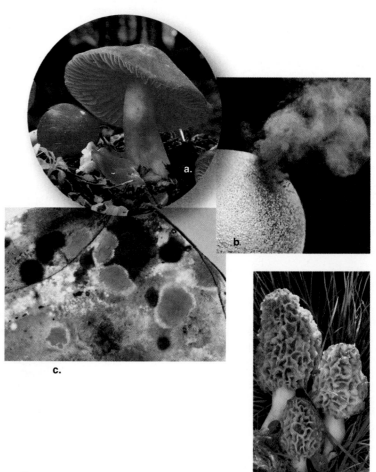

c.

d.

Figure 13.11 Body of a fungus.
a. The body of a fungus is called a mycelium. **b.** A mycelium contains many individual chains of cells, and each chain is called a hypha.

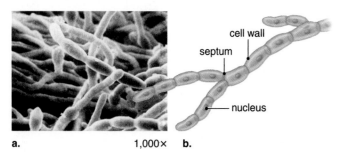

a.   1,000×   b.

cell wall

septum

nucleus

Black Bread Mold

In keeping with its name, black bread mold grows on bread and any other type of bakery goods. Notice in Figure 13.12 sporangia at the tips of aerial hyphae produce spores in both the asexual and sexual life cycles. A zygospore is diploid (2n); otherwise, all structures in the asexual and sexual life cycles of bread mold are haploid (n).

Figure 13.12 Black bread mold.
The mycelium of this mold, which uses sporangia to produce windblown spores, lives on bread.

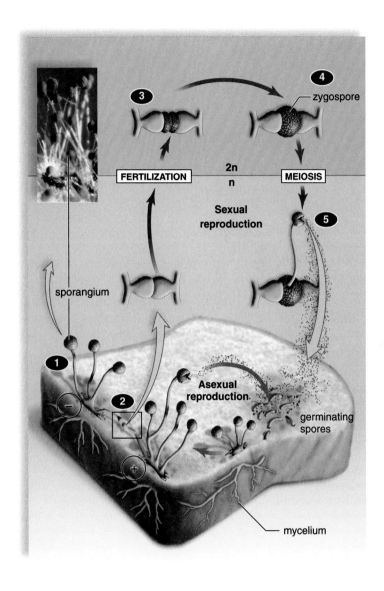

Observation: Black Bread Mold

1. If available, examine bread that has become moldy. Do you recognize black bread mold on the bread? _____
2. Obtain a petri dish that contains living black bread mold. Observe with a dissecting microscope. Identify the mycelium and a sporangium.
3. View a prepared slide of *Rhizopus*, using both a dissecting microscope and the low-power setting of a light microscope. The absence of cross walls in the hyphae is an identifying feature of zygospore fungi. *Label the mycelium and zygospore in Figure 13.13b.*

Figure 13.13 Microscope slides of black bread mold.
a. Asexual life cycle. **b.** Sexual life cycle.

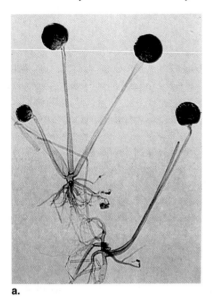

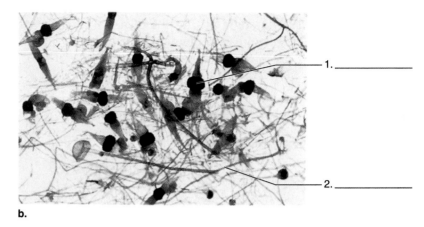

1._____

2._____

b.

a.

Club Fungi

Club fungi are just as familiar as black bread mold to most laypeople because they include the mushrooms. A gill mushroom consists of a stalk and a terminal cap with gills on the underside (Fig. 13.14). The cap, called a basidiocarp, is a fruiting body that arises following the union of + and − hyphae. The gills bear basidia, club-shaped structures where nuclei fuse, and meiosis occurs during spore production. The spores are called basidiospores.

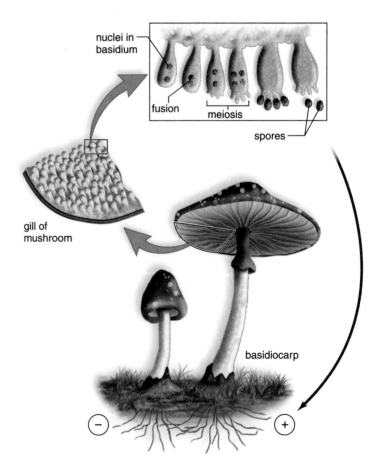

nuclei in basidium

fusion

meiosis

spores

gill of mushroom

basidiocarp

(−)   (+)

Figure 13.14 Sexual reproduction produces mushrooms.
Fusion of + and − hyphae tips results in hyphae that form the mushroom (a fruiting body). The nuclei fuse in clublike structures attached to the gills of a mushroom, and meiosis produces spores.

Observation: Mushrooms

1. Obtain an edible mushroom—for example, *Agaricus*—and identify as many of the following structures as possible:

   a. **Stalk:** The upright portion that supports the cap

   b. **Annulus:** A membrane surrounding the stalk where the immature (button-shaped) mushroom was attached

   c. **Cap:** The umbrella-shaped basidiocarp of the mushroom

   d. **Gills:** On the underside of the cap, radiating lamellae on which the basidia are located

   e. **Basidia:** On the gills, club-shaped structures where basidiospores are produced

   f. **Basidiospores:** Spores produced by basidia

2. View a prepared slide of a cross section of *Coprinus*. Using all three microscope objectives, look for the gills, basidia, and basidiospores.

3. Can you see individual hyphae in the gills? _____

4. Are the basidiospores inside or outside of the basidia? _____

5. What type of nuclear division does the zygote undergo to produce the basidiospores? _____

   _____

6. Can you suggest a reason for some of the basidia having fewer than four basidiospores? _____

   _____

7. What happens to the basidiospores after they are released? _____

   _____

Fungi and Human Diseases

Fungi cause a number of human diseases. Oral thrush is a yeast infection of the mouth common in newborns and AIDS patients (Fig. 13.15a). Ringworm is a group of related diseases caused by the fungi *tinea*. The fungal colony grows outward, forming a ring of inflammation (Fig. 13.15b). Athlete's foot is a form of tinea that affects the foot, mainly causing itching and peeling of the skin between the toes (Fig. 13.15c).

Figure 13.15 Human fungal diseases.
a. Thrush, or oral candidiasis, is characterized by the formation of white patches on the tongue.
b. Ringworm and **(c)** athlete's foot are caused by *Tinea* spp.

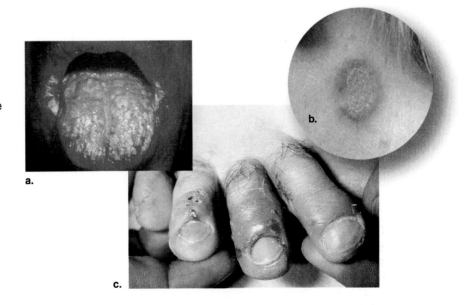

a.

b.

c.

1. What role do bacteria and fungi play in ecosystems? _____

   _____

2. What type of semisolid medium is used to grow bacteria? _____

3. What is the scientific name for spherical bacteria? _____

4. It is sometimes said that diatoms live in what kind of "houses"? _____

5. What type of nutrition do algae have? _____

6. Name a colonial alga studied today. _____

7. Gram-positive bacteria have a thick layer of what substance in their cell walls? _____

   _____

8. What color are Gram-negative bacteria following Gram staining? _____

9. Once called the blue-green algae, cyanobacteria are now classified as what? _____

10. What do you call the projection that allows amoeboids to move and feed? _____

11. What do you call the multinucleate stage of the plasmodial slime mold? _____

12. The stalk and cap of a mushroom that rise above the substratum are termed what? _____

13. What type of nutrition do fungi have? _____

14. What do fungi produce during both sexual and asexual reproduction? _____

15. Why aren't all the organisms studied today in the domain Eukarya? _____

   _____

   _____

16. In general, how does sexual reproduction differ from asexual reproduction among fungi? _____

   _____

   _____

14

Diversity: Plants

Learning Outcomes

14.1 Evolution and Diversity of Plants
- List the four main events in the evolution of plants. 164
- Associate each of these events with a major group of plants. 164–165
- Describe the plant life cycle and the concept of the dominant generation. 165
- Explain why you would expect the sporophyte to be the dominant generation in a plant adapted to a land existence. 165

14.2 Seedless Plants
- Describe the appearance and function of both generations in the life cycle of a moss. 166–167
- Describe the appearance and function of both generations in the life cycle of a fern. 168–169
- Contrast the adaptations of the moss and fern to a land environment. 169

14.3 Seed Plants
- Describe the appearance and function of both generations in the life cycle of a pine tree. 171–173
- Describe the appearance and function of both generations in the life cycle of a flowering plant. 174–176
- Contrast the adaptations of the pine tree and the flowering plant to a land environment. 177

Introduction

Your study of plant evolution in this laboratory will emphasize four groups of plants: the mosses; the ferns; the gymnosperms, represented by the pine tree; and the angiosperms, the flowering plants. While each of these groups of plants is successfully adapted to living on land, adaptation to the land environment is best demonstrated by the flowering plants (Fig. 14.1). The number and kinds of flowering plants is much greater than that of all the other groups of plants.

Adaptation to a land environment includes the ability to prevent excessive loss of water into the atmosphere; to obtain and transport water and nutrients to all parts of the plant; to support a large body against the pull of gravity; and to reproduce without dependence on external water.

With regard to human beings, consider that skin protects us from drying out, blood transports water and nutrients about the body, the skeleton supports us, males have a penis for delivering flagellated sperm to the female, and the embryo and fetus are protected from drying out within the uterus of the female.

Figure 14.1 The flowering plants.
The flowering plants provide us with much of our food including grains, potatoes, and beans. Here, we see the flower and the fruit of a watermelon plant.

14.1 Evolution and Diversity of Plants

In Figure 14.2, note the four events that mark the evolution of plants. Protection of the embryo is first seen in mosses (Fig. 14.3). The next event, evolution of vascular tissue, is first seen in ferns. Vascular tissue not only allows transport of water and nutrients, it also provides an internal skeleton that opposes the force of gravity. Conifers are the first seed plants. The last event, evolution of flowers, allows the seeds to be protected within a fruit. Fruits aid dispersal of seeds.

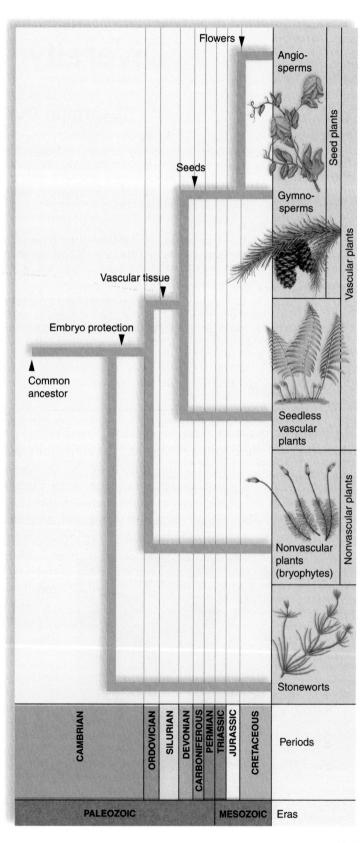

Figure 14.2 **Evolution of plants.**
The evolution of plants is marked by four significant events: protection of the embryo, evolution of vascular tissue, evolution of the seed, and evolution of the flower.

Figure 14.3 Representatives of the four major groups of plants.

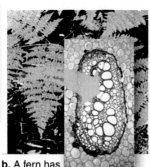

a. In mosses, the embryo is protected by a special structure.

b. A fern has vascular tissue.

c. In a conifer, seeds disperse offspring.

d. In flowering plants, the seeds are enclosed in fruits.

Alternation of Generations

All plants have a life cycle known as alternation of generations (Fig. 14.4). In this life cycle, there are two mature stages, known as the sporophyte and the gametophyte. The **sporophyte,** the 2n generation, produces spores, by the process of meiosis, in structures called sporangia (sing., sporangium). A **spore** is a haploid reproductive cell that produces a new generation that is also haploid. Spores develop into the gametophyte. The **gametophyte,** the n generation, produces gametes that later fuse to form a zygote. A zygote develops into the sporophyte, completing the cycle.

In plants, one generation is **dominant,** meaning that it lasts longer and is the generation we refer to as the plant. The gametophyte is dominant in mosses, and the sporophyte is dominant in ferns, gymnosperms, and angiosperms. In plants with dominant sporophytes, the sporophyte has vascular (transport) tissue that conducts water from the roots to the leaves. Is it beneficial for a sporophyte, the generation that has vascular tissue, to be dominant? _____

Why? _____

Figure 14.4 Alternation of generations.
In the alternation of generations life cycle of plants, the sporophyte is the 2n generation that produces spores by meiosis. The gametophyte is the n generation that produces the gametes. When the gametes fuse, a new sporophyte comes into being.

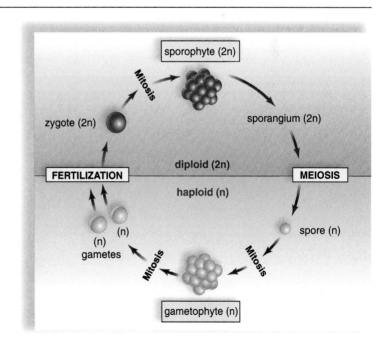

14.2 Seedless Plants

Mosses and ferns are both seedless plants. Mosses and their relatives, called the bryophytes, are low-lying plants, called the nonvascular plants because they lack vascular tissue. Ferns, characterized by large leaves, do have vascular tissue, but even so, share other characteristics with the bryophytes. For example, they are both seedless plants.

Mosses

The bryophytes (mosses and liverworts) were the first plants to live on land. The gametophyte is dominant in these nonvascular plants. The gametophyte produces eggs within archegonia and swimming sperm in antheridia (Fig. 14.5). The bryophytes are dependent on external moisture to ensure fertilization because the sperm must swim to the egg. However, the zygote developing within the archegonia is protected from drying out. The bryophytes have another adaptation to life on land in that the spores, produced by the dependent sporophyte, are windblown. The spores disperse the gametophyte.

Figure 14.5 Life cycle of the moss.

In mosses, the gametophyte consists of leafy shoots that produce flagellated sperm within antheridia and eggs within archegonia. Following fertilization, the sporophyte, consisting of a stalk and capsule (a sporangium), is dependent on the gametophyte.

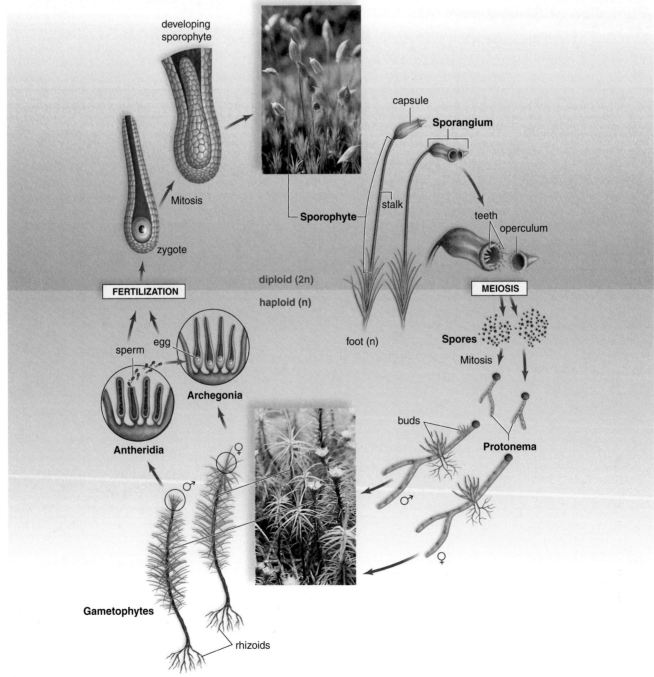

Life Cycle of Mosses

Study the life cycle of the moss (see Fig. 14.5) and find the gametophyte. Examine a living gametophyte or a plastomount of this generation. Describe its appearance. _____ _____ Considering that this is the generation we refer to as the "moss," what generation is dominant in mosses? _____ Describe the sporophyte. _____

Observation: Moss Life Cycle

1. Study the prepared slide of the female shoot head (top of female shoot) and the male shoot head (top of the male shoot). Find an **archegonium** in a female shoot head and locate an egg in at least one of these. Find an **antheridium** in a male shoot head. An antheridium is filled with many flagellated sperm. When sperm produced by the antheridia swim in a film of water to the eggs in the archegonia, zygotes result. A zygote develops into a new sporophyte.

2. Examine the plastomount of a shoot with a sporophyte attached. The sporophyte is dependent on the female shoot. Why female? _____

3. Examine a slide of a longitudinal section through the sporophyte of the moss. Identify the stalk and sporangium. What is being produced in the sporangium? _____ By what process? _____

The Life Cycle of a Moss

1. Which generation is haploid? _____ Which is diploid? _____
   Which generation is dominant in mosses? _____ Which generation is dependent? _____

2. Is there any evidence of vascular tissue in the moss sporophyte? _____

3. When spores germinate, what generation begins to develop? _____

4. Why is it proper to say that, in the moss, spores disperse the plant? _____ _____

5. By what means are spores disseminated? _____

Adaptation of Mosses to the Land Environment

Which of these is an indication that mosses are well adapted to life on land? Write "yes" if the feature is an adaptation to land and "no" if the feature is not an adaptation to living on land.

Lack of vascular tissue. _____

Body covered by a cuticle, a protection against drying out. _____

Flagellated sperm. _____

Egg and embryo protected by female shoot. _____

Spores are windblown. _____

Ferns

All the other plants to be studied in this laboratory are vascular plants. Ferns are seedless vascular plants in which the dominant sporophyte possesses vascular tissue. The windblown spores develop into an independent gametophyte that is water dependent because it lacks vascular tissue and also because it produces flagellated sperm. The sperm must swim from the antheridia to the archegonia, where the eggs are produced. The zygote, protected from drying out within an archegonium, develops directly into the sporophyte.

Figure 14.6 Fern life cycle.
The frond is the dominant sporophyte generation that produces windblown spores. A spore gives rise to an independent gametophyte. The prothallus produces gametes. When the gametes fuse, a new sporophyte begins to develop.

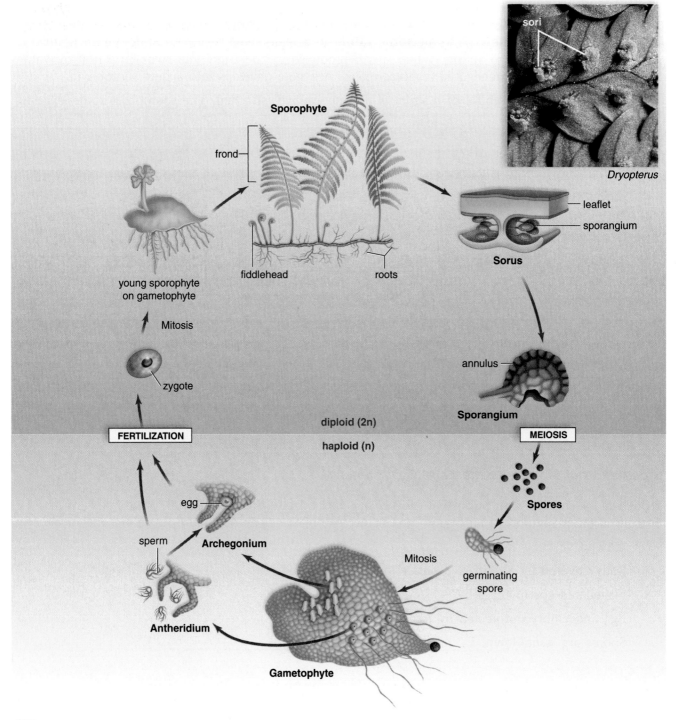

1. Study the life cycle of the fern (Fig. 14.6) and find the sporophyte and gametophyte. This large, complexly divided sporophyte leaf is known as a **frond.** Fronds arise from an underground stem. Examine a preserved specimen of a frond and on the underside, notice brownish **sori (sing., sorus),** each one a cluster of many sporangia (Fig. 14.7). What is being produced in the sporangia? _____ Considering that it is this generation that we call the fern, what generation is dominant in ferns? _____

2. Examine a prepared slide of a young sporangium cross section. Study the slide carefully and locate the fern leaf and a sorus. Within a sorus, find the sporangia and spores. Notice the shelflike structure that protects the sporangia until they are mature.

3. Examine a plastomount showing the fern life cycle. Notice a portion of the frond with sori and a small heart-shaped structure. The latter is the gametophyte generation of the fern. Most persons do not realize that this structure exists as a part of the fern life cycle. What is the function of this structure? _____

4. Examine a whole mount slide of a fern archegonia. What is being produced inside an archegonium? _____ If you focus up and down very carefully on an archegonium, you may be able to see an egg inside.

5. Examine a whole mount slide of fern antheridia. What is being produced inside the antheridia? _____ When sperm produced by the antheridia swim to the archegonia in a film of water, what results? _____ The latter develop into what generation? _____

The Life Cycle of a Fern

1. Is either generation in the fern dependent for any length of time on the other generation? _____

2. Which generation is dispersed in ferns? _____ How? _____

Adaptation of Ferns to a Land Environment

1. List one additional way in which the fern is adapted to life on land, when you compare it to the moss. _____

2. List one characteristic of the fern illustrating that sexual reproduction is not adapted to a land environment. _____

Figure 14.7 **Underside of frond leaflets showing sori.**

sorus

14.3 Seed Plants

Seed plants (gymnosperms and angiosperms) are further adapted to live and reproduce on land. The dominant sporophyte contains vascular tissue, which not only transports water but also serves as an internal skeleton, allowing these plants to oppose the force of gravity. For example, all of today's trees are seed plants.

Gymnosperms

The gymnosperms are usually evergreen trees in which the sporangia are found on **cones.** The four groups of living gymnosperms are **cycads, ginkgoes, gnetophytes,** and **conifers** (Fig. 14.8). All of these plants have ovules and subsequently develop seeds that are exposed on the surface of cone scales or analogous structures. (Because the seeds are not enclosed by fruit, gymnosperms are said to have "naked seeds.") Early gymnosperms were present in the swamp forests of the Carboniferous period, and they became dominant during the Triassic period. Today, living gymnosperms are classified into 780 species, the most plentiful being the conifers.

Conifers (phylum Coniferophyta) consist of about 575 species of trees, many of them evergreens such as pines, spruces, firs, cedars, hemlocks, redwoods, cypresses, yews, and junipers. Vast areas of northern temperate regions are covered in evergreen coniferous forests. The tough, needlelike leaves of pines conserve water because they have a thick cuticle and recessed stomata (openings for gas exchange).

Cycad, *Encephalartos humlis*
Female plant with large seed cones

Figure 14.8 **Gymnosperm diversity.**

Ginkgo, *Ginkgo biloba*
Female maidenhair tree with seeds

Gnetophyte, *Ephedra*
Branched shrub with scalelike leaves

Conifer, *Picea*
Spruce tree with pollen cones and seed cones

Life Cycle of Pine Trees

The life cycle of a pine tree (Fig. 14.9) illustrates that seed plants are no longer dependent on external water to ensure fertilization. In seed plants, the dominant sporophyte produces two types of spores, termed the microspore and the megaspore. Microspore mother cells produce **microspores** by meiosis and each develops into a male gametophyte generation, the pollen grain. During **pollination,** pollen grains are dispersed in the vicinity of female gametophytes.

A megaspore mother cell within an **ovule** produces **megaspores** by meiosis. Only one of these develops into a female gametophyte that produces an egg. The 2n zygote (a sporophyte) is still within the ovule, which becomes a seed. The ability of seeds to withstand harsh conditions until the environment is again favorable for growth largely accounts for the dominance of seed plants today.

Study the life cycle of the pine tree and describe the sporophyte. _____

Figure 14.9 Life cycle of pine.

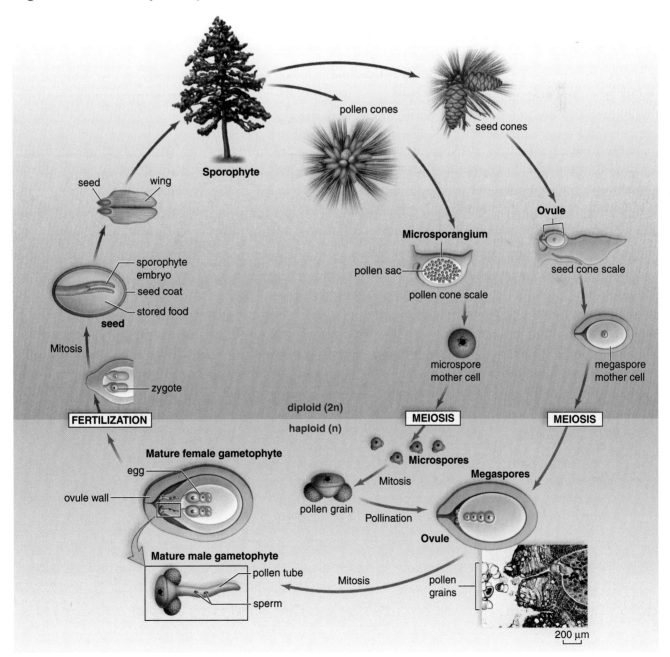

1. Observe the pollen and seed cones on display (Fig. 14.10). Compare their relative size and structure. Remove a single scale (sporophyll) from the male cone and from the seed cone, which has been heated so that the cone opens. Observe with a dissecting microscope. Note the two pollen sacs located on the lower surface of each scale from the pollen cone. Note also the two ovules that may have developed into seeds on the upper surface of each scale from the seed cone.

Figure 14.10 Pine cones.
a. The scales of pollen cones bear pollen sacs, where microspores become pollen grains. **b.** The scales of seed cones bear ovules that develop into winged seeds.

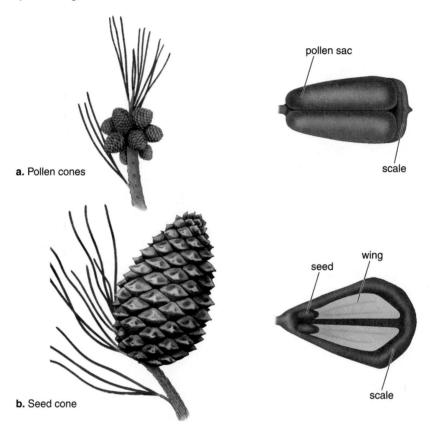

a. Pollen cones

pollen sac

scale

b. Seed cone

seed   wing

scale

2. Examine the prepared slide of a longitudinal section through a mature pollen cone and label Figure 14.11. On the lower surface of each scale are the pollen sacs in which the microspore mother cells produce microspores. Each microspore develops into a male gametophyte, a pollen grain. Under high power, focus on a pollen grain and note the external wings and interior cells. One of these will divide to produce two cells, one of which is the sperm nucleus and the other of which forms the pollen tube, through which a sperm travels to the egg.

Figure 14.11 Pine pollen cone.

Pollen cones bear **(a)** pollen sacs in which microspores develop into pollen grains. **b.** Enlargement of pollen grains.

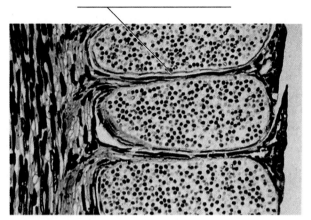

a. Longitudinal section through pine pollen cone, showing pollen sacs.

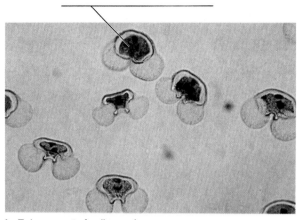

b. Enlargement of pollen grains.

3. Examine the prepared slide of a longitudinal section through a seed cone and label Figure 14.12. In some ovules, you will see a megaspore mother cell surrounded by nutritive cells. You may also be able to find pollen grains just outside or within the integuments of an ovule. At about the time the megaspore mother cell is undergoing meiosis, the scales swell and open to allow the wind-dispersed pollen grains to enter. The megaspore undergoes a series of mitotic divisions and develops into the female gametophyte that contains two archegonia, each of

   which encloses a single large egg. What generation is now within the ovule? _____

Figure 14.12 Seed cone.

Seed cones bear ovules, shown here in longitudinal section. Note pollen grains near the entrance.

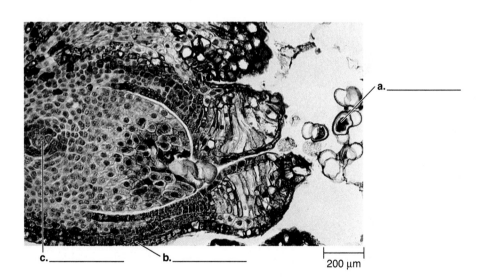

a. _____

c. _____   b. _____   ⊢——⊣ 200 μm

4. Following fertilization, a seed contains the embryonic plant and nutrient material within a seed coat. If available, examine pine seeds. The seeds of gymnosperms are windblown. In seed plants, seeds disperse the sporophyte. If you wish, dissect a seed and, with the help of a hand lens, attempt to find the embryo.

Angiosperms

The angiosperms are the flowering plants. The life cycle of a flowering plant (Fig. 14.13) is like that of the pine tree except for these innovations:

- The often brightly colored flower contains the pollen sacs and ovules. Pollen may be windblown or carried by animals (e.g., insects).
- Flowering plants practice **double fertilization.** A mature pollen grain contains two sperm; one fertilizes the egg, and the other joins with the two polar nuclei to form **endosperm** (3n), which serves as food for the developing embryo.
- Flowering plants have seeds enclosed within fruits. Fruits protect the seeds and aid in seed dispersal. Sometimes, animals eat the fruits, and after the digestion process, the seeds are deposited far away from the parent plant. The term *angiosperm* means "covered seeds." The seeds of angiosperms are found in fruits, which develop from parts of the flower.
  Which generation of a flowering plant bears flowers and fruits? _____

Figure 14.13 Flowering plant life cycle.
The parts of the flower involved in reproduction are the anthers of stamens and the ovules in the ovary of a carpel.

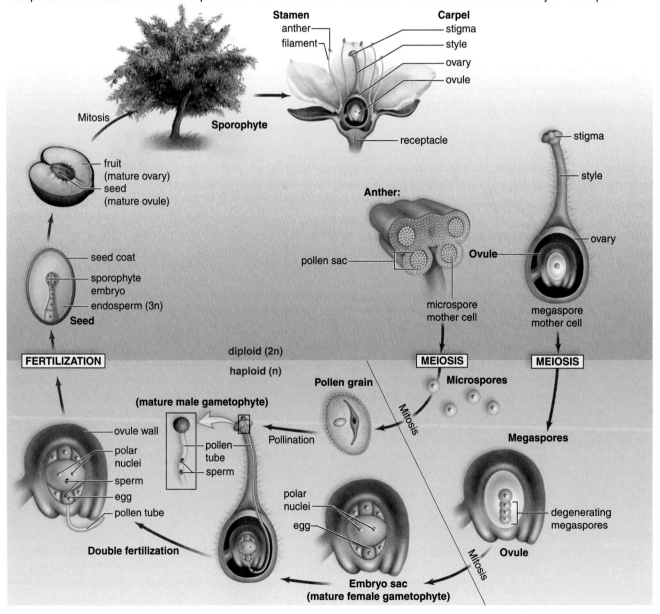

1. With the help of Figure 14.14, identify the following structures on a model of a flower:
   a. **Receptacle:** The portion of a stalk to which the flower parts are attached.
   b. **Sepals:** An outermost whorl of modified leaves, collectively termed the calyx. Sepals are green in most flowers. They protect a bud before it opens.
   c. **Petals:** Usually colored leaves that collectively comprise the corolla.
   d. **Stamen:** A swollen terminal **anther** and the slender supporting **filament.** The anther contains two pollen sacs, where microspores develop into microgametophytes (pollen grains).
   e. **Carpel:** A modified sporophyll consisting of a swollen basal ovary; a long, slender **style** (stalk); and a terminal **stigma** (sticky knob).
   f. **Ovary:** The enlarged part of the carpal that develops into a fruit.
   g **Ovule:** The structure within the ovary where a megaspore develops into a female gametophyte (embryo sac). The ovule becomes a seed.

2. Carefully inspect a fresh flower. What is the common name of your flower? _____

3. Remove the sepals and petals by breaking them off at the base. How many sepals and petals are there? _____

4. Are the stamens taller than the carpel? _____

5. Remove a stamen, and touch the anther to a drop of water on a slide. If nothing comes off in the water, crush the anther a little to release some of its contents. Place a coverslip on the drop, and observe with low- and high-power magnification. What are you observing?

   _____

6. Remove the carpel by cutting it free just below the base. Make a series of thin cross sections through the ovary. The ovary is hollow, and you can see nearly spherical bodies inside. What are these bodies? _____

7. Is your flower a monocot or a eudocot? _____

   _____

Figure 14.14 **Generalized flower.**
A flower has four main kinds of parts: sepals, petals, stamens, and a carpel. A stamen has an anther and filament. A carpel has a stigma, style, and ovary. An ovary contains ovules.

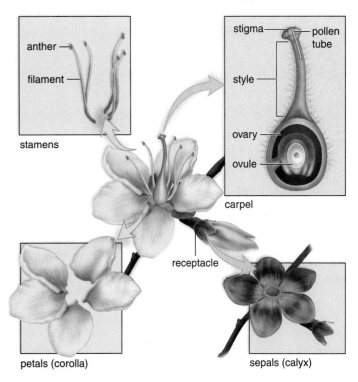

8. Remove a stamen and touch the anther to a drop of water on a slide. If nothing comes off in the water, crush the anther a little to squeeze out some of its contents. Place a coverslip on the drop and observe with low and high powers of the microscope. The somewhat spherical

cells with thick walls are pollen grains. How is pollen dispersed? _____
Flowering plants provide nectar to insects, whose mouthparts are adapted to acquiring the nectar from this particular species of plant. This is a mutualistic relationship. What does a

pollinator do for the plant? _____

9. Remove the carpel by cutting it free just below the base. Make a series of thin cross sections through the ovary. The ovary is hollow and in this cavity there are small, nearly spherical bodies much larger than pollen grains. These are ovules. Remove an ovule that may be still attached by a stalk to the ovary wall. Place the ovule on a drop of water on a slide; cover with a cover glass; press firmly on the top of the cover glass with a clean eraser so as to smear the ovule into a thin mass. Observe with the microscope. Do you see cells within the

ovule? _____

10. At maturity, an angiosperm seed contains the embryo and possibly some nutrient material in the form of endosperm covered by a seed coat. A fruit is a ripened ovary, together with any accessory flower parts that may be associated with the ovary. If available, examine a pea pod and an apple. Find the remnants of the flower parts still attached to the fruit. Open the pea pod and slice the apple. Which portion of each should be associated with the ovules? _____
_____ Which with the ovary? _____ Label flower remnants, fruit, and seed in the following diagram. Can you think of a biological advantage to producing fruits? _____ To producing a fleshy fruit?_____

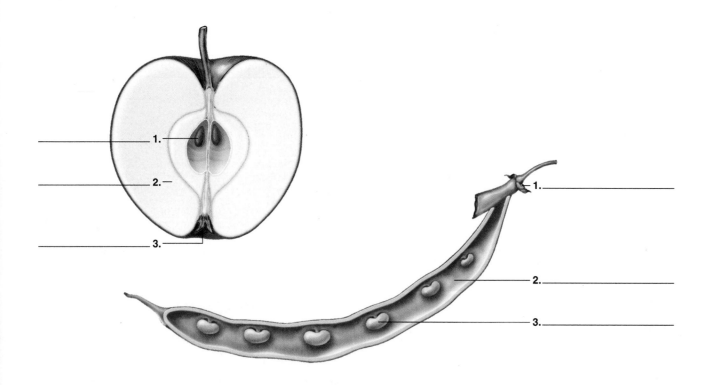

Compare the Life Cycle of a Pine Tree to That of a Flowering Plant

Complete the following table:

	Dominant Generation	Vascular Tissue (Present or Absent)	Dispersal of Sporophyte	Fruit (Present or Absent)
Conifer				
Flowering plant				

Adaptation of Seed Plants to a Land Environment

1. Do seed plants have vascular tissue that transports water and nutrients to all parts of the plant? _____

2. Can seed plants support a large body against the pull of gravity? _____

3. Do seed plants reproduce without dependence on external water? _____

Comparison of Moss, Fern, Pine, and Flowering Plants

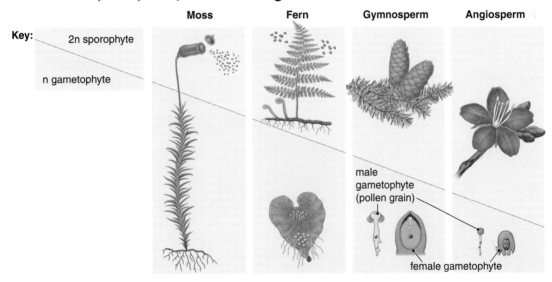

1. The preceding diagram tells you that the size of the _____ became progressively _____ as plants became adapted to live on land. Why is it suitable for this generation to be dependent on the sporophyte? _____

2. Which plants disperse spores? _____

3. Which plants have male and female gametophytes and disperse seeds? _____

4. Why is it appropriate to refer to "bees and plants" to explain sexual reproduction? _____

1. Is the sporophyte the haploid or the diploid generation? —————————————————

2. What generation is in a seed? ——————————————————————————————————

3. Which generation is the dependent one in mosses? —————————————————————

4. What structure is used for dispersal in seedless plants? ———————————————————

5. Which groups of plants have flagellated sperm? —————————————————————

6. Do bryophytes or ferns have vascular tissue? ——————————————————————

7. What generation is dominant in ferns, gymnosperms, and angiosperms? ——————————————

8. Which group of plants studied has vascular tissue but uses spores for dispersal? ———————————

9. What kind of spore becomes a pollen grain in seed plants? ——————————————————

10. The carpel should be associated with the development of which gametophyte, male or female?

    ———

11. The ovule becomes the seed, while the ovary becomes what? —————————————————

12. The gametophyte in seed plants is either dependent on or independent of the sporophyte? ——————

    ———

13. A pine tree, unlike a fern, is able to reproduce sexually in a dry environment. Explain. ———————

    ———

    ———

14. What is the difference between pollination and fertilization? ——————————————————

    ———

LABORATORY

15

Diversity: Animals

Learning Outcomes

Introduction

This laboratory concerns the animal kingdom. Animals are all multicellular and have varying degrees of motility. They are heterotrophic, and most digest their food in a digestive cavity.

One of the themes of today's laboratory will be the similarities and differences between the animals studied. The molluscs (e.g., clam) and arthropods (e.g., crayfish and grasshopper) are invertebrates, while the amphibians (e.g., frog) are vertebrates. Invertebrates lack a backbone of vertebrae that is present in vertebrates. All of the animals studied in today's laboratory have true tissues, bilateral symmetry, tube-within-a-tube body plan, and a coelom or body cavity. They differ in that the molluscs are nonsegmented, while the arthropods and vertebrates are segmented animals with jointed appendages. Jointed appendages are particularly adapted for locomotion on land. The insects are arthropods that are plentiful on land. Among the vertebrates, the amphibians live on land at least part of their lives, while the reptiles, birds, and mammals are primarily land animals (Fig. 15.1).

Figure 15.1 Animal diversity.
The chameleon, a reptile, and the fly, an arthropod, are both very well adapted to living on land. The skin of a reptile and the external skeleton of an arthropod resist drying out. They both breathe by taking in air and both have a suitable means of locomotion on land.

Figure 15.2 Evolution of animals.

The numbered features trace the increasing complexity of animals. These features are a part of the trends that occur during animal evolution. These trends are described in Table 15.1.

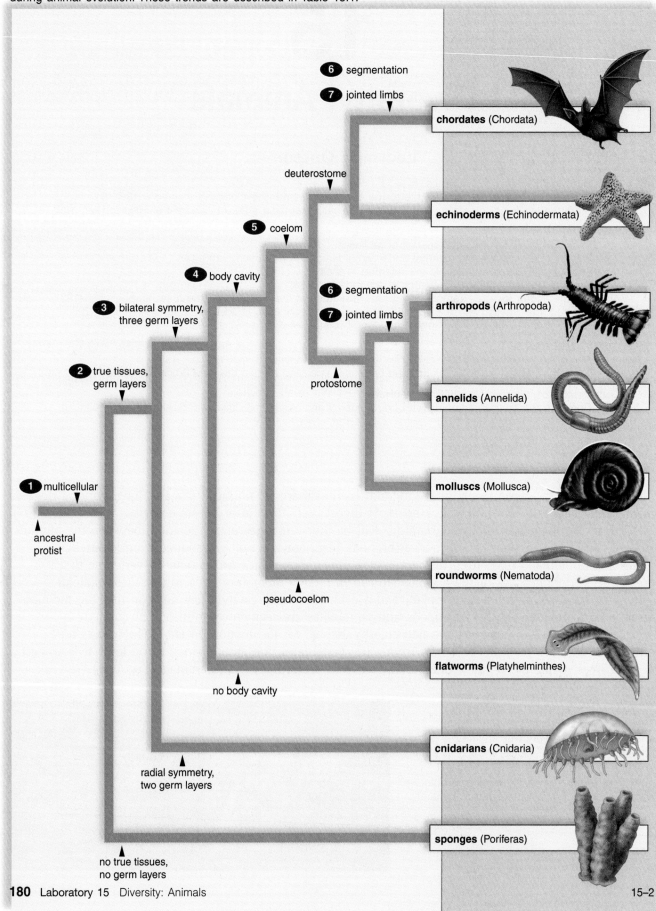

15.1 Evolution and Diversity of Animals

As we survey the evolution of animals in Figure 15.2, certain features (see numbers) allow us to trace an increasing complexity during the evolution of animals. All animals are multicellular, but thereafter the trends listed and described in Table 15.1 become apparent. *The levels of complexity found among the coelomate animals studied today have an asterik.*

Table 15.1	Anatomical Trends Observed During the Evolution of Animals
Level of Organization	
Cellular	Multicellular but cells not organized into tissues
Tissues	Cells organized into tissues
Organs	Tissues organized into organs
*Organ systems	Organs organized into systems
Germ Layers	
None	
Two	Two germ layers: the ectoderm (outside) and the endoderm (inside)
*Three	Three germ layers: the ectoderm, the endoderm, and the mesoderm (middle); presence of mesoderm allows for development of various internal organs
Symmetry	
None	
Radial	Any longitudinal cut through the midpoint yields equal halves; this design allows animals to reach out in all directions
*Bilateral	Only one longitudinal cut through the midpoint yields equal halves; such animals often have well-developed head regions (cephalization)
Body Plan	
None	
Sac plan	Mouth used for intake of nutrient molecules and exit of waste molecules
*Tube-within-a-tube	Separate openings (mouth and anus) for food intake and waste exit; allows for specialization of parts along the digestive canal
Coelom	
Acoelomate	Have no coelom (fluid-filled extracellular spaces)
Pseudocoelomate	Have false coelom; coelom incompletely lined with mesoderm
*Coelomate	Have true coelom, a fluid-filled body cavity completely lined with mesoderm
Segmentation	
Nonsegmented	No segmentation (repeating parts)
*Segmented	A repeating series of parts from anterior to posterior; allows for specialization of some sections for different functions
Jointed appendages	
No jointed appendages	
*Jointed appendages	Jointed appendages, which arose as an added feature to segmentation, permit a means of locomotion highly adapted to the land environment

15.2 Invertebrate Study

All of the animal groups shown in Figure 15.2 contain invertebrates. The vertebrates only occur among the chordates. Most types of invertebrates are adapted to living in the sea; the insects being the major exception to this statement. We will have an opportunity to contrast adaptations to the land environment with adaptations to the aquatic environment.

Molluscs

Most **molluscs** are marine, but there are also some freshwater and terrestrial molluscs (Fig. 15.2). Molluscs are not segmented; instead all molluscs have a three-part body. They have a ventral, muscular **foot** specialized for various means of locomotion; a **visceral mass** that includes the internal organs; and a **mantle,** a thin tissue that encloses the visceral mass and may secrete a shell.

Observation: Diversity of Molluscs

Most molluscs belong to one of four groups. The polyplacophores are grazing marine herbivores, such as chitons, with a body flattened and covered by a shell consisting of eight plates (Fig. 15.3). The bivalves are marine and freshwater sessile filter feeders, such as clams, with a body enclosed by a shell consisting of two valves. These animals have a hatchet-shaped foot but no head or radula. The cephalopods are marine active predators, such as squids. The shell may be reduced or absent; the head, which is anterior to an elongated visceral mass, bears a circle of tentacles and arms. The circulatory system is efficient, and the well-developed nervous system is accompanied by cephalization. The gastropods are marine, freshwater, and often terrestrial herbivores, such as snails, with a coiled shell that distorts body symmetry. The head has tentacles.

1. Examine mollusc specimens, and complete the first two columns of Table 15.2.
2. Examine the foot, and complete the third column of Table 15.2. Some molluscs have a broad, flat foot, others a hatchet-shaped foot; in still others the foot has become tentacles and arms that assist in the capture of food.
3. Indicate in the last column if cephalization is present or not present.

Figure 15.3 Three groups of molluscs.
Top row: Snails (*left*) and nudibranchs (*right*) are gastropods. Middle row: Octopuses (*left*) and nautiluses (*right*) are cephalopods. Bottom row: Scallops (*left*) and mussels (*right*) are bivalves.

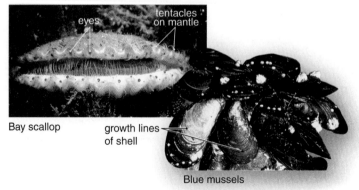

Table 15.2   Molluscan Diversity

Common Name of Specimen	Group	Description of Foot	Cephalization (Yes or No)

Anatomy of Clam

Clams are bivalved because they have right and left shells secreted by the mantle. Clams have no head, and they burrow in sand by extending a "hatchet" foot between the valves. Clams are filter feeders and feed on debris that enters the mantle cavity. Clams have an open circulatory system; the blood leaves the heart and enters sinuses (cavities) by way of anterior and posterior aortas. There are many different types of clams. The one examined here is the freshwater clam *Venus*.

Observation: Anatomy of Clam

External Anatomy

1. Examine the external shell (Fig. 15.4) of a preserved clam (*Venus*). The shell is an **exoskeleton.**
2. Find the posterior and anterior ends. The more pointed end of the **valves** (the halves of the shell) is the posterior end.
3. Determine the clam's dorsal and ventral regions. The valves are hinged together dorsally.
4. What is the function of a heavy shell? _____

Figure 15.4 External view of clam shell.

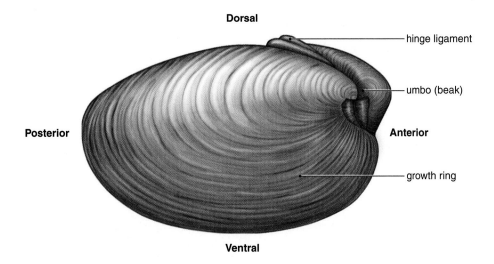

Internal Anatomy

1. Place the clam in the dissecting pan, with the **hinge ligament** and **umbo** (blunt dorsal protrusion) down. Carefully separate the **mantle** from the right valve by inserting a scalpel into the slight opening of the valves. What is a mantle? _____

   _____

2. Insert the scalpel between the mantle and the valve you just loosened.

3. The **adductor muscles** hold the valves together. Cut the adductor muscles at the anterior and posterior ends by pressing the scalpel toward the dissecting pan. After these muscles are cut, the valve can be carefully lifted away. What is the advantage of powerful adductor muscles? _____

   _____

4. Examine the inside of the valve you removed. Note the concentric lines of growth on the outside, the hinge teeth that interlock with the other valve, the adductor muscle scars, and the mantle line. The inner layer of the shell is mother-of-pearl.

5. Examine the rest of the clam (Fig. 15.5) attached to the other valve. Notice the mantle, which lies over the visceral mass and foot, and also the adductor muscles.

6. Bring the two halves of the mantle together. Explain the term *mantle cavity.* _____

   _____

7. Identify the **incurrent** (more ventral) and **excurrent siphons** at the posterior end (Fig. 15.5). Explain how water enters and exits the mantle cavity. _____

   _____

8. Cut away the free-hanging portion of the mantle to expose the **gills.** Does the clam have a respiratory organ? _____

   What type of respiratory organ? _____

9. A mucous layer on the gills entraps food particles brought into the mantle cavity, and the cilia on the gills convey these food particles to the mouth. Why is the clam called a filter feeder?

   _____

   _____

10. The nervous system is composed of three pairs of ganglia (located anteriorly, posteriorly, and in the foot), all connected by nerves. The clam does not have a brain. A ganglion contains a limited number of neurons, whereas a brain is a large collection of neurons in a definite head region.

11. Identify the **foot,** a tough, muscular organ for locomotion, and the **visceral mass,** which lies above the foot and is soft and plump. The visceral mass contains the digestive and reproductive organs.

12. Identify the **labial palps** that channel food into the open mouth.

13. Identify the **anus,** which discharges into the excurrent siphon.

14. Find the **intestine** by its dark contents. Trace the intestine forward until it passes into a sac, the clam's only evidence of a coelom.

15. Locate the **pericardial sac (pericardium)** that contains the heart. The intestine passes through the heart. The heart pumps blood into the aortas, which deliver it to blood sinuses in the tissues.

    A clam has an **open circulatory system.** Explain. _____

    _____

16. Cut the visceral mass and the foot into exact left and right halves, and examine the cut surfaces. Identify the greenish-brown digestive glands; the stomach, embedded in the digestive glands; and the intestine, which winds about in the visceral mass. Reproductive organs are also present.

Figure 15.5 Anatomy of a bivalve.

The mantle has been removed to reveal the internal organs. **a.** Drawing. **b.** Dissected specimen.

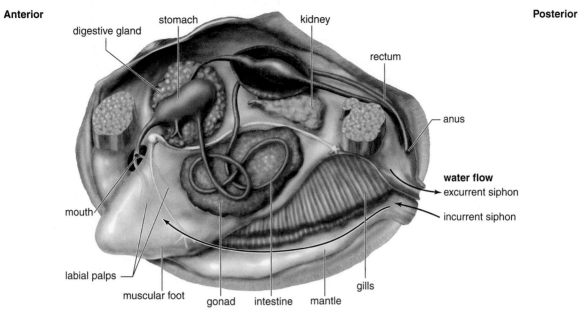

a Digestive system (green) of clam

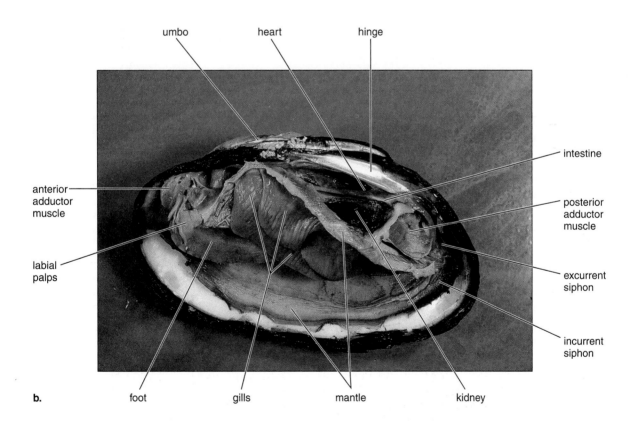

b.

Anatomy of Squid

Squids are cephalopods (means head-foot) because they have a well-defined head (Fig. 15.6). The head contains a brain and bears sense organs. The squid moves quickly by jet propulsion of water, which enters the mantle cavity by way of a space that circles the head. When the cavity is closed off, water exits by means of a funnel. Then the squid moves rapidly in the opposite direction.

The squid seizes fish with its tentacles; the mouth has a pair of powerful, beaklike jaws and a **radula,** a filelike organ containing rows of teeth. The squid has a closed circulatory system composed of vessels and three hearts, one of which pumps blood to all the internal organs, while the other two pump blood to the gills located in the mantle cavity.

Observation: Anatomy of Squid

1. Examine a preserved squid.
2. Refer to Figure 15.6 for help in locating the beaklike jaws, which are encircled by the tentacles and arms.
3. Locate the head with its sense organs, notably the large, well-developed eye.
4. Find the funnel, where water exits from the mantle cavity, causing the squid to move backward.
5. If the squid has been dissected, note the heart, gills, and blood vessels.

Clam Anatomy Compared with Squid Anatomy

1. Compare clam anatomy with squid anatomy by completing Table 15.3.
2. Explain how both clams and squids are adapted to their way of life.

_____

_____

_____

Table 15.3	Comparison of Clam and Squid	
	Clam	Squid
Feeding mode		
Skeleton		
Circulation		
Cephalization		
Locomotion		
Nervous system		

Figure 15.6 Anatomy of a squid.

The squid is an active predator and lacks the external shell of a clam. It captures fish with its tentacles and bites off pieces with its jaws. A strong contraction of the mantle forces water out the funnel, resulting in "jet propulsion." **a.** Drawing. **b.** Dissected specimen.

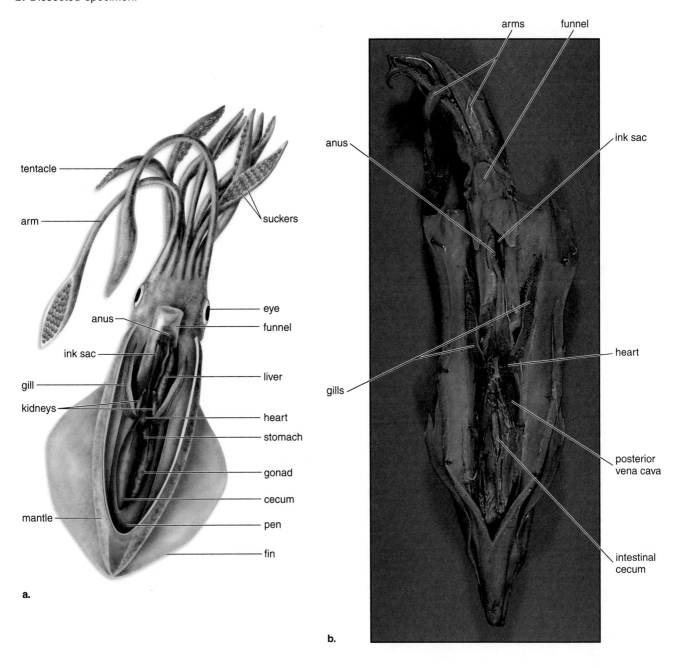

a.

b.

Arthropods

The **arthropods** are segmented and have paired, jointed appendages and a hard exoskeleton that contains chitin. The chitinous exoskeleton consists of hardened plates separated by thin, membranous areas that allow movement of the body segments and appendages.

Today we will study two main groups of arthropods: the **crustaceans** (e.g., crayfish, lobsters, shrimps, and crabs) and the **insects** (e.g., grasshoppers, ants, beetles, and butterflies). Other familiar arthropods are the arachnids (e.g., spiders, scorpions, and ticks) and also the millipedes and centipedes. The millipedes generally have two pairs of legs per segment, as exemplified by the pillbug that you studied in Laboratory 1. The **centipedes** have one pair of legs per segment and, compared to millipedes, they have a flattened body. They also have poison fangs directly under the head.

Observation: Diversity of Arthropods

1. Examine various specimens of arthropods (Fig. 15.7), and complete Table 15.4.
2. In the last column of Table 15.4, note the number and type of appendages attached to the thorax and/or abdomen.

Table 15.4	Arthropod Diversity	
Common Name of Specimen	**Group**	**Appendages (Attached to Body)**

Figure 15.7 Crustacean diversity.

a. A copepod uses its long antennae for floating and its feathery maxillae for filter feeding. **b.** Shrimp and **(c)** crabs are decapods—they have five pairs of walking legs. Marine shrimp feed on codepods. **d.** Barnacles have no abdomen and a reduced head; the thoracic legs project through a shell to filter feed. The gooseneck barnacles attach to an object such as ships and buoys by a long stalk.

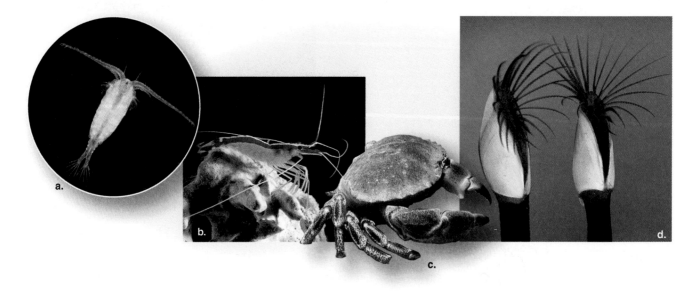

Anatomy of Crayfish

Crayfish belong to a group of arthropods called crustaceans (Fig. 15.8). Crayfish are adapted to an aquatic existence. They are known to be scavengers, but they also prey on other invertebrates. The mouth is surrounded by appendages modified for feeding, and there is a well-developed digestive tract. Dorsal, anterior, and posterior arteries carry hemolymph (blood plus lymph) to tissue spaces (hemocoel) and sinuses. In contrast to vertebrates, there is a ventral solid nerve cord.

Observation: Anatomy of Crayfish

External Anatomy

1. Obtain a preserved crayfish, and place it in a dissecting pan.
2. Identify the chitinous **exoskeleton.** With the help of Figure 15.8, identify the cephalothorax and abdomen. Together, the head and thorax are called the **cephalothorax;** the cephalothorax is

Figure 15.8 Anatomy of a crayfish.
a. Drawing of external anatomy (male). **b.** Dissection of internal anatomy (female).

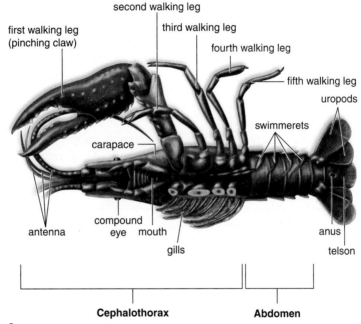

second walking leg

first walking leg (pinching claw)

third walking leg

fourth walking leg

fifth walking leg

uropods

swimmerets

carapace

antenna

compound eye mouth

gills

anus

telson

Cephalothorax   Abdomen

a.

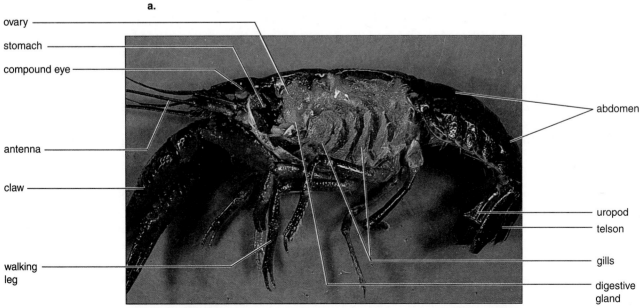

ovary

stomach

compound eye

antenna

claw

walking leg

abdomen

uropod

telson

gills

digestive gland

b.

covered by the **carapace.** Has specialization of segments occurred? _____ Explain. _____

_____

3. Find the **antennae,** which project from the head. At the base of each antenna, locate a small, raised nipple containing an opening for the **green glands,** the organs of excretion. Crayfish excrete a liquid nitrogenous waste.
4. Locate the **compound eyes,** composed of many individual units for sight. Do crayfish demonstrate cephalization? _____ Explain. _____
5. Identify the six pairs of appendages around the mouth for handling food.
6. Find the five pairs of walking legs attached to the cephalothorax. The most anterior pair is modified as pincerlike claws.
7. Locate the five pairs of **swimmerets** on the abdomen. In the male, the anterior two pairs are stiffened and folded forward. They are claspers that aid in the transfer of sperm during mating.
8. In the female, identify the **seminal receptacles,** a swelling located between the bases of the third and fourth pairs of walking legs. Sperm from the male are deposited in the seminal receptacles. In the male, identify the opening of the sperm duct located at the base of the fifth walking leg.

    What sex is your specimen? _____

9. Examine the opposite sex also.
10. Find the last abdominal segment, which bears a pair of broad, fan-shaped **uropods** that, together with a terminal extension of the body, form a tail. Has specialization of appendages occurred? _____ Explain. _____

Internal Anatomy

1. Place the crayfish in the dissecting pan.
2. Cut away the lateral surface of the carapace with scissors to expose the **gills** (Fig. 15.8*b*). Observe that the gills occur in distinct, longitudinal rows. How many rows of gills are there in your specimen? _____ The outer row of gills is attached to the base of certain appendages. Which ones? _____

    These outer gills are the **podobranchia** ("foot gills"). How many podobranchia do you find in your specimen? _____

3. Carefully separate the gills with a probe or dissecting needle, and locate the inner row(s) of gills. These inner gills are the **arthrobranchia** ("joint gills") and are attached to the chitinous membrane that joins the appendages to the thorax. How many rows of arthrobranchia do you find in the specimen? _____

4. Remove a gill with your scissors by cutting it free near its point of attachment, and place it in a watch glass filled with water. Observe the numerous gill filaments arranged along a central axis.
5. Carefully cut away the dorsal surface of the carapace with scissors and a scalpel. The epidermis that adheres to the exoskeleton secretes the exoskeleton. Remove any epidermis adhering to the internal organs.
6. Identify the diamond-shaped heart lying in the middorsal region. A crayfish has an open circulatory system. Carefully remove the heart.
7. Locate the **gonads** anterior to the heart in both the male and female. The gonads are tubular structures bilaterally arranged in front of the heart and continuing behind it as a single mass. In the male, the testes are highly coiled, white tubes.
8. Find the **mouth;** the short, tubular **esophagus;** and the two-part **stomach,** with the attached **digestive gland,** that precedes the intestine.
9. Identify the **green glands,** two excretory structures just anterior to the stomach, on the ventral segment wall.
10. Remove the thoracic contents previously identified.
11. Identify the **brain** in front of the esophagus. The brain is connected to the ventral nerve cord by a pair of nerves that pass around the esophagus.

12. Remove the animal's entire digestive tract, and float it in water. Observe the various parts, especially the connections of the digestive gland to the stomach.

13. Cut through the stomach, and notice in the anterior region of the stomach wall the heavy, toothlike projections, called the **gastric mill,** that grind up food. Do you see any grinding stones ingested by the crayfish? _____

If possible, identify what your specimen had been eating. _____

Anatomy of Grasshopper

The grasshopper is an insect (Fig. 15.9). All insects have a head, a thorax, and an abdomen. Their appendages always include three pairs of jointed legs and usually two antennae as sensory organs. In insects with wings, such as the grasshopper, wings are attached to the thorax. Respiration is by a highly branched internal system of tubes, called tracheae. Insects are adapted to life on land.

Observation: Anatomy of Grasshopper

External Anatomy

1. Obtain a preserved grasshopper (*Romalea*), and study its external anatomy with the help of Figure 15.10. Identify the head, thorax, and abdomen.

2. Locate the leathery forewings and the inner, membranous hind wings attached to the thorax. Then, Identify the three pairs of legs attached to the thorax. Which pair of legs is used for jumping? _____ How many segments does each leg have? _____

Figure 15.9 Insect diversity.

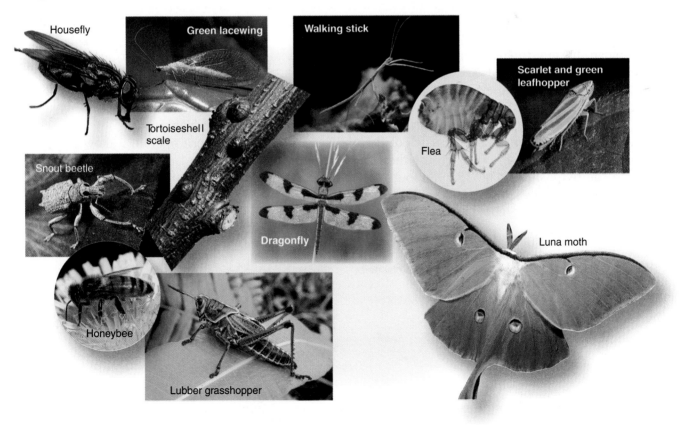

Figure 15.10 External anatomy of a female grasshopper, *Romalea*.
a. The legs and wings are attached to the thorax. b. The head has mouthparts of various types.

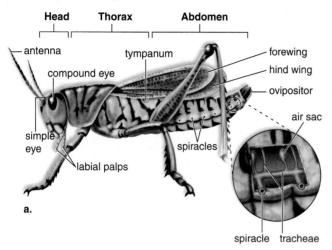

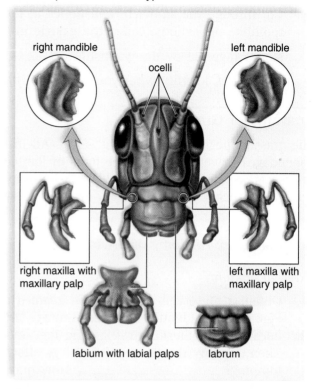

3. Is locomotion in the grasshopper adapted to land?_____

   Explain. _____

   _____

4. Use a hand lens or dissecting microscope to examine the grasshopper's special sense organs of the head. Identify the **antennae** (a pair of long, jointed feelers); the **compound eyes;** and the three, dotlike **simple eyes.**
5. Remove the **mouthparts** by grasping them with forceps and pulling them out. Arrange them in order on an index card, and compare them with Figure 15.10*b*. These mouthparts are used for chewing and are quite different from those of a piercing and sucking insect.
6. Identify the **tympana** (sing., **tympanum**), one on each side of the first abdominal segment (Fig. 15.10*a*). The grasshopper detects sound vibrations with these membranes.
7. Locate the **spiracles,** along the sides of the abdominal segments. These openings allow air to enter the tracheae, which constitute the respiratory system.
8. Find the **ovipositors** (Fig. 15.11*a*), four curved and pointed processes projecting from the hind end of the female. These are used to dig a hole in which eggs are laid. The male has **claspers** that are used during copulation (Fig. 15.11*b*).

Figure 15.11 Grasshopper genitalia.
a. Females have an ovipositor, and (b) males have claspers.

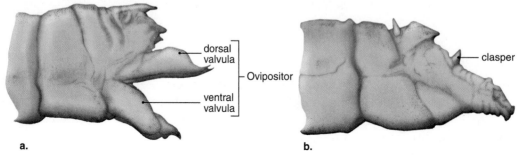

Internal Anatomy

1. Detach the wings and legs of the grasshopper identified in number 2, page 191. Then turn the organism on its side, and use scissors to carefully cut through the exoskeleton (above the spiracles) along the full length (from the head to the posterior end) of the animal. Repeat this procedure on the other side.
2. Cut crosswise behind the head so that you can remove a strip of the exoskeleton. If necessary, reach in with a probe to loosen the muscle attachments and membranes.
3. Pin the insect to the dissecting pan, dorsal side up. Cover the specimen with water to keep the tissues moist.
4. Identify the heart (Fig. 15.12) and aorta just beneath the portion of exoskeleton you removed. A grasshopper has an open circulatory system. Remove the heart and adjacent tissues.
5. Locate the **fat body,** a yellowish fatty tissue that covers the internal organs. Carefully remove it.
6. Find the **tracheae,** the respiratory system of insects. Using the dissecting microscope, look for glistening white tubules, which deliver air to the muscles.
7. Identify the reproductive organs that lie on either side of the digestive tract in the abdomen. If your specimen is a male, look for the testis, a coiled, elongated cord containing many tubules. If your specimen is a female, look for the ovary, essentially a collection of parallel, tapering tubules containing cigar-shaped eggs.
8. Locate the digestive tract and, in sequence, the **crop,** a large pouch for storing food (a grasshopper eats grasses); the **gastric ceca,** digestive glands attached to the stomach; the stomach and the intestine, which continues to the anus; and **Malpighian tubules,** excretory organs attached to the intestine. Insects secrete a dry solid nitrogenous waste. Is this an

   adaptation to life on land? _____ Explain. _____

   _____

9. Work the digestive tract free, and move it to one side. Now identify the **salivary glands** that extend into the thoracic cavity.
10. Remove the internal organs. Now identify the ventral **nerve cord,** thickened at intervals by ganglia.
11. Remove one side of the exoskeleton covering the head. Identify the brain, anterior to the esophagus.

Figure 15.12 Internal anatomy of a female grasshopper.
The digestive system of a grasshopper shows specialization of parts.

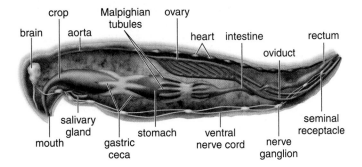

Conclusions

Compare the adaptations of a crayfish with those of a grasshopper by completing Table 15.5. Put a star beside each item that indicates an adaptation to life in the water (crayfish) and to life on land (grasshopper). How many did you identify? _____ Check with your instructor to see if you identified the maximum number of adaptations.

Table 15.5	Comparison of Crayfish and Grasshopper	
	Crayfish	Grasshopper
Locomotion		
Respiration		
Nervous system		
Reproductive features		
Sense organs		

Insect Metamorphosis

Metamorphosis means a change, usually a drastic one, in form and shape. Some insects undergo what is called *complete metamorphosis,* in which case they have three stages of development: the larval stages, the pupa stage, and finally the adult stage. Metamorphosis occurs during the pupa stage, when the animal is enclosed within a hard covering. The animals best known for metamorphosis are the butterfly and the moth, whose larval stage is called a caterpillar and whose pupa stage is the cocoon; the adult is the butterfly or moth (Fig. 15.13a). Grasshoppers undergo *incomplete metamorphosis,* a gradual change in form rather than a drastic change. The immature stages of the grasshopper are called nymphs rather than larvae, and they are recognizable as grasshoppers even though they differ somewhat in shape and form (Fig. 15.13b).

If available, examine life cycle displays or plastomounts that illustrate complete and incomplete metamorphosis.

Observation: Insect Metamorphosis

Observe any insects available, and state in Table 15.6 whether they have complete metamorphosis or incomplete metamorphosis.

Table 15.6	Insect Metamorphosis
Common Name of Specimen	**Complete or Incomplete Metamorphosis**

Figure 15.13 Insect metamorphosis.

During **(a)** complete metamorphosis, a series of larvae leads to pupation. The adult hatches out of the pupa. During **(b)** incomplete metamorphosis, a series of nymphs leads to a full-grown grasshopper.

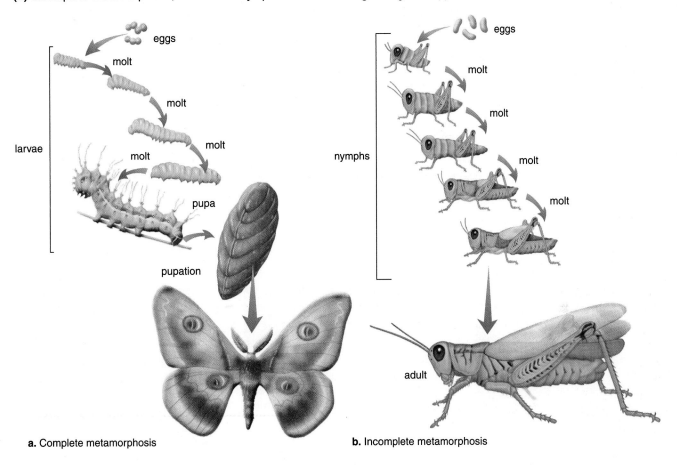

a. Complete metamorphosis

b. Incomplete metamorphosis

Conclusions

- With reference to Figure 15.13, what stage is missing when an insect does not have complete metamorphosis? _____ What happens during this stage? _____

- What form, the larvae or the adult, disperses offspring in flying insects? _____ How is this a benefit? _____

  _____

- In insects that undergo complete metamorphosis, the larvae and the adults use different food sources and habitats. Why might this be a benefit? _____

  _____

- With reference to insects that undergo incomplete metamorphosis, which form, the nymphs or the adult, have better developed wings? _____ What is the benefit of wings to an insect?

  _____

  _____

15.3 Vertebrate Study

Like arthropods, **vertebrates** are segmented, and specialization of parts has occurred. Arthropods have an exoskeleton, while vertebrates have an endoskeleton, but they both have jointed appendages. In vertebrates, two pairs of appendages are characteristic. The vertebrate brain is more complex than that of arthropods and is enclosed by a skull. Among vertebrates, a high degree of cephalization is the rule. All organ systems are present and efficient.

Diversity of Vertebrates

Like arthropods, there are both aquatic and terrestrial vertebrates (as shown in Fig. 15.14). Fishes are aquatic; adult amphibians may be terrestrial, but most must return to an aquatic habitat to reproduce; and reptiles are terrestrial, although some reptiles, such as sea turtles, are secondarily adapted to life in the water. Mammals are adapted to a wide variety of habitats, including air (e.g., bats), sea (e.g., whales), and land (e.g., humans).

Figure 15.14 Vertebrate groups.

a. Cartilaginous fishes: blue shark, *Prionace glauca*

b. Boneyfishes: blueback butterflyfish *Chaetodon plebius*

c. Amphibians: northern leopard frog, *Rana pipiens*

d. Reptiles: Pearl River redbelly turtle, *Pseudemys* sp.

e. Birds: scissor-tailed flycatcher, *Aves tyrannidae*

f. Mammals: grey fox, *Urocyon cinereoargenteus*

Anatomy of Frog

Frogs are amphibians, a group of animals in which metamorphosis occurs. Metamorphosis includes a change in structure, as when an aquatic tadpole becomes a frog with lungs and limbs (Fig. 15.15). Amphibians were the first vertebrates to be adapted to living on land; however, they typically return to the water to reproduce. *Place a check in the margin below for every adaptation to a land environment.*

Observation: External Anatomy of Frog

1. Place a preserved frog (*Rana pipiens*) in a dissecting tray.
2. Identify the bulging eyes, which have nonmovable upper and lower lids but can be covered by a **nictitating membrane** that moistens the eye.
3. Locate the **tympanum** behind each eye (Fig. 15.15). What is the function of a tympanum? _____

   _____

4. Examine the external **nares** (sing., **naris,** or **nostril**) (Fig. 15.15). Insert a probe into an external naris, and observe that it protrudes from one of the paired, small openings, the internal

   nares, inside the mouth cavity. What is the function of the nares? _____
5. Identify the paired limbs. The bones of the forelimbs and hindlimbs are the same as in all tetrapods, in that the first bone articulates with a girdle and the limb ends in phalanges. The hind feet have five phalanges, and the forefeet have only four phalanges. Which pair of limbs

   is longest? _____

   How does a frog locomote on land? _____

   _____

   What is a frog's means of locomotion in the water? _____

   _____

Figure 15.15 **External frog anatomy.**

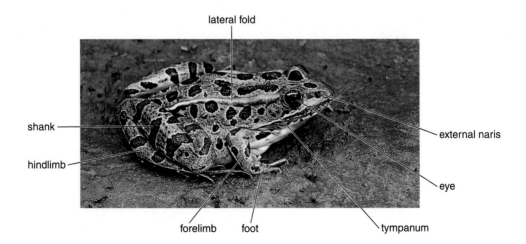

Mouth

1. Open your frog's mouth very wide (Fig. 15.16), cutting the angles of the jaws if necessary.
2. Identify the tongue attached to the lower jaw's anterior end.
3. Find the **auditory (eustachian) tube** opening in the angle of the jaws. These tubes lead to the ears. Auditory tubes equalize air pressure in the ears.
4. Examine the **maxillary teeth** located along the rim of the upper jaw. Another set of teeth—**vomerine teeth**—is present just behind the midportion of the upper jaw.
5. Locate the **glottis,** a slit through which air passes into and out of the **trachea,** the short tube from glottis to lungs. What is the function of a glottis? _____

   _____

6. Identify the **esophagus,** which lies dorsal and posterior to the glottis and leads to the stomach.

    If available, examine life cycle displays or plastomounts that illustrate complete and incomplete metamorphosis.

Figure 15.16 **Mouth cavity of a frog.**
a. Drawing. **b.** Dissected specimen.

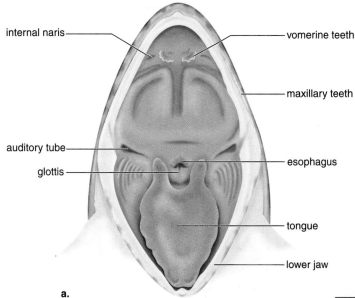

internal naris — — vomerine teeth

— maxillary teeth

auditory tube —
glottis — — esophagus

— tongue

— lower jaw

a.

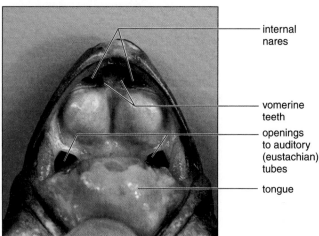

— internal nares

— vomerine teeth
— openings to auditory (eustachian) tubes
— tongue

b.

Opening the Frog

1. Place the frog ventral side up in the dissecting pan. Lift the skin with forceps, and use scissors to make a large, circular cut to remove the skin from the abdominal region as close to the limbs as possible. Cut only skin, not muscle.
2. Now, remove the muscles by cutting through them in the same circular fashion. At the same time, cut through any bones you encounter. A vein, called the abdominal vein, will be slightly attached to the internal side of the muscles.
3. Identify the **coelom,** or body cavity.
4. If your frog is female, the abdominal cavity is likely to be filled by a pair of large, transparent **ovaries,** each containing hundreds of black and white eggs. Gently lift the left ovary with forceps, and find its place of attachment. Cut through the attachment, and remove the ovary in one piece.

Respiratory System and Liver

1. Insert a probe into the glottis, and observe its passage into the trachea. Enlarge the glottis by making short cuts above and below it. When the glottis is spread open, you will see a fold on either side; these are the vocal cords used in croaking.
2. Identify the **lungs,** two small sacs on either side of the midline and partially hidden under the liver (Fig. 15.17). Trace the path of air from the external nares to the lungs. _____

   _____
3. Locate the **liver,** the large, prominent, dark brown organ in the midventral portion of the trunk (Fig. 15.17). Between the right half and left half of the liver, find the **gallbladder.**

Circulatory System

1. Lift the liver gently. Identify the **heart,** covered by a membranous covering (the **pericardium**). With forceps, lift the covering, and gently slit it open. The heart consists of a single, thick-walled **ventricle** and two (right and left) anterior, thin-walled **atria.**
2. Locate the three large veins that join beneath the heart to form the **sinus venosus.** (To lift the heart, you may have to snip the slender strand of tissue that connects the atria to the pericardium.) Blood from the sinus venosus enters the right atrium. The left atrium receives blood from the lungs.
3. Find the **conus arteriosus,** a single, wide arterial vessel leaving the ventricle and passing over the right atrium. Follow the conus arteriosus forward to where it divides into three branches on each side. The middle artery on each side is the **systemic artery,** which fuses behind the heart to become the **dorsal aorta.** The dorsal aorta transports blood through the body cavity and gives off many branches. The **posterior vena cava** begins between the two kidneys and returns blood to the sinus venosus. Which vessel lies above (dorsal to) the other? _____

   _____

Digestive Tract

1. Identify the **esophagus,** a very short connection between the mouth and the stomach. Lift the left liver lobe, and identify the stomach, whitish and J-shaped. The **stomach** connects with the esophagus anteriorly and with the small intestine posteriorly.
2. Find the **small intestine** and the **large intestine,** which enters the **cloaca** (see Figs. 15.18 and 15.19). The cloaca lies beneath the pubic bone and is a general receptacle for the intestine, the reproductive system, and the urinary system. It opens to the outside by way of the anus. Trace the path of food in the digestive tract from the mouth to the cloaca. _____

   _____

Accessory Glands

1. You identified the liver and gallbladder previously. Now try to find the **pancreas,** a yellowish organ near the stomach and intestine.

Figure 15.17 Internal organs of a female frog, ventral view.

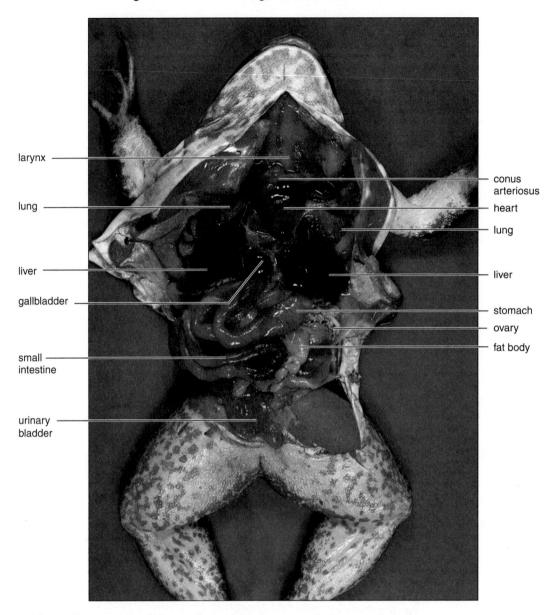

larynx

lung

liver

gallbladder

small intestine

urinary bladder

conus arteriosus

heart

lung

liver

stomach

ovary

fat body

2. Locate the **spleen,** a small, pea-shaped body near the stomach.

Urogenital System

1. Identify the **kidneys,** long, narrow organs lying against the dorsal body wall (Fig. 15.18).
2. Locate the **testes** in a male frog (Fig. 15.18). Testes are yellow, oval organs attached to the anterior portions of the kidneys. Several small ducts, the **vasa efferentia,** carry sperm into kidney ducts that also carry urine from the kidneys. **Fat bodies,** which store fat, are attached to the testes.
3. Locate the ovaries in a female frog. The ovaries are attached to the dorsal body wall (Fig. 15.19). Fat bodies are also attached to the ovaries. Highly coiled **oviducts** lead to the cloaca. The ostium (opening) of the oviduct is dorsal to the liver.
4. Find the **mesonephric ducts**—thin, white tubes that carry urine from the kidney to the cloaca. In female frogs, you will have to remove the left ovary to see the mesonephric ducts.
5. Locate the **cloaca.** You will need to split through the bones of the pelvic girdle in the midventral line and carefully separate the bones and muscles to find the cloaca.
6. Identify the urinary bladder attached to the ventral wall of the cloaca. In frogs, urine backs up into the bladder from the cloaca.

7. Explain the term *urogenital system.* _____

   _____

8. The cloaca receives material from (1) _____,

   (2) _____, and (3) _____.

Figure 15.18 **Urogenital system of a male frog.**
a. Drawing. **b.** Dissected specimen.

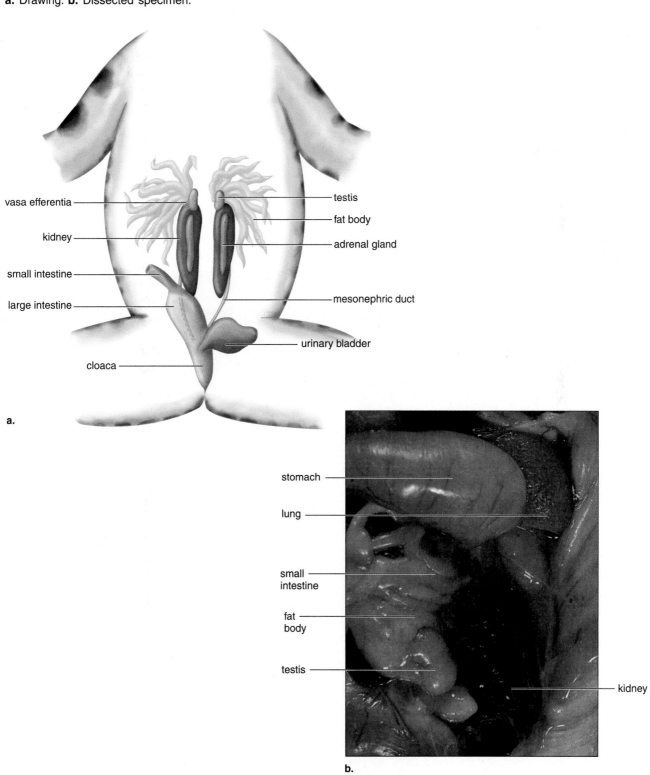

vasa efferentia

kidney

small intestine

large intestine

cloaca

testis

fat body

adrenal gland

mesonephric duct

urinary bladder

a.

stomach

lung

small
intestine

fat
body

testis

kidney

b.

Figure 15.19 Urogenital system of a female frog.
a. Drawing. **b.** Dissected specimen.

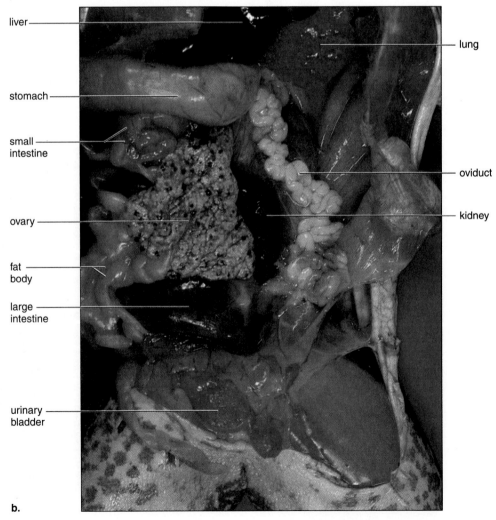

ostium

oviduct

kidney

ovary

large intestine

urinary bladder

fat body

adrenal gland

mesonephric duct

uterus

cloaca

a.

liver

stomach

small intestine

ovary

fat body

large intestine

urinary bladder

lung

oviduct

kidney

b.

15.4 Comparison of Invertebrates with Vertebrates

Vertebrates are segmented, and specialization of parts has occurred. Arthropods have an exoskeleton, while vertebrates have an endoskeleton, but they both have jointed appendages. In vertebrates, two pairs of appendages are characteristic. The vertebrate brain is more complex than that of arthropods and is enclosed by a skull. Among vertebrates, a high degree of cephalization is the rule. All organ systems are present and efficient.

Complete Table 15.7 with a "yes" or "no" to compare the crayfish with the frog.

Table 15.7	Comparison of an Invertebrate (Crayfish) with a Vertebrate (Frog)				
	Coelom	Exoskeleton	Endoskeleton	Jointed Appendages	Location of Nerve Cord (ventral/dorsal)
Crayfish					
Frog					

1. Jointed appendages and an exoskeleton are characteristic of what group of animals? _____

2. Crayfish belong to what group of arthropods? _____

3. A clam belongs to what group of molluscs? _____

4. A visceral mass, foot, and mantle are characteristic of what group of animals? _____

5. In a clam, which structure secretes the shell? _____

6. The arthropods have what feature not seen in any other invertebrate group? _____

7. Contrast the respiratory organ of a crayfish with that of a grasshopper. _____

8. In a frog, the glottis allows air to enter the _____

9. In a frog, the esophagus allows food to enter the _____

10. In a frog, the cloaca receives material from the intestine, the kidneys, and the _____

11. Compare respiratory organs in the crayfish and the grasshopper. How are these suitable to the habitat of each? _____

12. For each of the following characteristics, name an animal with the characteristic, and state the characteristic's advantages:

    a. Presence of a head region _____
    _____

    b. Jointed appendages _____
    _____

16

Organization of Flowering Plants

Learning Outcomes

Introduction

Despite their great diversity in size and shape, flowering plants all have three vegetative organs that function in growth and nutrition but have nothing to do with reproduction. The vegetative organs are the root, the stem, and the leaf. Roots anchor a plant and absorb water and minerals from the soil. A stem usually supports the leaves so that they are exposed to sunlight. Leaves carry on photosynthesis, and thereby produce the nutrients that sustain a plant.

Each of these organs contains various tissues, arranged differently depending on whether a flowering plant is a monocot or a eudicot. The arrangement of tissues is distinctive enough that you should be able to identify the plant as a monocot or eudicot when examining a slide of a root, stem, or leaf.

Another way to group flowering plants is according to whether they are herbaceous or woody. All flowering trees are woody; their stems contain wood. Many flowering garden plants and all grasses are herbaceous (nonwoody). Herbaceous plants have only **primary growth,** which increases their height. Woody plants have both primary and **secondary growth.** Secondary growth increases the girth of a tree.

16.1 External Anatomy of a Flowering Plant

Figure 16.1 shows that a plant has a root system and a shoot system. The **root system** consists of a primary root and any and all of its lateral (side) roots. The **shoot system** consists of the stem and leaves.

Observation: A Living Plant

Root System

Observe the root system of a living plant if the root system is exposed. Does the plant have a taproot system—that is, a main root many times larger than the lateral roots?

_____ Or does the plant have a fibrous root system—that is, all the roots approximately the same size? _____ What is the primary function of the root system?

_____

Shoot System

What is the primary function of the shoot system? _____

The Stem

1. Observe the **stem.** Locate a **node** and an **internode.**
2. Measure the length of the internode in the middle of the stem. Does the internode get larger or smaller toward the apex of the stem? _____ Toward the roots? _____ Based on the fact that a stem elongates as it grows, explain your observation. _____

    _____

    _____

3. Where is the **terminal bud** of a stem?

    _____

    Where is the **axillary bud?**

    _____

The Leaves

1. Describe the **blade.** _____
2. Describe the **petiole.** _____

Figure 16.1 Organization of a plant.
Roots, stems, and leaves are vegetative organs.

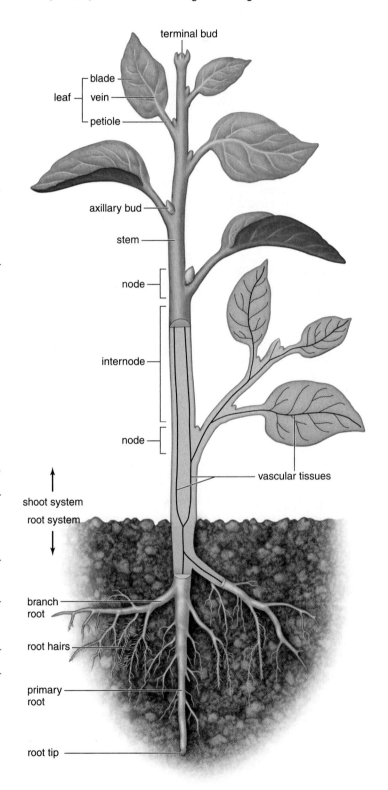

Figure 16.2 Monocots versus eudicots.

The five features illustrated here are used to distinguish monocots from eudicots.

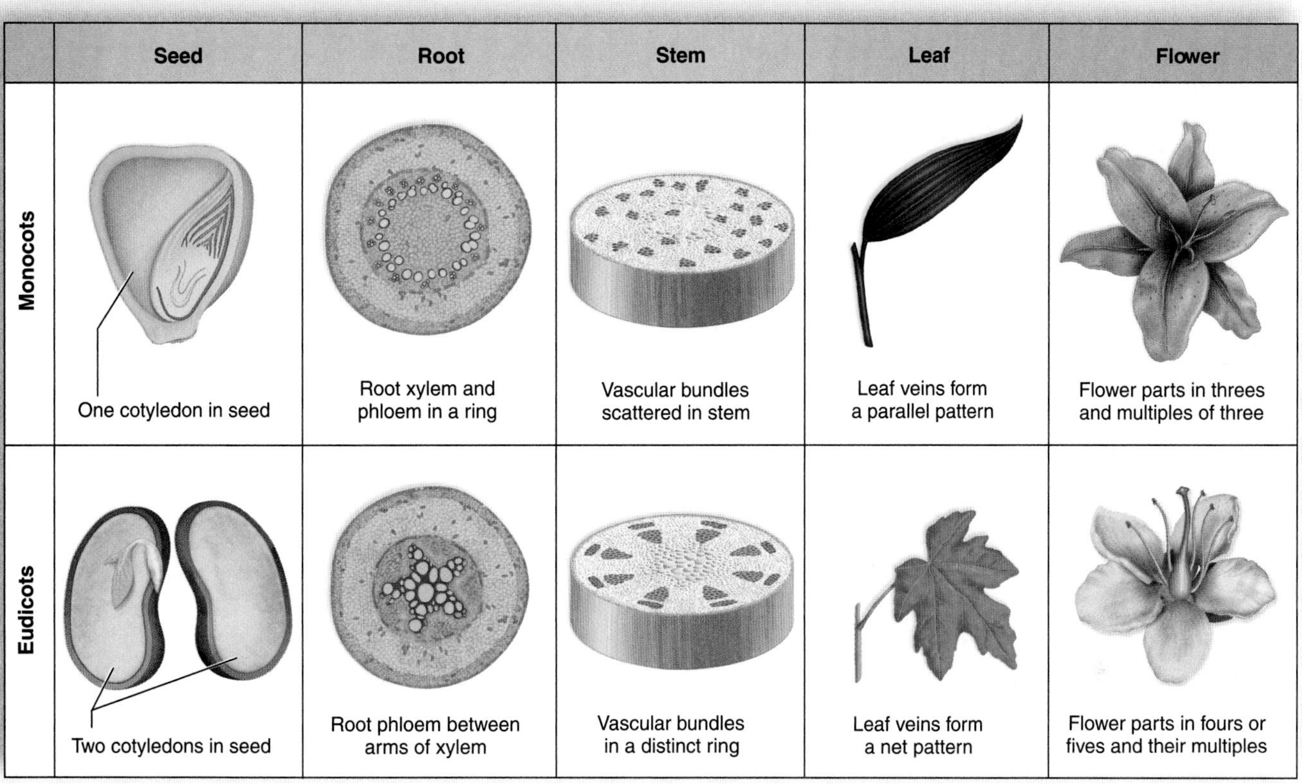

	Seed	Root	Stem	Leaf	Flower
Monocots	One cotyledon in seed	Root xylem and phloem in a ring	Vascular bundles scattered in stem	Leaf veins form a parallel pattern	Flower parts in threes and multiples of three
Eudicots	Two cotyledons in seed	Root phloem between arms of xylem	Vascular bundles in a distinct ring	Leaf veins form a net pattern	Flower parts in fours or fives and their multiples

Monocots Versus Eudicots

Flowering plants are classified into two groups: **monocots** and **eudicots.** In this laboratory, you will be studying the differences between monocots and eudicots with regard to the roots, stems, and leaves, as noted in Figure 16.2.

Experimental Procedure: Monocot Versus Eudicot

1. Observe the leaves of the plant you are studying. Based on Figure 16.2, is this plant a monocot or a eudicot? _____ Explain. _____
   _____

2. Observe any other available types of leaves, and note in Table 16.1 the name of the plant and whether it is a monocot or a eudicot.

Table 16.1	Monocots Versus Eudicots	
Name of Plant	**Organization of Leaf Veins**	**Monocot or Eudicot?**
1.		
2.		
3.		
4.		

16.2 Major Tissues of Vegetative Organs

Root tips and shoot tips contain **apical meristem** that continually divides by mitosis to produce new cells. These cells become one of the three specialized tissue systems in plants: epidermal tissue, ground tissue, or vascular tissue.

Epidermal tissue contains the epidermal cells that make up the **epidermis,** the outer covering of nonwoody plants. In each part of the plant, the epidermis is specialized for a particular function (Table 16.2).

Ground tissue forms the bulk of the plant and includes the cortex, pith, and mesophyll. Ground tissue cells may photosynthesize, may store the products of photosynthesis, or may merely lend support. **Parenchyma** and **sclerenchyma cells** are in ground tissue. The blue cells in the adjoining figure are nonliving when they reach functional maturity.

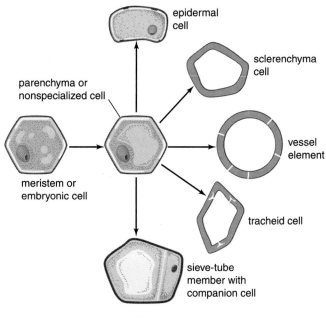

Vascular tissue consists of xylem and phloem. **Xylem** contains hollow cells (vessel elements and tracheids) that transport water and minerals, while **phloem** contains sieve-tube members that transport organic nutrients. Sieve-tube members contain cytoplasm but no nucleus; however, each has a companion cell that does contain a nucleus.

Table 16.2	Major Tissues of Vegetative Organs		
Tissue Type	**Roots**	**Stems**	**Leaves**
Epidermis (epidermal tissue)	Root hairs absorb water and minerals.	Protect inner tissues	Protect against cuticle-H_2O loss Stomata carry on gas exchange.
Cortex (ground tissue)	Store products of photosynthesis	Carry on photosynthesis, if green	(Not present)
Xylem and phloem (vascular tissue)	Transport water and nutrients	Transport water and nutrients	Transport water and nutrients
Pith (ground tissue)	Store products of photosynthesis	Store products of photosynthesis	(Not present)
Mesophyll (ground tissue) spongy layer*	(Not present)	(Not present)	Gas exchange/ photosynthesis
palisade layer*			Photosynthesis

*Discussed later in the laboratory.

16.3 Root System

The **root system** anchors the plant in the soil, absorbs water and minerals from the soil, and stores the products of photosynthesis received from the leaves.

Anatomy of a Eudicot Root Tip

Primary growth increases the length of a plant, while **secondary growth** increases its girth. The focus here is on the primary growth of roots. Note the location of the root apical meristem in Figure 16.3. As primary growth occurs, root cells enter zones that correspond to various stages of differentiation and specialization.

Observation: Anatomy of Root Tip

1. Examine a model and/or a slide of a root tip (Fig. 16.3).
2. Identify the **root cap** (dead cells at the tip of a plant that provide protection as the root grows).
3. Locate the **zone of cell division.** Meristem is found in this zone. As mentioned previously, meristematic tissue is composed of embryonic cells that continually divide, providing new cells for root growth.
4. Find the **zone of elongation.** In this zone, rows of newly produced cells elongate as they begin to grow larger.
5. Identify the **zone of maturation.** In this zone, the cells become differentiated into particular cell types. When epidermal cells differentiate, they produce root hairs. Also noticeable are the cells that make up the xylem and phloem of vascular tissue.

Figure 16.3 Eudicot root tip.
In longitudinal section, the root cap is followed by the zone of cell division, zone of elongation, and zone of maturation.

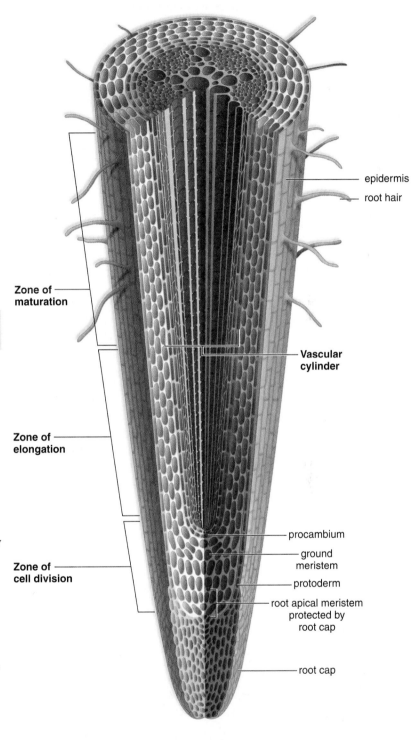

epidermis

root hair

Zone of maturation

Vascular cylinder

Zone of elongation

procambium

ground meristem

Zone of cell division

protoderm

root apical meristem protected by root cap

root cap

Anatomy of Eudicot Roots

As noted in Figure 16.2, eudicot and monocot roots differ in their arrangement of tissues, particularly the vascular tissue. You will observe the organization of a eudicot root.

Observation: Anatomy of Eudicot Root

1. Obtain a prepared cross-section slide of a buttercup (*Ranunculus*) root. Use both low power and high power to identify the **epidermis** (the outermost layer of small cells that gives rise to root hairs). The epidermis protects inner tissues and absorbs water and minerals.
2. Locate the **cortex,** which consists of several layers of thin-walled cells (Fig. 16.4*a* and *b*). In Figure 16.4*b*, note the many stained starch grains in the cortex cells. The cortex functions in food storage.
3. Find the **endodermis,** a single layer of cells whose walls are thickened by a layer of waxy material known as the **Casparian strip.** (It is as though these cells are glued together with a waxy glue.) Because of the Casparian strip, the only access to the xylem is through the living endodermal cells. Therefore, the endodermis regulates what materials enter a plant through the root.

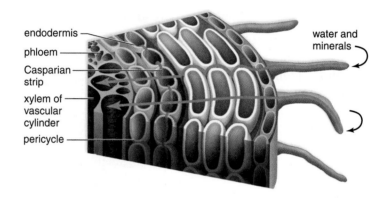

4. Identify the **pericycle,** a layer one or two cells thick just inside the endodermis. Branch roots originate from this tissue.

Figure 16.4 Eudicot root cross section.

The vascular cylinder of a dicot root contains the vascular tissue. Xylem is typically star-shaped, and phloem lies between the points of the star. **a.** Drawing. **b.** Micrograph.

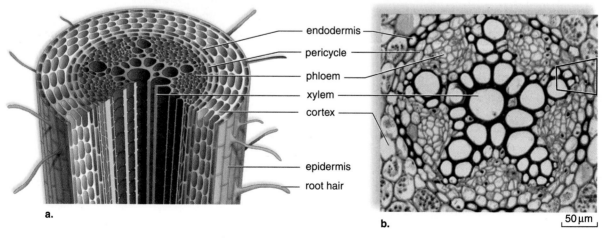

5. Locate the **xylem** in the central portion of the root. Xylem has several "arms" that extend like the spokes of a wheel. This tissue conducts water and minerals from the roots to the stem.

6. Find the **phloem,** located between the arms of the xylem. Phloem conducts organic nutrients from the leaves to the roots and other parts of the plant.

Conclusion

- Trace the path of water through all the tissues of a eudicot root from a root hair to the xylem. In parenthesis, state a function for each tissue mentioned:

_____

_____

Root Diversity

There are essentially two types of roots: **taproots** and **fibrous roots** (Fig. 16.5a and b). In a taproot system, the main root is many times larger than the branch roots and often serves as a food storage organ. In a fibrous root system, all of the roots are approximately the same size.

Observation: Root Diversity

1. Compare the root system of a carrot, or other specimens provided by your instructor, to the root system of a dandelion. Identify the type of root for each.

   a. Carrot _____

   b. Dandelion _____

2. Taproots in particular function in food storage. Examine the taproots on display, and name one or two in which the taproot is enlarged for storage. _____

   _____

Figure 16.5 Root diversity.
a. Carrots have a taproot. **b.** Grass has a fibrous root system.

a. Taproot system   b. Fibrous root system

16.4 Stems

Stems are usually found aboveground and provide support for leaves and flowers. Stems that do not contain wood are called **herbaceous,** or nonwoody, stems. Usually, monocots remain herbaceous throughout their lives. Some eudicots, such as those that live a single season, are also herbaceous. Other eudicots, such as trees, become woody as they mature.

Anatomy of Herbaceous Stems

Herbaceous stems undergo primary growth, which results in an increase in length, but they do not undergo secondary growth, which results in an increase in girth. Activity of the **apical meristem** located in the terminal bud (see Fig. 16.1) results in primary growth of the shoot system.

> *Observation: Anatomy of Eudicot and Monocot Herbaceous Stems*

Herbaceous Eudicot Stem

1. Examine a prepared slide of a herbaceous eudicot stem (Fig. 16.6), and identify the **epidermis** (the outer protective layer).
2. Locate the **cortex,** which may photosynthesize or store nutrients.
3. Find the **vascular bundle,** which transports water and organic nutrients. The vascular bundles in a herbaceous eudicot stem occur in a ring pattern. Label the vascular bundle in Figure 16.6.
4. Identify the **pith,** which stores organic nutrients.
5. Just as in the eudicot root, which tissue (xylem or phloem) is closer to the stem surface? _____

Figure 16.6 **Eudicot herbaceous stem.**
The vascular bundles are in a definite ring in this photomicrograph of a eudicot herbaceous stem. Complete the labeling as directed by the Observation. **a.** Drawing. **b.** Micrograph.

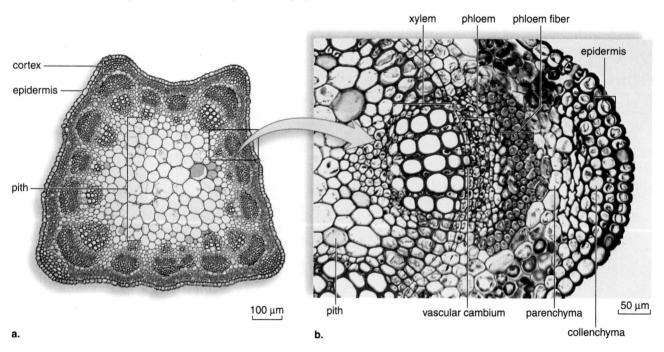

Herbaceous Monocot Stem

1. Examine a prepared slide of a herbaceous monocot stem (Fig. 16.7). Locate the same four tissues that you found in the herbaceous eudicot stem.

2. The vascular bundles in a herbaceous monocot stem are said to be scattered. Explain. _____

_____

_____

Figure 16.7 Monocot stem.

The vascular bundles, one of which is enlarged, are scattered in this photomicrograph of a monocot herbaceous stem.

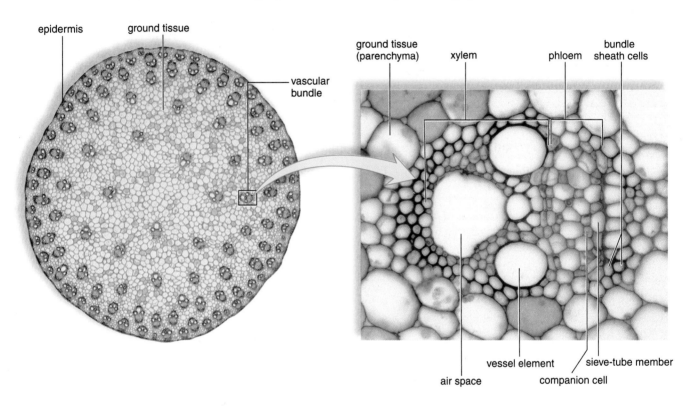

Conclusion

- The organization of a eudicot stem allows you to easily locate a centrally placed pith, but not the organization of a monocot stem. Explain:

_____

_____

Anatomy of Woody Stems

Woody stems undergo both primary growth (increase in length) and secondary growth (increase in girth). When *primary growth* occurs, the apical meristem within a terminal bud is active. When *secondary growth* occurs, the vascular cambium is active. **Vascular cambium** is meristem tissue, which produces new xylem and phloem called **secondary xylem** and **phloem** each year. The buildup of secondary xylem year after year is called **wood**.

Observation: Anatomy of Winter Twig

1. A winter twig typically shows several years' past primary growth. Examine several examples of winter twigs (Fig. 16.8), and identify the **terminal bud scales** located at the tip of the twig. This is where new primary growth will originate.
2. Locate a **terminal bud scale scar.** These scars encircle the twig and indicate where the terminal bud was located in previous years. The distance between two adjacent terminal bud scale scars equals one year's primary growth.
3. Find a **leaf scar.** This is where a leaf was attached to the stem.
4. Note the **vascular bundle scars.** Complete this sentence: Vascular bundle scars appear where the vascular bundles _____.
5. Identify a **node.** This is the region where you find leaf scars and bundle scars.
6. Locate an **axillary bud.** This is where new branch growth can occur.

Figure 16.8 **External structure of a winter twig.**
Counting the terminal bud scale scars tells the age of a particular branch.

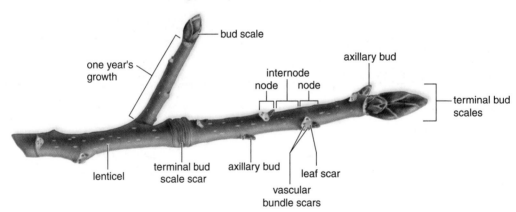

Conclusion

- The twig in Figure 16.8 increased in length by how much due to primary growth?

_____

_____

1. Examine a prepared slide of a cross section of a woody stem (Fig. 16.9), and identify the **bark** (the dark outer area), which contains **cork,** a protective outer layer; **cortex,** which stores nutrients; and **phloem,** which transports organic nutrients.
2. Locate the **vascular cambium** at the inner edge of the bark, between the bark and the wood. Vascular cambium is meristem tissue whose activity accounts for secondary growth that causes increased girth of a tree. Secondary phloem (disappears) and secondary xylem (builds up) are produced by vascular cambium each growing season.
3. Find the **wood,** which contains annual rings. An **annual ring** is the amount of xylem added to the plant during one growing season. Rings appear to be present because spring wood has large xylem vessels and looks light in color, while summer wood has much smaller vessels and appears much darker. How old is the stem you are observing? _____ Are all the rings the same width? _____
4. Identify the **pith,** a ground tissue that stores organic nutrients and may disappear.
5. Locate **rays,** groups of small, almost cuboid cells that extend out from the pith laterally.

Figure 16.9 **Woody eudicot stem cross section.**
Because xylem builds up year after year, it is possible to count the annual rings to determine the age of a tree. This tree is three years old. **a.** Drawing. **b.** Photomicrograph.

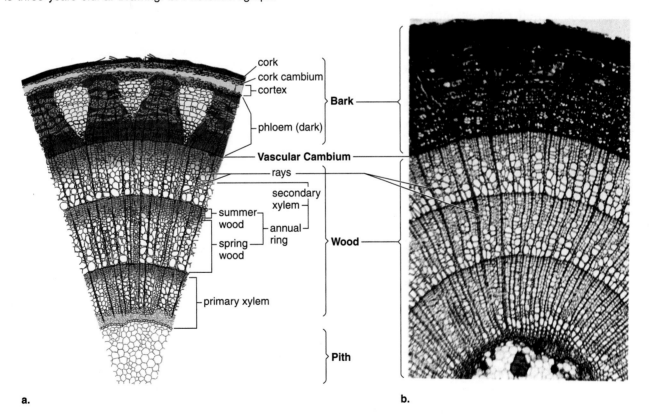

cork
cork cambium
cortex
} **Bark**

phloem (dark)

Vascular Cambium

rays

secondary xylem
summer wood
annual ring
spring wood
} **Wood**

primary xylem

Pith

a.

b.

16.5 Leaves

A **leaf** is the organ that produces food for the plant by carrying on photosynthesis. Leaves are generally broad and quite thin. An expansive surface facilitates the capture of solar energy and gas exchange. Water and nutrients are transported to the cells of a leaf by leaf veins, extensions of the vascular bundles from the stem.

Anatomy of Leaves

Observation: Anatomy of Leaves

1. Examine a prepared slide of a leaf cross section. With the help of Figure 16.10, identify the waxy **cuticle,** the outermost layer that protects the leaf and prevents water loss.
2. Locate the **upper epidermis** and **lower epidermis,** single layers of cells at the upper and lower surfaces. Trichomes are hairs that grow from the upper epidermis and help protect the leaf from insects and water loss.
3. Find the leaf **vein.** This transports water and organic nutrients. If you are examining a monocot leaf, the leaf veins will all be circular. If you are examining a eudicot leaf, the leaf veins may appear circular or oval, because they will be at various angles. Explain why. _____

   _____

   _____

4. Identify the **palisade mesophyll,** located near the upper epidermis. These cells contain chloroplasts and carry on most of the plant's photosynthesis. Label the palisade mesophyll in Figure 16.10.
5. Locate the **spongy mesophyll,** located near the lower epidermis. These cells have air spaces that facilitate the exchange of gases across the plasma membrane. Label the spongy mesophyll in Figure 16.10.
6. Find the **stoma** (pl., **stomata**), openings through which gas exchange occurs. These are more numerous in the lower epidermis. Each has two guard cells that regulate opening and closing.

Figure 16.10 Leaf anatomy.

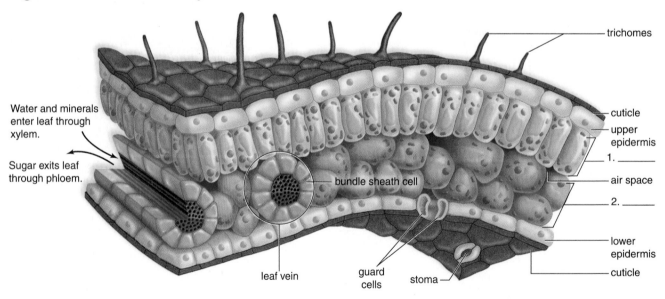

Water and minerals enter leaf through xylem.

Sugar exits leaf through phloem.

bundle sheath cell

leaf vein

guard cells

stoma

trichomes

cuticle

upper epidermis

1. _____

air space

2. _____

lower epidermis

cuticle

Leaf Diversity

Most leaves consist of a flattened **blade** and a stalk called the **petiole.** Stipules, a pair of appendages, are sometimes present where the petiole attaches to the stem.

Observation: Leaf Diversity

1. Examine a variety of leaf specimens. Notice whether each leaf is simple or compound. A **simple leaf** has a single blade, while a **compound leaf** has a blade divided into leaflets. There is one **axillary bud** at the base of the petiole of each leaf regardless of whether it is simple or compound.
2. In **palmately compound leaves,** the leaflets are attached at one point at the end of the petiole. In **pinnately compound leaves,** the leaflets are attached at intervals along the petiole. *Determine whether each leaf is palmately compound or pinnately compound and complete the labels in Figure 16.11*
3. The leaves can alternate their position on the stem or they can be opposite one another. Some leaves occur in a whorl at a node. *Determine the arrangement of the leaves on the stem and complete the labels in Figure 16.11.*

Figure 16.11 Classification of leaves.

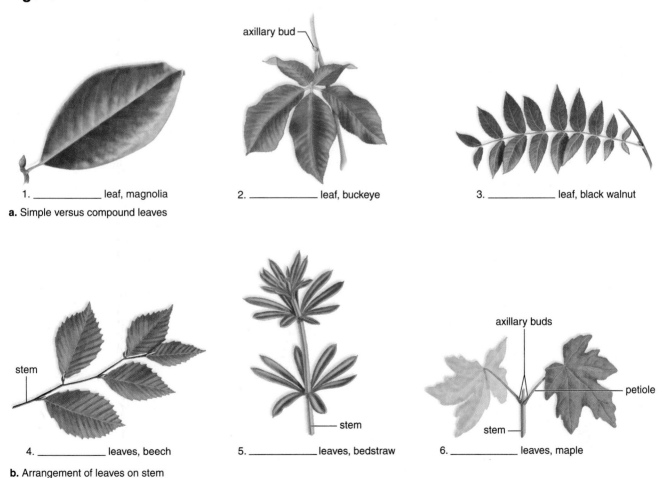

1. _____ leaf, magnolia   2. _____ leaf, buckeye   3. _____ leaf, black walnut

a. Simple versus compound leaves

4. _____ leaves, beech   5. _____ leaves, bedstraw   6. _____ leaves, maple

b. Arrangement of leaves on stem

1. Given the information in this laboratory, how would you distinguish between a monocot plant and a eudicot plant based on their external anatomy? _____

    _____

2. What is meristem tissue? _____ How is this tissue different from all other types of plant tissue? _____

    _____

3. In which zone of a eudicot root would you expect to find vascular tissue? Why? _____

    _____

4. In a eudicot root, what structural feature allows the endodermis to regulate the entrance of water and materials into the vascular cylinder, where xylem and phloem are located? _____

    _____

    _____

5. Characterize the root of a carrot. _____

6. How would you microscopically distinguish a eudicot stem from a monocot stem? _____

    _____

7. Distinguish between primary and secondary growth of a woody stem, and explain how each arises.

    _____

8. Contrast how you could determine one year's growth by looking at a winter twig with how you determine one year's growth in a cross section of a tree stem. _____

9. Contrast the manner in which water reaches the inside of a leaf with the manner in which carbon dioxide reaches the inside of a leaf. _____

    _____

10. How would you recognize the epidermis of a root versus the epidermis of a leaf? _____

    _____

17

Reproduction in Plants

Introduction

Flowers are the reproductive structures of angiosperms. **Pollination** involves the transport of pollen from an **anther,** where pollen is produced, to the **carpel,** where a pollen grain germinates and forms a **pollen tube.** The tube grows through the **style** and into the **ovary,** where there are **ovules,** each containing an egg. One sperm from the pollen tube fertilizes an egg in the ovule. The ovule becomes a **seed** containing an embryo enclosed within a fruit. The **fruit** develops from the ovary and also, at times, from accessory structures. Fruits assist the dispersal of angiosperm seeds. When animals eat fleshy fruits, they may ingest the seeds

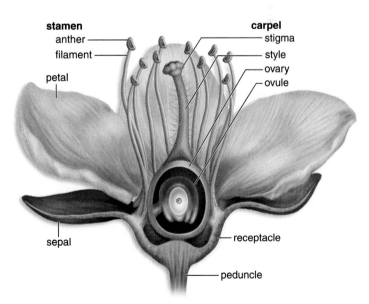

and defecate them sometime later. Coconuts are dispersed by oceanic currents. Some lightweight fruits with wings and some with seed hairs are dispersed by wind. Seeds contain an embryonic plant. When seeds germinate, a new plant begins to develop.

17.1 Flowers

Angiosperms are divided into two groups: monocots and eudicots. **Monocots** have flower parts in threes or multiples thereof. **Eudicots** have flower parts in fours or fives or multiples thereof. Complete flowers contain stamens, carpels, sepals, and petals. Flowers that lack one of these basic parts are called incomplete.

Observation: A Flower

1. Examine a flower model or living flower. Use Figure 17.1 to help you identify:
   a. **Sepals:** The outermost set of modified leaves, collectively termed the **calyx.** Sepals are green in most flowers.
   b. **Petals:** The inner leaves that collectively constitute the **corolla.** Petals often have a design and color that attract specific pollinators, such as bees and butterflies.
   c. **Stamen:** A swollen terminal **anther,** where pollen is produced, and the slender **filament** that supports it.
   d. **Carpel:** This structure consists of a swollen basal **ovary**, a long slender **style** (stalk), and a terminal **stigma** (sticky knob).
   e. **Ovary:** The enlarged part of the carpel that develops into a fruit.

2. Is the flower you are examining a monocot or eudicot? _____

   Explain. _____

Life Cycle of Flowering Plants

Figure 17.1 describes the life cycle of flowering plants. Use the figure to identify the major steps in this life cycle.

1. The parts of the flower involved in reproduction are the _____ and the _____.

2. The anther at the top of the stamen has _____, which contain numerous microsporocytes that undergo meiosis to produce _____

   _____. Each microspore becomes a _____.

3. The carpel contains an ovary that encloses _____. Within an ovule, a

   megasporocyte undergoes meiosis to produce four _____. One

   megaspore develops into a(n) _____.

4. After pollination, the generative cell of a pollen grain divides to produce two _____.
   A pollen grain develops a pollen tube that passes down the style of the carpel.

5. During double fertilization, one sperm from the pollen tube fertilizes the egg within the embryo

   sac, and the other joins with two _____ nuclei.

6. The fertilized egg becomes an _____, and the joining of polar nuclei and sperm becomes

   the 3n _____. A seed contains the three parts labeled in Figure 17.1. In
   angiosperms, seeds are enclosed by fruits.

Figure 17.1 Flowering plant life cycle.

In flowering plants, meiosis produces microspores that develop into male gametophytes (pollen grains) and megaspores that develop into female gametophytes (embryo sacs).

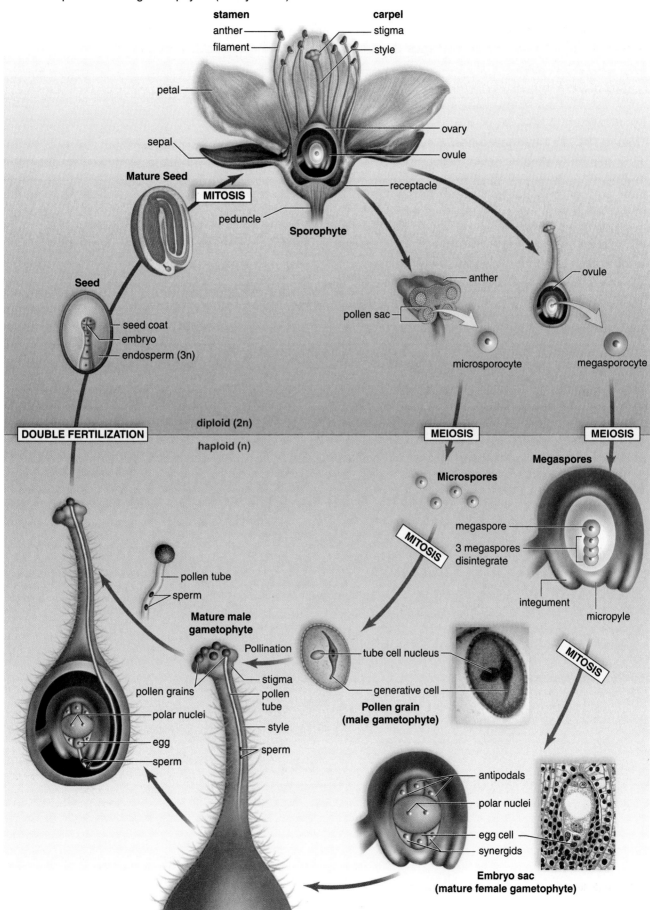

Development of Eudicot Embryo

Stages in the development of a eudicot embryo are shown in Figure 17.2. During development, the **suspensor** anchors the embryo and transfers nutrients to it from the mature plant. The **cotyledons** store nutrients that the embryo uses as nourishment. An embryo consists of the **epicotyl,** which becomes the leaves; the **hypocotyl,** which becomes the stem; and the **radicle,** which becomes the roots.

Figure 17.2 **Development of a eudicot embryo.**
Embryogenesis consists of these stages, described in the text.

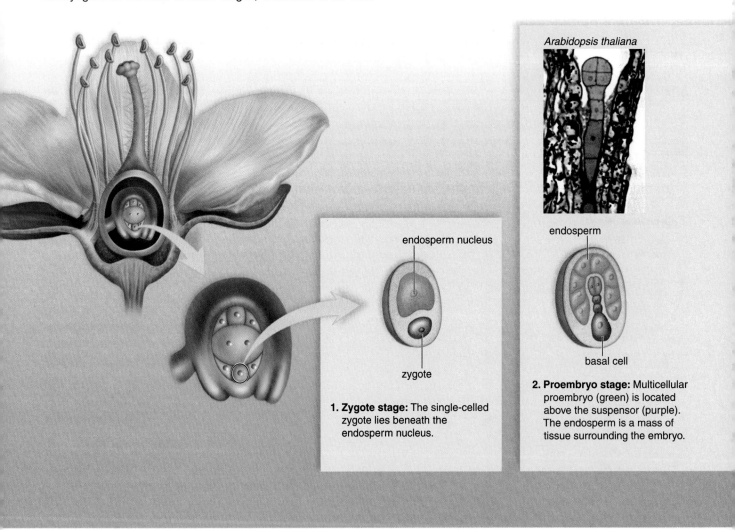

Arabidopsis thaliana

endosperm nucleus

zygote

1. Zygote stage: The single-celled zygote lies beneath the endosperm nucleus.

endosperm

basal cell

2. Proembryo stage: Multicellular proembryo (green) is located above the suspensor (purple). The endosperm is a mass of tissue surrounding the embryo.

Observation: Development of the Embryo

Study prepared slides and identify the stages described in Figure 17.2.

List the stages you were able to identify. _____

_____

_____

_____

_____

A. thaliana

endosperm

3. Globular stage: As cell division continues, the proembryo (green) becomes globe-shaped. The stalklike suspensor (purple) anchors the embryo.

A. thaliana

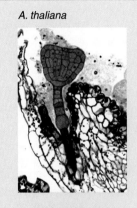

cotyledons appearing

4. Heart stage: The embryo becomes heart-shaped as the cotyledons begin to appear.

Capsella

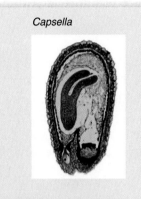

shoot apical meristem   bending cotyledons

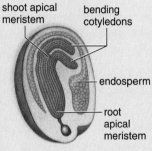

endosperm

root apical meristem

5. Torpedo stage: The embryo becomes torpedo-shaped as the cotyledons enlarge. The endosperm lessens, and tissues become differentiated.

Capsella

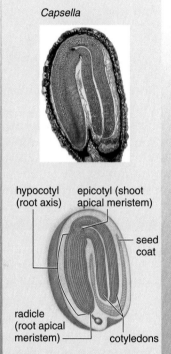

hypocotyl (root axis)   epicotyl (shoot apical meristem)

seed coat

radicle (root apical meristem)   cotyledons

6. Mature embryo stage: The embryo consists of the epicotyl (represented here by the shoot apex), the hypocotyl, and the radicle (which contains the root apex).

17.2 Fruits

A fruit is derived from an ovary or from an ovary and closely associated tissues, whereas "vegetables" are not derived from floral tissues and would never contain seeds. As an ovary develops into a fruit, the ovarian wall thickens and becomes a **pericarp,** the outer covering of a fruit.

Kinds of Fruits

Biologists have divided **simple fruits** (derived from a single ovary) into various types according to various characteristics. Simple fruits can be classified as drupes, pepos, hesperidia, berries, pomes, legumes, follicles, samaras, drupaceous nuts, nuts, grains, or achenes.

In addition to simple fruits, there are also **aggregate fruits,** which develop from a number of ovaries within a single flower. Examples include blackberries, raspberries, and strawberries. **Multiple fruits** develop from a number of ovaries of several flowers. Examples are pineapples, mulberries, and figs.

Observation: Keying Simple Fruits

Record your observations in Table 17.1. Use Figure 17.3 to learn the difference between fleshy and dry fruits. Then use Table 17.2 to identify various types of simple fruits. In Table 17.2 you must choose between 1a or 1b to get started. Thereafter, you continue by following the "Go to" instructions until you reach the fruit type for the fruit you are keying.

Table 17.1	Identification of Simple Fruits		
Common Name	Fleshy or Dry?	Eaten as a Vegetable, Fruit, Other?	Type of Fruit (from Key in Table 17.2)
1			
2			
3			
4			
5			
6			
7			
8			
9			
10			

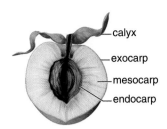

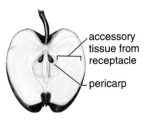

a. peach (drupe)   **b.** apple—an accessory fruit (pome)   **c.** Pea (legume)   **d.** Maple (samara)

Figure 17.3 **Fleshy fruits versus dry fruits.**
Fleshy fruits are represented by **(a)** peach and **(b)** apple. Dry fruits are represented by **(c)** pea, which is dehiscent because it splits open; and **(d)** maple, which is an indehiscent fruit.

Table 17.2	A Dichotomous Key for Simple Fruit Types

1*a*.	Fruit is fleshy; pericarp is soft.	Go to 2*a*/2*b*.
1*b*.	Fruit is dry; pericarp is dry.	Go to 6*a*/6*b*.
2*a*.	Fruit has a single seed inside a hard and stony pit.	DRUPE
	Examples: plum, cherry.	
2*b*.	Fruit contains several seeds.	Go to 3*a*/3*b*.
3*a*.	Fruit has a firm rind.	Go to 4*a*/4*b*.
3*b*.	Fruit does not have a firm rind.	Go to 5*a*/5*b*.
4*a*.	Fruit has a firm rind and is not segmented.	PEPO
	Examples: squash, cucumber, watermelon.	
4*b*.	Fruit has a firm rind and is segmented.	HESPERIDIUM
	Examples: lemon, lime, orange.	
5*a*.	Entire wall of fruit is fleshy, and seeds may be eaten.	BERRY
	Examples: tomato, grape.	
5*b*.	Fruit has a papery core.	POME
	Examples: apple, pear.	
6*a*.	Dry fruit is dehiscent (splits open).	Go to 7*a*/7*b*.
6*b*.	Dry fruit is indehiscent (does not split open).	Go to 8*a*/8*b*.
7*a*.	Fruit splits at two seams.	LEGUME
	Examples: pea, soybean, locust.	
7*b*.	Fruit splits at one seam.	FOLLICLE
	Examples: milkweed, larkspur.	
8*a*.	Fruit has one or more wings.	SAMARA
	Examples: maple, elm, ash.	
8*b*.	Fruit does not have wings.	Go to 9*a*/9*b*.
9*a*.	Pericarp has three complete layers.	DRUPACEOUS NUT
	Examples: coconut, hickory.	
9*b*.	Pericarp does not have three complete layers.	Go to 10*a*/10*b*.
10*a*.	Fruit is relatively large; pericarp is thick and stony; seed separates from ovarian wall.	NUT
	Examples: walnut, oak.	
10*b*.	Fruit is relatively small; pericarp is thin; seed is at least partially attached to ovarian wall.	Go to 11*a*/11*b*.
11*a*.	Pericarp is completely fused to seed coat.	GRAIN
	Examples: wheat, corn, oats.	
11*b*.	Pericarp attaches to seed coat at only one point.	ACHENE
	Examples: sunflower, dandelion.	

17.3 Seeds

The seeds of flowering plants develop from ovules. A seed contains an embryonic plant, stored food, and a seed coat. Monocot seeds have one **cotyledon** (seed leaf); eudicots have two cotyledons.

Observation: Eudicot and Monocot Seeds

Bean Seed

1. Obtain a presoaked bean seed (eudicot). Carefully dissect it, using Figure 17.4 to help you identify:

   a. **Seed coat:** The outer covering. Remove the seed coat with your fingernail.

   b. **Cotyledons:** Food storage organs. The endosperm was absorbed by the cotyledons during development. What is the function of these cotyledons? _____

   _____

   c. **Epicotyl:** The small portion of the embryo located above the attachment of the cotyledons. The first true leaves **(plumules)** develop from the epicotyl.

   d. **Hypocotyl:** The small portion of embryo located below the attachment of the cotyledons. The lower end develops into the embryonic root, or **radicle.**

Figure 17.4 **Eudicot seed structure and germination.**

a. A seed contains an embryo, stored food, and a seed coat as exemplified by a bean seed. A eudicot seed has two cotyledons. Following **(b)** germination, a seedling grows to become a mature plant.

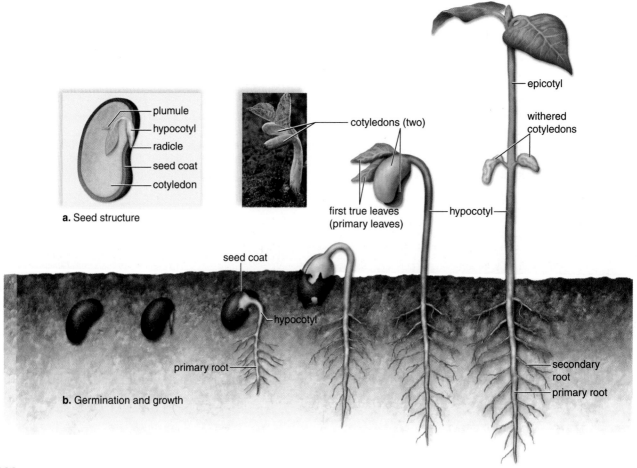

2. Observe seedlings in various stages of development. Which organ emerges first from the seed—the plumule or the radicle? _____

Of what advantage is this to the plant? _____

_____

3. The hypocotyl is the first part to emerge from the soil. What is the advantage of the hypocotyl pulling the plumule up out of the ground instead of pushing it up through the ground? _____

_____

4. Do cotyledons stay beneath the ground? _____

Corn Kernel

1. Obtain a presoaked corn kernel (monocot). Lay the seed flat, and with a razor, carefully slice it in half. A corn kernel is a fruit, and the seed coat is tightly attached to the pericarp (Figure 17.5).
2. Identify the cotyledon, plumule, and radicle. In addition, identify the:
   a. **Endosperm:** Stored food for the embryo. Passes into the cotyledon as the seedling grows.
   b. **Coleoptile:** A sheath that covers the emerging leaves.
3. Examine corn seedlings in various stages of development. Does the cotyledon of a corn seed stay beneath the ground? _____

Figure 17.5 Monocot seed structure and germination.

a. A monocot seed has only one cotyledon as exemplified by a corn kernel. A corn kernel is a fruit—the seed is covered by a pericarp. Following **(b)** germination, a corn seedling grows to become a mature plant.

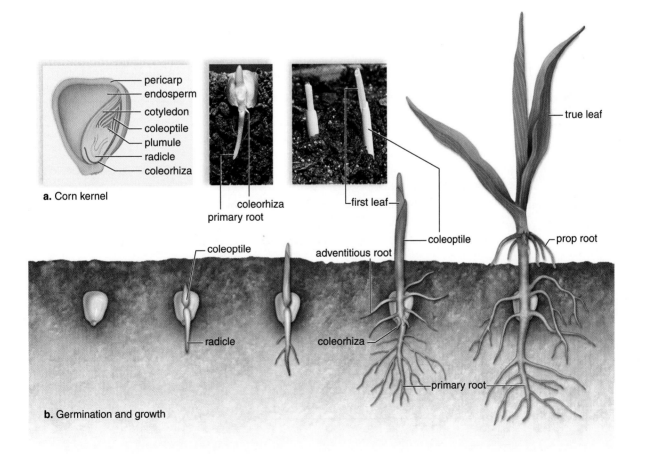

17.4 Seed Germination

Mature seeds contain an embryo that does not resume growth until after germination, which requires the proper environmental conditions. Mature seeds are dry, and for germination to begin, the dry tissues must take up water in a process called **imbibition**. After water has been imbibed, enzymes break down the food source into small molecules that can provide energy or be used as building blocks until the seedling is ready to photosynthesize.

Explain why the cotyledons of a bean seedling shrivel as the seedling grows. _____

_____

Experimental Procedure: Effect of Acid Rain

In certain parts of the United States, rain has a much lower pH than normal. This is known to have detrimental effects on the sustainability of forests. Would you predict that acid rain also affects

germination of seeds? _____

 Your instructor has placed 20 sunflower seeds in each of five containers with water of increasing acidity: 0% vinegar (tap water), 1% vinegar, 5% vinegar, 20% vinegar, and 100% vinegar.

1. What hypothesis do you propose regarding the effect of these solutions on the germination of

 sunflower seeds? _____

 _____

2. Count the number of germinated sunflower seeds in each container, and complete Table 17.3.
3. Record the pH of each container as directed by your instructor.

4. Do the data support or falsify your hypothesis? _____ Explain. _____

 _____

5. You learned in Laboratory 5 that each enzyme has an optimum pH. Explain why acid rain is

 expected to inhibit metabolism, and therefore, seedling development. _____

 _____

Table 17.3	Effect of Increasing Acidity on Germination of Sunflower Seeds		
Concentration of Vinegar	pH	Number of Seeds That Germinated	Percent Germination
0%			
1%			
5%			
20%			
100%			

1. Relate the parts of a carpel to the germination, growth, and function of a pollen tube.

   _____

   _____

   _____

2. How can you tell a monocot flower from a eudicot flower?

   _____

   _____

3. Fruits ordinarily contains seeds. Explain.

   _____

   _____

4. Botanists identify a string bean, tomato, okra, and cucumber as fruits. Explain.

   _____

   _____

5. How can you tell a monocot seed from a eudicot seed?

   _____

   _____

6. Name three general parts of a seed.

   _____

7. Relate the plumule and radicle to parts of an adult plant.

   _____

   _____

8. Name two growth pattern differences between monocot and eudicot seeds after germination.

   _____

   _____

   _____

9. Why do you expect acidic conditions to affect the ability of seeds to germinate?

   _____

18

Animal Organization and Structure

Learning Outcomes

18.1 Tissue Level of Organization
- Identify slides and models of various types of epithelium. 234–236
- Tell where particular types of epithelium are located in the body, and state a function. 234–236
- Identify slides and models of various types of connective tissue. 237–240
- Tell where particular connective tissues are located in the body, and state a function. 237–240
- Identify slides and models of three types of muscular tissue. 241–242
- Tell where each type of muscular tissue is located in the body, and state a function. 241–242
- Identify a slide and model of a neuron. 243
- Tell where nervous tissue is located in the body, and state a function. 243

18.2 Organ Level of Organization
- Identify a slide of the intestinal wall and the layers in the wall. State a function for each tissue. 244
- Identify a slide of skin and the two regions of skin. State a function for each region of skin. 245

Introduction

Humans, as well as all living things, are made up of **cells.** Groups of cells that have the same structural characteristics and perform the same functions are called **tissues.** Figure 18.1 shows the four categories of tissues in the human body. An **organ** is composed of different types of tissues, and various organs form **organ systems.** Humans thus have the following levels of biological organization: cells ⟶ tissues ⟶ organs ⟶ organ systems.

The photomicrographs of tissues in this laboratory were obtained by viewing prepared slides with a light microscope. Preparation required the following sequential steps:

1. **Fixation:** The tissue is immersed in a preservative solution to maintain the tissue's existing structure.
2. **Embedding:** Water is removed with alcohol, and the tissue is impregnated with paraffin wax.
3. **Sectioning:** The tissue is cut into extremely thin slices by an instrument called a microtome. When the section runs the length of the tissue, it is called a longitudinal section (l.s.); when the section runs across the tissue, it is called a cross section (c.s.).
4. **Staining:** The tissue is immersed in dyes that stain different structures. The most common dyes are hematoxylin and eosin stains (H & E). They give a differential blue and red color to the basic and acidic structures within the tissue. Other dyes are available for staining specific structures.

Figure 18.1 The major tissues in the human body.

The many kinds of tissues in the human body are grouped into four types: epithelial tissue, muscular tissue, nervous tissue, and connective tissue.

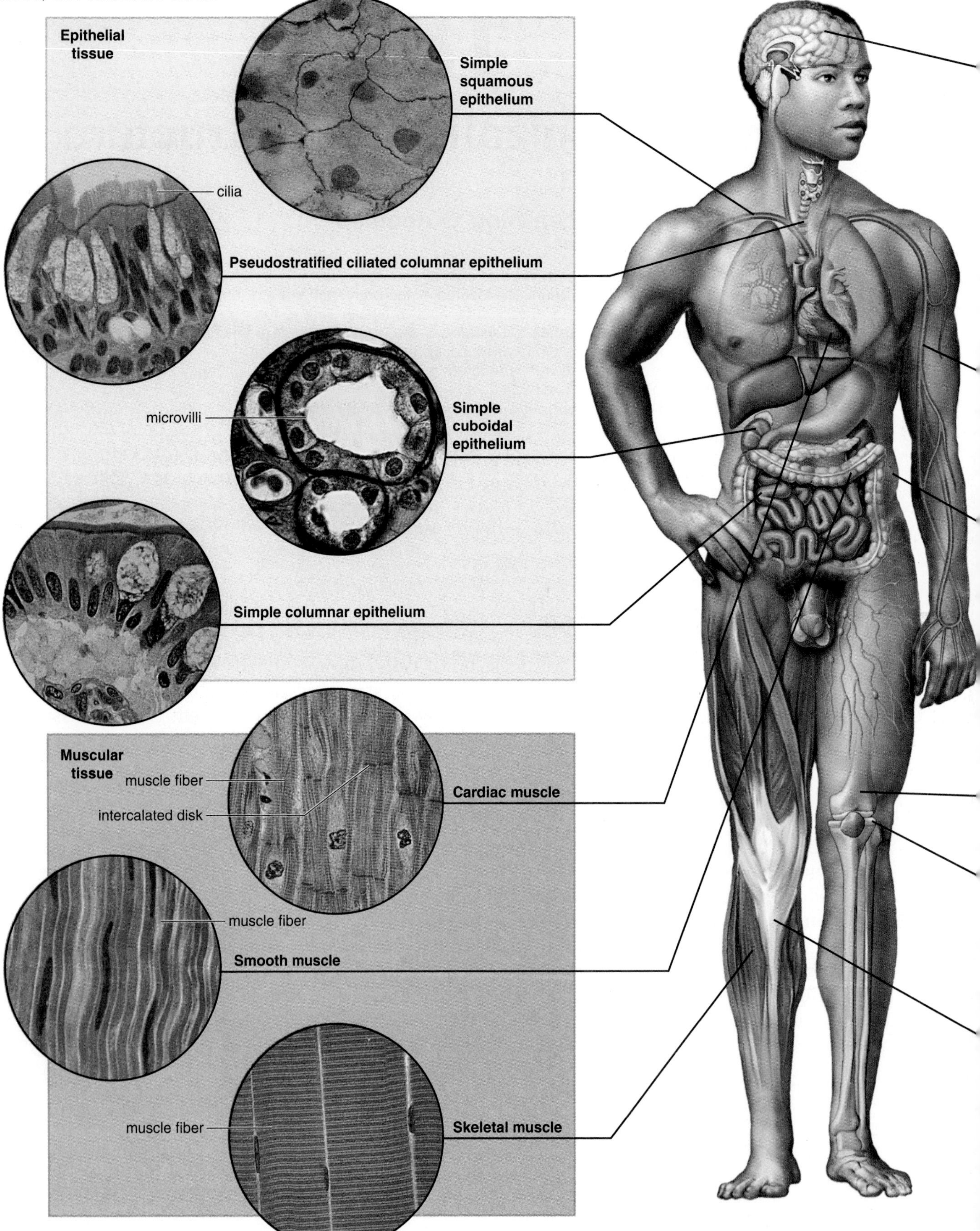

Epithelial tissue

cilia

Simple squamous epithelium

Pseudostratified ciliated columnar epithelium

microvilli

Simple cuboidal epithelium

Simple columnar epithelium

Muscular tissue

muscle fiber

intercalated disk

Cardiac muscle

muscle fiber

Smooth muscle

muscle fiber

Skeletal muscle

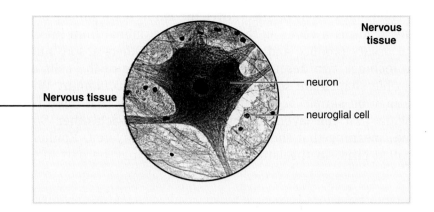

Nervous tissue

Nervous tissue

neuron

neuroglial cell

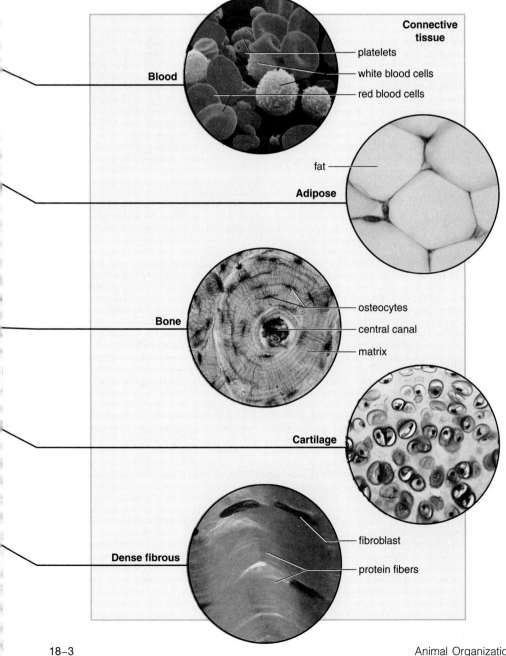

Connective tissue

Blood

platelets

white blood cells

red blood cells

fat

Adipose

osteocytes

Bone

central canal

matrix

Cartilage

fibroblast

Dense fibrous

protein fibers

18.1 Tissue Level of Organization

Epithelial tissue (epithelium) forms a continuous layer, or sheet, over the entire body surface and most of the body's inner cavities. Externally, it forms a covering that protects the animal from infection, injury, and drying out. Some epithelial tissues produce and release secretions. Others absorb nutrients.

The name of an epithelial tissue includes two descriptive terms: the shape of the cells and the number of layers. The three possible shapes are *squamous, cuboidal,* and *columnar.* With regard to layers, an epithelial tissue may be simple or stratified. **Simple** means that there is only one layer of cells; **stratified** means that cell layers are placed on top of each other. Some epithelial tissues are **pseudostratified,** meaning that they only appear to be layered. Epithelium may also have hairlike extensions called **cilia.** In the latter case, "ciliated" may be part of the tissue's name.

Observation: Simple and Stratified Squamous Epithelium

Simple Squamous Epithelium

Simple squamous epithelium is a single layer of thin, flat, many-sided cells, each with a central nucleus. It lines internal cavities, the heart, and all the blood vessels. It also lines parts of the urinary, respiratory, and male reproductive tracts.

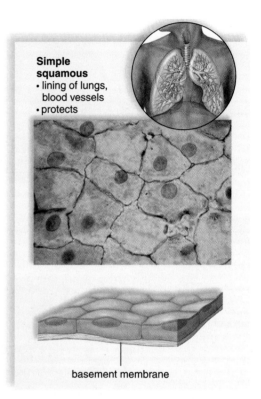

Simple squamous
• lining of lungs, blood vessels
• protects

basement membrane

1. Study a model or diagram of simple squamous epithelium.

   What does squamous mean? _____

2. Examine a prepared slide of squamous epithelium. Under low power, note the close packing of the flat cells. What shapes are the cells?

   _____

3. Under high power, examine an individual cell, and identify the plasma membrane, cytoplasm, and nucleus.

Stratified Squamous Epithelium

As would be expected from its name, stratified squamous epithelium consists of many layers of cells. The innermost layer produces cells that are first cuboidal or columnar in shape, but as the cells push toward the surface, they become flattened.

The outer region of the skin, called the epidermis, is stratified squamous epithelium. As the cells move toward the surface, they flatten, begin to accumulate a protein called **keratin,** and eventually die. Keratin makes the outer layer of epidermis tough, protective, and able to repel water.

The linings of the mouth, throat, anal canal, and vagina are stratified epithelium. The outermost layer of cells surrounding the cavity is simple squamous epithelium. In these organs, this layer of cells remains soft, moist, and alive.

1. Either now or when you are studying skin in Section 18.2, examine a slide of skin and find the portion of the slide that is stratified squamous epithelium.

2. Approximately how many layers of cells make up this portion of skin? _____

3. Which layers of cells best represent squamous epithelium? _____

Observation: Simple Cuboidal Epithelium

Simple cuboidal epithelium is a single layer of cube-shaped cells, each with a central nucleus. It is found in tubules of the kidney and in the ducts of many glands, where it has a protective function. It also occurs in the secretory portions of some glands—that is, where the tissue produces and releases secretions.

1. Study a model or diagram of simple cuboidal epithelium.
2. Examine a prepared slide of simple cuboidal epithelium. Move the slide until you locate cube-shaped cells that line a lumen (cavity). Are these cells ciliated? _____

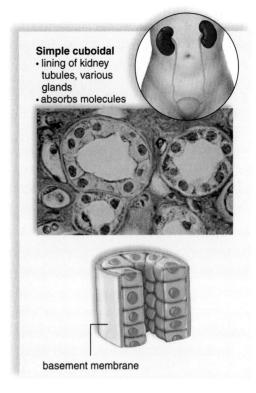

Simple cuboidal
- lining of kidney tubules, various glands
- absorbs molecules

basement membrane

Observation: Simple Columnar Epithelium

Simple columnar epithelium is a single layer of tall, cylindrical cells, each with a nucleus near the base. This tissue, which lines the digestive tract from the stomach to the anus, protects, secretes, and allows absorption of nutrients.

1. Study a model or diagram of simple columnar epithelium.
2. Examine a prepared slide of simple columnar epithelium. Find tall and narrow cells that line a lumen. Under high power, focus on an individual cell. Identify the plasma membrane, the cytoplasm, and the nucleus. Epithelial tissues are attached to underlying tissues by a basement membrane composed of extracellular material containing protein fibers.
3. The tissue you are observing contains mucus-secreting cells. Search among the columnar cells until you find a **goblet cell,** so named because of its goblet-shaped, clear interior. This region contains mucus, which may be stained a light blue. In the living animal, the mucus is discharged into the gut cavity and protects the lining from digestive enzymes.

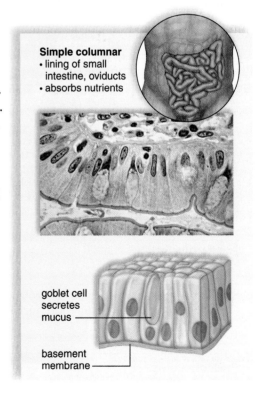

Simple columnar
- lining of small intestine, oviducts
- absorbs nutrients

goblet cell secretes mucus

basement membrane

Observation: Pseudostratified Ciliated Columnar Epithelium

Pseudostratified ciliated columnar epithelium appears to be layered, while actually all cells touch the basement membrane. Many cilia are located on the free end of each cell. In the human trachea, the cilia wave back and forth, moving mucus and debris up toward the throat so that it cannot enter the lungs. Smoking destroys these cilia, but they will grow back if smoking is discontinued.

1. Study a model or diagram of pseudostratified ciliated columnar epithelium.
2. Examine a prepared slide of pseudostratified ciliated columnar epithelium. Concentrate on the part of the slide that resembles the model. Identify the cilia.

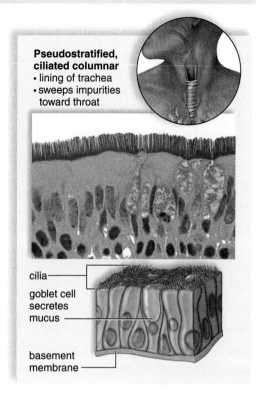

Pseudostratified, ciliated columnar
- lining of trachea
- sweeps impurities toward throat

cilia

goblet cell secretes mucus

basement membrane

Summary of Epithelial Tissue

Complete Table 18.1 to summarize your study of epithelial tissue.

Table 18.1	Epithelial Tissue		
Type	**Structure**	**Function**	**Location**
Simple squamous			Walls of capillaries, lining of blood vessels, air sacs of lungs, lining of internal cavities
Stratified squamous	Innermost layers are cuboidal or columnar; outermost layers are flattened	Protection, repel water	
Simple cuboidal		Secretion, absorption	
Simple columnar	Columnlike—tall, cylindrical nucleus at base		Lining of uterus, tubes of digestive tract
Pseudostratified ciliated columnar		Protection, secretion, movement of mucus and sex cells	

Connective Tissue

Connective tissue joins different parts of the body together. There are four general classes of connective tissue: connective tissue proper, bone, cartilage, and blood. All types of connective tissue consist of cells surrounded by a matrix (a noncellular material that varies from solid to semifluid to fluid) that usually contains fibers. Elastic fibers are composed of a protein called elastin. Collagenous fibers contain the protein collagen.

Observation: Connective Tissue

There are several different types of connective tissue. We will study loose fibrous connective tissue, dense fibrous connective tissue, adipose tissue, bone, cartilage, and blood. **Loose fibrous connective tissue** supports epithelium and also many internal organs, such as muscles, blood vessels, and nerves. Its presence allows organs to expand. **Dense fibrous connective tissue** contains many collagenous fibers packed together, as in tendons, which connect muscles to bones, and in ligaments, which connect bones to other bones at joints.

1. Examine a slide of loose fibrous connective tissue, and compare it to the figure below (*left*).

    What is the function of loose fibrous connective tissue? _____

2. Examine a slide of dense fibrous connective tissue, and compare it to the figure below (*right*).

    What two kinds of structures in the body contain dense fibrous connective tissue? _____

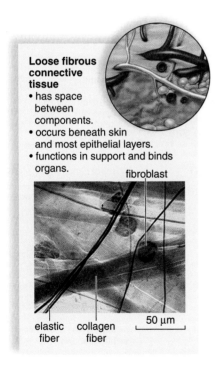

Loose fibrous connective tissue
• has space between components.
• occurs beneath skin and most epithelial layers.
• functions in support and binds organs.

fibroblast

elastic fiber   collagen fiber   50 μm

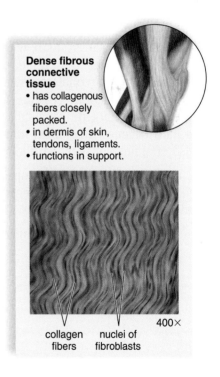

Dense fibrous connective tissue
• has collagenous fibers closely packed.
• in dermis of skin, tendons, ligaments.
• functions in support.

collagen fibers   nuclei of fibroblasts   400×

Observation: Adipose Tissue

In **adipose tissue,** the cells have a large, central, fat-filled vacuole that causes the nucleus and cytoplasm to be at the perimeter of the cell. Adipose tissue occurs beneath the skin, where it insulates the body, and around internal organs, such as the kidneys and heart. It cushions and helps protect these organs.

1. Examine a prepared slide of adipose tissue. Why is the nucleus pushed to one side? _____

   _____

2. State a location for adipose tissue in the body.

   _____

   _____

   What are two functions of adipose tissue at this location?

   _____

   _____

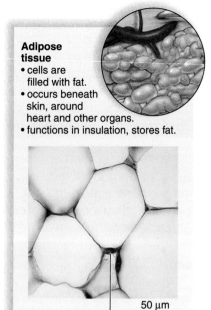

Adipose tissue
- cells are filled with fat.
- occurs beneath skin, around heart and other organs.
- functions in insulation, stores fat.

50 µm

nucleus

Observation: Compact Bone

Compact bone is found in the bones that make up the skeleton. It consists of **osteons** (Haversian system) with a central canal, and concentric rings of spaces called **lacunae,** connected by tiny crevices called canaliculi. The central canal contains a nerve and blood vessels, which service bone. The lacunae contain bone cells called **osteocytes,** whose processes extend into the canaliculi. Separating the lacunae is a matrix that is hard because it contains minerals, notably calcium salts. The matrix also contains collagenous fibers.

1. Study a model or diagram of compact bone. Then look at a prepared slide and identify the central canal, lacunae, and canaliculi.

2. What is the function of the central canal and canaliculi?

   _____

   _____

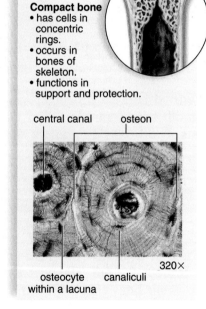

Compact bone
- has cells in concentric rings.
- occurs in bones of skeleton.
- functions in support and protection.

central canal   osteon

320×

osteocyte   canaliculi
within a lacuna

Observation: Hyaline Cartilage

In **hyaline cartilage,** cells called **chondrocytes** are found in twos or threes in lacunae. The lacunae are separated by a flexible matrix containing weak collagenous fibers.

1. Study the diagram and photomicrograph of hyaline cartilage in the figure at right. Then study a prepared slide of hyaline cartilage, and identify the matrix, lacunae, and chondrocytes.
2. Compare compact bone and hyaline cartilage. Which of these types of connective tissue is more organized? _____

   Why? _____
3. Which of these two types of connective tissue lends more support to body parts? _____

   _____

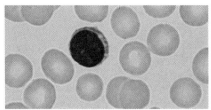

Hyaline cartilage
- has cells in lacunae.
- occurs in nose and walls of respiratory passages; at ends of bones, including ribs.
- functions in support and protection.

chondrocyte within lacunae   matrix   50 µm

Observation: Blood

Blood is a connective tissue in which the matrix is an intercellular fluid called **plasma. Red blood cells** (erythrocytes) have a biconcave (indented on both sides) appearance and lack a nucleus. These cells carry oxygen combined with the respiratory pigment hemoglobin. **White blood cells** (leukocytes) have a nucleus and are typically larger than the more numerous red blood cells. These cells fight infection.

1. Study a prepared slide of human blood. With the help of Figure 18.2, identify the numerous red blood cells and the less numerous but larger white blood cells, which appear faint because of the stain.
2. Try to identify a neutrophil, the most common type of white blood cell. A neutrophil has a multilobed nucleus. Try to identify a lymphocyte, the next most common type of white blood cell. A lymphocyte is the smallest of the white blood cells, with a spherical or slightly indented nucleus.

Figure 18.2 Blood cells.

Red blood cells are more numerous than white blood cells. White blood cells can be separated into five distinct types. If you have blood work done that includes a complete blood count (CBC), the doctor is getting a count of each of these types of WBCs. (*a–e:* Magnification 1,050×)

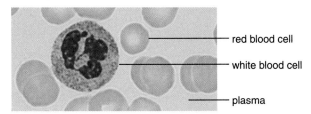

red blood cell

white blood cell

plasma

a. Neutrophil

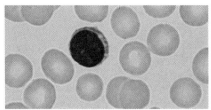

b. Lymphocyte

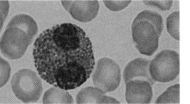

c. Eosinophil

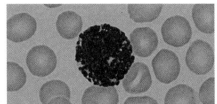

d. Basophil

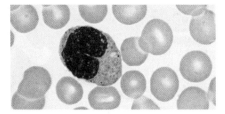

e. Monocyte

Summary of Connective Tissue

1. Complete Table 18.2 to summarize your study of connnective tissue.

Table 18.2	Connective Tissue		
Type	**Structure**	**Function**	**Location**
Loose fibrous connective tissue			Between the muscles; beneath the skin; beneath most epithelial layers
Dense fibrous connective tissue		Binds organs together, binds muscle to bones, binds bone to bone	
Adipose			Beneath the skin; around the kidney and heart; in the breast
Compact bone		Support, protection	
Hyaline cartilage	Cells in lacunae		Nose; ends of bones; rings in walls of respiratory passages; between ribs and sternum
Blood	Red and white cells floating in plasma		Blood vessels

2. Working with others in a group, decide how the structure of each connective tissue suits its function.

   Loose fibrous connective tissue _____

   Dense fibrous connective tissue _____

   Apidose tissue _____

   Compact bone _____

   Hyaline cartilage _____

   Blood _____

Muscular Tissue

Muscular (contractile) tissue is composed of cells called muscle fibers. Muscular tissue has the ability to contract, and contraction usually results in movement. The body contains skeletal, cardiac, and smooth muscle.

Observation: Skeletal Muscle

Skeletal muscle occurs in the muscles attached to the bones of the skeleton. The contraction of skeletal muscle is said to be **voluntary** because it is under conscious control. Skeletal muscle is striated; it contains light and dark bands. The striations are caused by the arrangement of contractile filaments (actin and myosin filaments) in muscle fibers. Each fiber contains many nuclei, all peripherally located.

1. Study a model or diagram of skeletal muscle, and note that striations are present. You should see several muscle fibers, each marked with striations.
2. Examine a prepared slide of skeletal muscle. The striations may be difficult to make out, but bringing the slide in and out of focus may help.

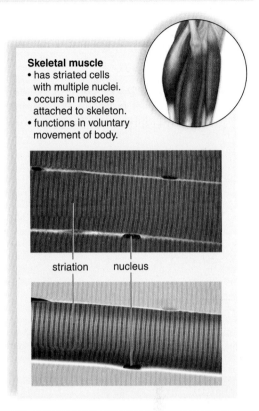

Skeletal muscle
- has striated cells with multiple nuclei.
- occurs in muscles attached to skeleton.
- functions in voluntary movement of body.

striation   nucleus

Observation: Cardiac Muscle

Cardiac muscle is found only in the heart. It is called **involuntary** because its contraction does not require conscious effort. Cardiac muscle is striated in the same way as skeletal muscle. However, the fibers are branched and bound together at **intercalated disks,** where their folded plasma membranes touch. This arrangement aids communication between fibers.

1. Study a model or diagram of cardiac muscle, and note that striations are present.
2. Examine a prepared slide of cardiac muscle. Find an intercalated disk. What is the function of cardiac muscle?

_____

_____

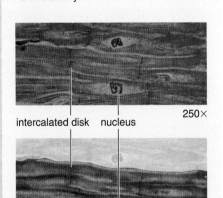

Cardiac muscle
- has branching, striated cells, each with a single nucleus.
- occurs in the wall of the heart.
- functions in the pumping of blood.
- is involuntary.

intercalated disk   nucleus   250×

Smooth muscle is sometimes called **visceral muscle** because it makes up the walls of the internal organs, such as the intestines and the blood vessels. Smooth muscle is involuntary because its contraction does not require conscious effort.

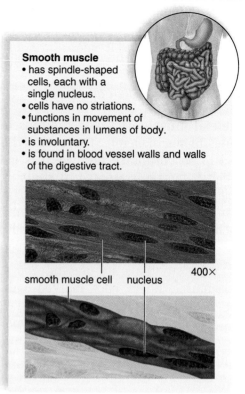

Smooth muscle
- has spindle-shaped cells, each with a single nucleus.
- cells have no striations.
- functions in movement of substances in lumens of body.
- is involuntary.
- is found in blood vessel walls and walls of the digestive tract.

smooth muscle cell   nucleus   400×

1. Study a model or diagram of smooth muscle, and note the shape of the cells and the centrally placed nucleus. Smooth muscle has spindle-shaped cells. What does

   spindle-shaped mean? _____

   _____

2. Examine a prepared slide of smooth muscle. Distinguishing the boundaries between the different cells may require you to take the slide in and out of focus.

Summary of Muscular Tissue

1. Complete Table 18.3 to summarize your study of muscular tissue.

Table 18.3	Muscular Tissue		
Type	Striations (Yes or No)	Branching (Yes or No)	Conscious Control (Yes or No)
Skeletal			
Cardiac			
Smooth			

2. How does it benefit an animal that skeletal muscle is voluntary while cardiac and smooth muscle are involuntary? _____

   _____

Nervous Tissue

Nervous tissue is found in the brain, spinal cord, and nerves. Nervous tissue receives and integrates incoming stimuli before conducting nerve impulses which control the glands and muscles of the body. Nervous tissue is composed of two types of cells: **neurons** that transmit messages and **neuroglia** that support and nourish the neurons. Motor neurons, which take messages from the spinal cord to the muscles, are often used to exemplify typical neurons. Motor neurons have several **dendrites,** processes that take signals to a **cell body,** where the nucleus is located, and an **axon** that takes nerve impulses away from the cell body.

Observation: Nervous Tissue

1. Study a model or diagram of a neuron, and then examine a prepared slide. Most likely, you will not be able to see neuroglial cells because they are much smaller than neurons.
2. Identify the dendrites, cell body, and axon in Figure 18.3 and label the micrograph. Long axons are called nerve fibers.
3. Explain the appearance and function of the parts of a motor neuron:

   a. Dendrites _____

   _____

   b. Cell body _____

   _____

   c. Axon _____

   _____

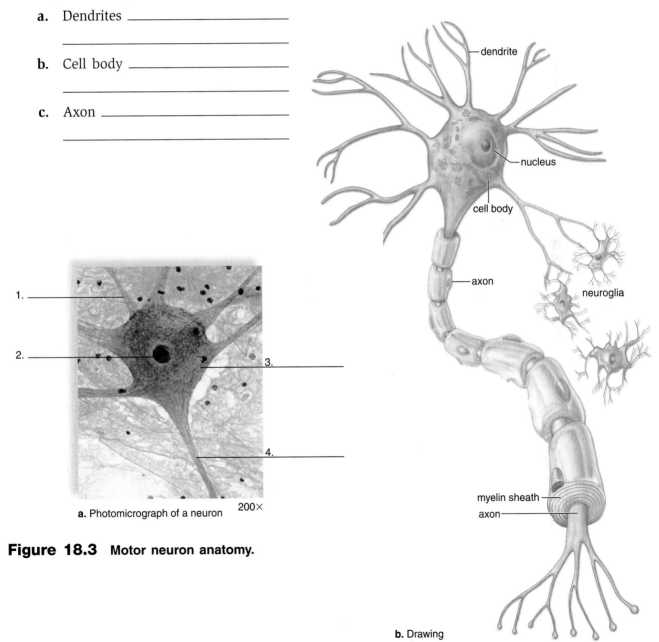

a. Photomicrograph of a neuron   200×

Figure 18.3 **Motor neuron anatomy.**

18.2 Organ Level of Organization

Organs are structures composed of two or more types of tissue that work together to perform particular functions. You may tend to think that a particular organ contains only one type of tissue. For example, muscular tissue is usually associated with muscles and nervous tissue with the brain. However, muscles and the brain also contain other types of tissue—for example, loose connective tissue and blood. Here we will study the compositions of two organs—the intestine and the skin.

Intestine

The **intestine,** a part of the digestive system, processes food and absorbs nutrient molecules.

Observation: Intestinal Wall

Study a slide of a cross section of intestinal wall. With the help of Figure 18.4, identify the following layers:

1. **Mucosa** (mucous membrane layer): This layer, which lines the central lumen (cavity), is made up of columnar epithelium overlying a layer of connective tissue. The epithelium is glandular—that is, it secretes mucus from goblet cells and digestive enzymes from the rest of the epithelium. The membrane is arranged in deep folds (fingerlike projections) called **villi,** which increase the small intestine's absorptive surface.
2. **Submucosa** (submucosal layer): This connective tissue layer contains nerve fibers, blood vessels, and lymphatic vessels. The products of digestion are absorbed into these blood and lymphatic vessels.
3. **Muscularis** (smooth muscle layer): Circular muscular tissue and then longitudinal muscular tissue are found in this layer. Rhythmic contraction of these muscles causes **peristalsis,** a wavelike motion that moves food along the intestine.
4. **Serosa** (serous membrane layer): In this layer, a thin sheet of connective tissue underlies a thin, outermost sheet of squamous epithelium. This membrane is part of the **peritoneum,** which lines the entire abdominal cavity.

Figure 18.4 The intestinal wall.
A cross section reveals the various layers of the intestinal wall, noted to the right of this photomicrograph. (Magnification 25×)

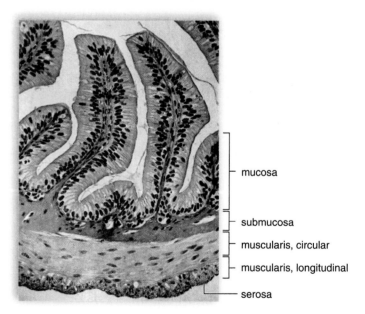

mucosa

submucosa

muscularis, circular

muscularis, longitudinal

serosa

Skin

The skin covers the entire exterior of the human body. Skin functions include protection, water retention, sensory reception, body temperature regulation, and vitamin D synthesis.

Observation: Skin

Study a model or diagram and also a prepared slide of the skin. With the help of the figure below, identify the two skin regions and the subcutaneous layer from the exterior surface down:

1. **Epidermis:** This region is composed of stratified squamous epithelial cells. The outer cells of the epidermis are nonliving and create a waterproof covering that prevents excessive water loss. These cells are always being replaced because an inner layer of the epidermis is composed of living cells that constantly produce new cells.
2. **Dermis:** This region is a connective tissue containing blood vessels, nerves, sense organs, and the expanded portions of oil (sebaceous) and sweat glands and hair follicles.

    List the structures you can identify on your slide: _____

    _____

3. **Subcutaneous layer:** This is a layer of loose connective tissue and adipose tissue that lies beneath the skin proper and serves to insulate and protect inner body parts.

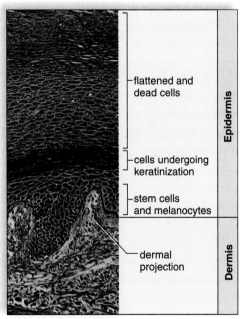

Photomicrograph of skin

1. Name four major types of tissues, and state a general function for each.

   _____

   _____

   _____

   _____

2. Describe the shapes of three types of epithelial tissue, and state a function for each.

   _____

   _____

   _____

3. Describe the appearance of three types of muscular tissue, and state a function for each.

   _____

   _____

   _____

4. What is meant by the term *involuntary muscle?* _____

   _____

5. List the three major parts of a neuron, and state a function for each part in a motor neuron.

   _____

   _____

   _____

6. Why are certain types of connective tissue called support tissues? _____

   _____

7. Describe how you would recognize a slide of compact bone. _____

8. The outer region of skin consists of stratified squamous epithelium. Define these terms:

   a. Stratified _____

   b. Squamous _____

   c. Epithelium _____

9. Identify the tissues a.–d. below as epithelial tissue, muscular tissue, nervous tissue, or connective tissue. On the same line, name the structure(s) shown in the circles.

   a. _____

   b. _____

   c. _____

   d. _____

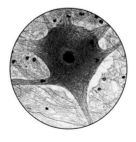

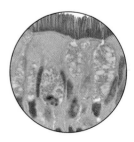

a. _____   b. _____   c. _____   d. _____

19

Basic Mammalian Anatomy I

Learning Outcomes

Introduction

In this laboratory, you will dissect a fetal pig. Both pigs and humans are mammals; therefore, you will be studying mammalian anatomy. The period of pregnancy, or gestation, in pigs is approximately 17 weeks (compared with an average of 40 weeks in humans). The piglets used in class will usually be within 1 to 2 weeks of birth.

The pigs may have a slash in the right neck region, indicating the site of blood drainage. A red latex solution may have been injected into the **arterial system,** and a blue latex solution may have been injected into the **venous system** of the pigs. If so, when a vessel appears red, it is an artery, and when a vessel appears blue, it is a vein.

Caution: Exercise caution when handling dissecting instruments. Wear protective latex gloves when handling preserved animal organs.

19.1 External Anatomy

Mammals are characterized by the presence of mammary glands and hair. Mammals also occur in two distinct sexes, called males and females, often distinguishable by their external **genitals,** the reproductive organs.

Both pigs and humans are placental mammals, which means that development occurs within the uterus of the mother. An **umbilical cord** stretches externally between the fetal animal and the **placenta,** where carbon dioxide and organic wastes are exchanged for oxygen and organic nutrients.

Pigs and humans are tetrapods—that is, they have four limbs. Pigs walk on all four of their limbs; in fact, they walk on their toes, and their toenails have evolved into hooves. In contrast, humans walk only on the feet of their hindlimbs.

Observation: External Anatomy

Body Regions and Limbs

1. Place your animal in a dissecting pan, and observe the following body regions: the rather large head; the short, thick neck; the cylindrical trunk with two pairs of appendages (forelimbs and hindlimbs); and the short tail (Fig. 19.1a). The tail is an extension of the vertebral column.
2. Examine the four limbs, and feel for the joints of the digits, wrist, elbow, shoulder, hip, knee, and ankle.
3. Determine which parts of the forelimb correspond to your upper arm, elbow, lower arm, wrist, and hand.
4. Do the same for the hindlimb, comparing it with your leg.
5. The pig walks on its toenails, which would be like a ballet dancer on "tiptoe." Notice how your heel touches the ground when you walk. Where is the heel of the pig? _____

Umbilical Cord

1. Locate the umbilical cord arising from the ventral (toward the belly) portion of the abdomen.
2. Note the cut ends of the umbilical blood vessels. If they are not easily seen, cut the umbilical cord near the end and observe this new surface.
3. What is the function of the umbilical cord? _____
   _____

Nipples and Hair

1. Locate the small **nipples,** the external openings of the **mammary glands.** The nipples are *not* an indication of sex, since both males and females possess them. How many nipples does your pig have? _____ When is it advantageous for a pig to have so many nipples?
   _____

2. Can you find hair on your pig? _____ Where? _____
   _____

Directional Terms for Dissecting Fetal Pig

Anterior: toward the head end	**Ventral: toward the belly**
Posterior: toward the hind end	**Dorsal: toward the back**

Figure 19.1 External anatomy of the fetal pig.

a. Body regions and limbs. **b, c.** The sexes can be distinguished by the external genitals.

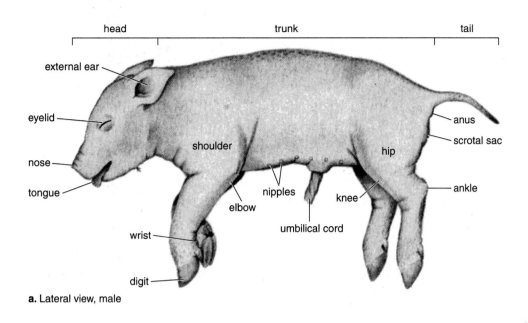

a. Lateral view, male

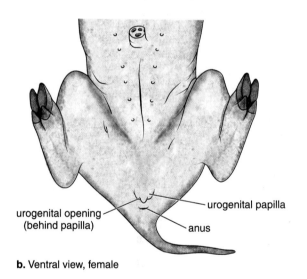

b. Ventral view, female

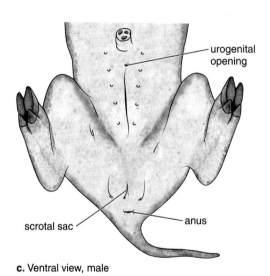

c. Ventral view, male

Anus and External Genitals

1. Locate the **anus** under the tail. The anus is an opening for what system in the body? _____
2. In females, locate the **urogenital opening,** just anterior to the anus, and a small, fleshy **urogenital papilla** projecting from the urogenital opening (Fig. 19.1*b*).
3. In males, locate the urogenital opening just posterior to the umbilical cord. The duct leading to it runs forward from between the legs in a long, thick tube, the **penis,** which can be felt under the skin. In males, the urinary system and the genital system are always joined (Fig. 19.1*c*).
4. You are responsible for identifying pigs of both sexes. What sex is your pig? _____
   Be sure to look at a pig of the opposite sex that another group of students is dissecting.

19.2 Oral Cavity and Pharynx

The **oral cavity** is the space in the mouth that contains the tongue and the teeth. The **pharynx** is dorsal to the oral cavity and has three openings: The **glottis** is an opening through which air passes on its way to the **trachea** (the windpipe) and lungs. The **esophagus** is a portion of the digestive tract that leads through the neck and thorax to the stomach. The **nasopharynx** leads to the nasal passages.

Observation: Oral Cavity and Pharynx

Oral Cavity

1. Insert a sturdy pair of scissors into one corner of the specimen's mouth, and cut posteriorly for approximately 4 cm. Repeat on the opposite side.
2. Place your thumb on the tongue at the front of the mouth, and gently push downward on the lower jaw. This will tear some of the tissue in the angles of the jaws so that the mouth will remain partly open (Fig. 19.2).
3. Note small, underdeveloped teeth in both the upper and lower jaws. Other embryonic, nonerupted teeth may also be found within the gums. The teeth are used to chew food.
4. Examine the tongue, which is partly attached to the lower jaw region but extends posteriorly and is attached to a bony structure at the back of the oral cavity (Fig. 19.2). The tongue manipulates food for swallowing.
5. Locate the hard and soft palates (Fig. 19.2). The **hard palate** is the ridged roof of the mouth that separates the oral cavity from the nasal passages. The **soft palate** is a smooth region posterior to the hard palate. An extension of the soft palate—the **uvula**—hangs down into the throat in humans. (A pig does not have a uvula.)

Figure 19.2 Oral cavity of the fetal pig.
The roof of the oral cavity contains the hard and soft palates, and the tongue lies above the floor of the oral cavity.

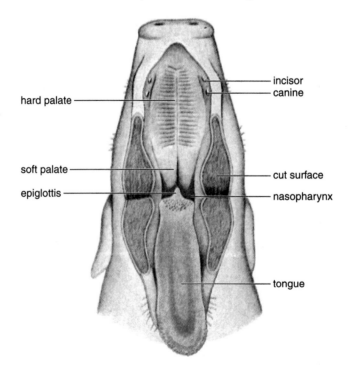

Pharynx

1. Push down on the tongue until you open the jaws far enough to see a slightly pointed flap of tissue pointing dorsally (toward the back) (Fig. 19.2). This flap is the **epiglottis,** which covers the glottis. The **glottis** leads to the trachea (Fig. 19.3*a*).
2. Posterior and dorsal to the glottis, find the opening into the **esophagus,** a tube that takes food to the stomach. Note the proximity of the glottis and the opening to the esophagus. Each time the pig—or a human—swallows, the epiglottis instantly closes to keep food and fluids from going into the lungs via the trachea.
3. Insert a blunt probe into the glottis, and note that it enters the trachea. Remove the probe, insert it into the esophagus, and note the position of the esophagus beneath the trachea.
4. Make a midline cut in the soft palate from the epiglottis to the hard palate. Then make two lateral cuts at the edge of the hard palate.
5. Posterior to the soft palate, locate the openings to the nasal passages.
6. Explain why it is correct to say that the air and food passages cross in the pharynx.

_____

_____

Figure 19.3 **Air and food passages in the fetal pig.**
The air and food passages cross in the pharynx. **a.** Drawing. **b.** Dissection of specimen.

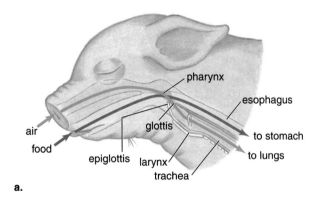

a.

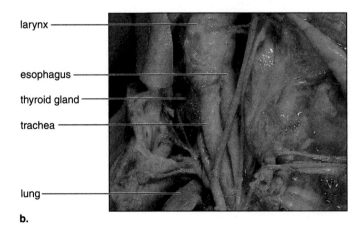

b.

19.3 Thoracic and Abdominal Incisions

First, prepare your pig according to the following directions, and then make thoracic and abdominal incisions so that you will be able to study the internal anatomy of your pig.

Preparation of Pig for Dissection

1. Place the fetal pig on its back in the dissecting pan.
2. Tie a cord around one forelimb, and then bring the cord around underneath the pan to fasten back the other forelimb.
3. Spread the hindlimbs in the same way.
4. With scissors always pointing up (never down), make the following incisions to expose the thoracic and abdominal cavities. The incisions are numbered on Figure 19.4 to correspond with the following steps.

Thoracic Incisions

1. Cut anteriorly up from the **diaphragm,** a structure that separates the thoracic cavity from the abdominal cavity, until you reach the hairs in the throat region.
2. Make two lateral cuts, one on each side of the midline incision anterior to the forelimbs, taking extra care not to damage the blood vessels around the heart.
3. Make two lateral cuts, one on each side of the midline just posterior to the forelimbs and anterior to the diaphragm, following the ends of the ribs. Pull back the flaps created by these cuts to expose the **thoracic cavity.** List the organs you find in the thoracic cavity.

Abdominal Incisions

1. With scissors pointing up, cut posteriorly from the diaphragm to the umbilical cord.
2. Make a flap containing the umbilical cord by cutting a semicircle around the cord and by cutting posteriorly to the left and right of the cord.
3. Make two cuts, one on each side of the midline incision posterior to the diaphragm. Examine the diaphragm, attached to the chest wall by radially arranged muscles. The central region of the diaphragm, called the **central tendon,** is a membranous area.
4. Make two more cuts, one on each side of the flap containing the umbilical cord and just anterior to the hindlimbs. Pull back the side flaps created by these cuts to expose the **abdominal cavity.**
5. Lifting the flap with the umbilical cord requires cutting the **umbilical vein.** Before cutting the umbilical vein, tie a thread on each side of where you will cut to mark the vein for future reference.
6. Rinse out your pig as soon as you have opened the abdominal cavity. If you have a problem with excess fluid, obtain a disposable plastic pipette to suction off the liquid.
7. Anatomically, the diaphragm separates what two cavities?

---

8. List the organs you find in the abdominal cavity.

Figure 19.4 **Ventral view of the fetal pig indicating incisions.**
These incisions are to be made preparatory to dissecting the internal organs. They are numbered here in the order they should be done.

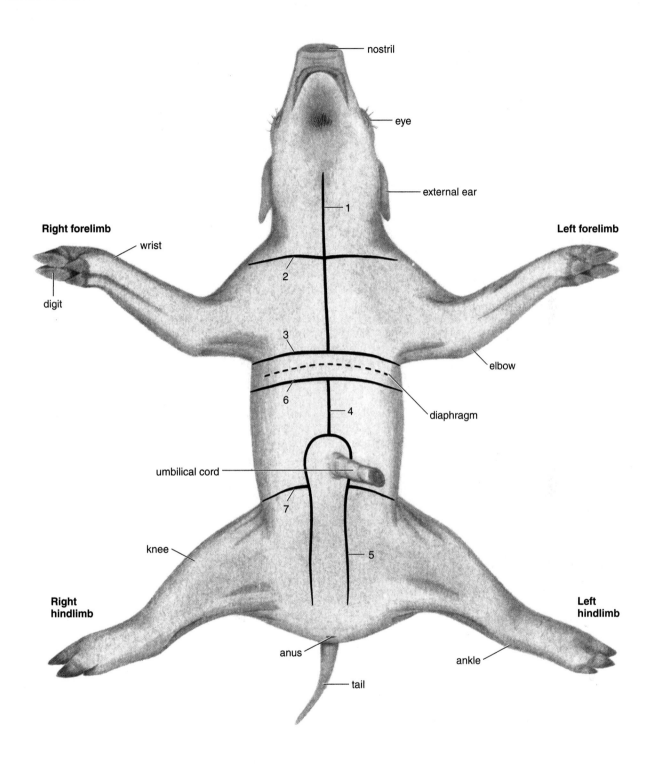

19.4 Neck Region

You will locate several organs in the neck region. Use Figures 19.3*b* and 19.5 as a guide, but *keep all the flaps on your pig* so you can close the thoracic and abdominal cavities at the end of the laboratory session.

The **thymus gland** is a part of the lymphatic system. Certain white blood cells called T (for thymus) lymphocytes mature in the thymus gland and help us fight disease. The **larynx,** or voice box, sits atop the **trachea,** or windpipe. The esophagus is a portion of the digestive tract that leads to the stomach. The **thyroid gland** secretes hormones that travel in the blood and act upon other body cells. These hormones (e.g., thyroxine) regulate the rate at which metabolism occurs in cells.

Observation: Neck Region

Thymus Gland

1. Move the skin apart in the neck region just below the hairs mentioned earlier. If necessary, cut the body wall laterally to make flaps. You will most likely be viewing exposed muscles.
2. *Cut through and clear away muscle* to expose the thymus gland, a diffuse gland that lies among the muscles. Later you will notice that the thymus flanks the thyroid and overlies the heart. The thymus is particularly large in fetal pigs, since their immune systems are still developing.

Larynx, Trachea, and Esophagus

1. Probe down into the deeper layers of the neck. Medially (toward the center), beneath several strips of muscle, you will find the hard-walled larynx and the trachea, parts of the respiratory passage to be examined later. Dorsal to the trachea, find the esophagus.
2. Open the mouth and insert a probe into the glottis and esophagus from the pharynx to better understand the orientation of these two organs.

Thyroid Gland

Locate the thyroid gland just posterior to the larynx, lying ventral to (on top of) the trachea.

19.5 Thoracic Cavity

As previously mentioned, the body cavity of mammals, including human beings, is divided by the diaphragm into the thoracic cavity and the abdominal cavity. The heart and lungs are in the thoracic cavity (Figs. 19.5 and 19.6). The **heart** is a pump for the cardiovascular system, and the **lungs** are organs of the respiratory system where gas exchange occurs.

Observation: Thoracic Cavity

Heart and Lungs

1. If you have not yet done so, fold back the chest wall flaps. To do this, you will need to tear the thin membranes that divide the thoracic cavity into three compartments: the left **pleural cavity** containing the left lung, the right pleural cavity containing the right lung, and the **pericardial cavity** containing the heart.
2. Examine the lungs. Locate the four lobes of the right lung and the three lobes of the left lung. The trachea, dorsal to the heart, divides into the **bronchi,** which enter the lungs. Later, when the heart is removed, you will be able to see the trachea and bronchi.
3. Trace the path of air from the nasal passages to the lungs.

_____

_____

Figure 19.5 Internal anatomy of the fetal pig.

The major organs are featured in this drawing. In the fetal pig, a red color tells you a vessel is an artery, and a blue color tells you it is a vein. (It does not tell you whether this vessel carries O$_2$-rich or O$_2$-poor blood.) Contrary to this drawing, *keep all the flaps on your pig* so you can close the thoracic and abdominal cavities at the end of the laboratory session.

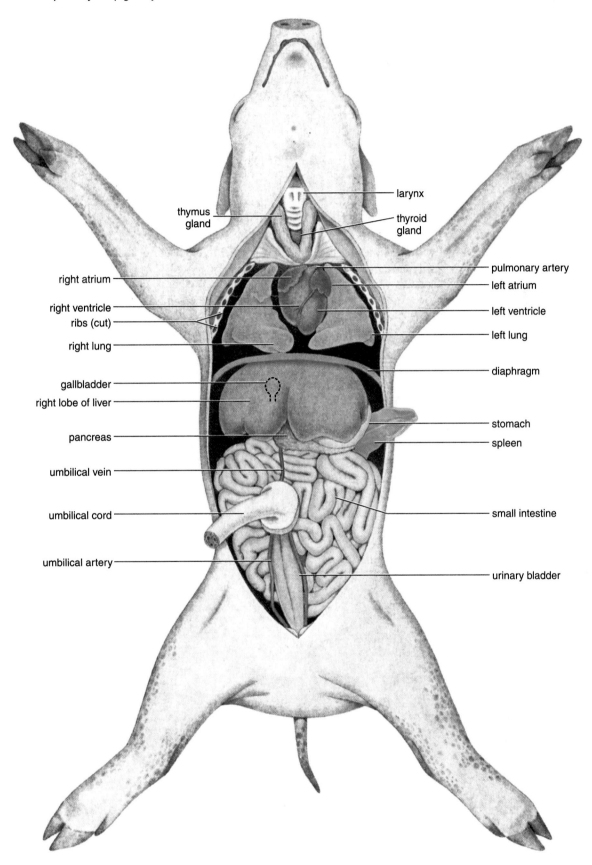

19.6 Abdominal Cavity

The abdominal wall and organs are lined by a membrane called **peritoneum,** consisting of epithelium supported by connective tissue. Double-layered sheets of peritoneum, called **mesenteries,** project from the body wall and support the organs.

The **liver,** the largest organ in the abdomen (Fig. 19.6), performs numerous vital functions, including (1) disposing of worn-out red blood cells, (2) producing bile, (3) storing glycogen, (4) maintaining the blood glucose level, and (5) producing blood proteins.

The abdominal cavity also contains organs of the digestive tract, such as the stomach, small intestine, and large intestine. The **stomach** (see Fig. 19.5) stores food and has numerous gastric glands that secrete gastric juice, which digests protein. The **small intestine** is the part of the digestive tract that receives secretions from the pancreas and gallbladder. Besides being an area for the digestion of all components of food—carbohydrate, protein, and fat—the small intestine absorbs the products of digestion: glucose, amino acids, glycerol, and fatty acids. The **large intestine** is the part of the digestive tract that absorbs water and prepares feces for defecation at the anus.

The **gallbladder** stores and releases bile, which aids the digestion of fat. The **pancreas** (see Fig. 19.5) is both an exocrine and an endocrine gland. As an exocrine gland, it produces and secretes pancreatic juice, which digests all the components of food in the small intestine. Both bile and pancreatic juice enter the duodenum by way of ducts. As an endocrine gland, the pancreas secretes the hormones insulin and glucagon into the bloodstream. Insulin and glucagon regulate blood glucose levels.

The **spleen** (see Fig. 19.5) is a lymphoid organ in the lymphatic system that contains both white and red blood cells. It purifies blood and disposes of worn-out red blood cells.

Observation: Abdominal Cavity

Liver

1. If your particular pig is partially filled with dark, brownish material, take your animal to the sink and rinse it out. This material is clotted blood. Consult your instructor before removing any red or blue latex masses, since they may enclose organs you will need to study.
2. Locate the liver, a large, brown organ. Its anterior surface is smoothly convex and fits snugly into the concavity of the diaphragm.
3. Name several functions of the liver. _____

_____

Stomach and Spleen

1. Push aside and identify the stomach, a large sac dorsal to the liver on the left side.
2. Locate the point near the midline of the body where the **esophagus** penetrates the diaphragm and joins the stomach.
3. Find the spleen, a long, flat, reddish organ attached to the stomach by mesentery.
4. The stomach is a part of what system? _____

   What is its function? _____

5. The spleen is a part of what system? _____

   What is its function? _____

Figure 19.6 Internal anatomy of the fetal pig.

Most of the major organs are shown in this photograph. The stomach has been removed. The spleen, gallbladder, and pancreas are not visible. *Do not* remove any organs or flaps from your pig.

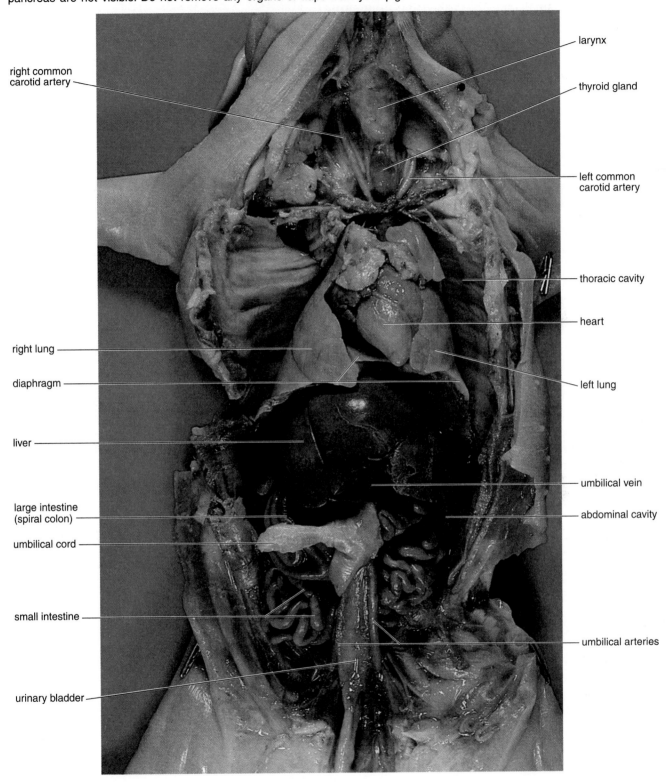

right common carotid artery

larynx

thyroid gland

left common carotid artery

thoracic cavity

heart

right lung

left lung

diaphragm

liver

umbilical vein

large intestine (spiral colon)

abdominal cavity

umbilical cord

small intestine

umbilical arteries

urinary bladder

Small Intestine

1. Look posteriorly where the stomach makes a curve to the right and narrows to join the anterior end of the small intestine called the **duodenum.**
2. From the duodenum, the small intestine runs posteriorly for a short distance and is then thrown into an irregular mass of bends and coils held together by a common mesentery.
3. The small intestine is a part of what system? _____

   What is its function? _____

Gallbladder and Pancreas

1. Locate the **bile duct,** which runs in the mesentery stretching between the liver and the duodenum. Find the gallbladder, embedded in the liver on the underside of the right lobe. It is a small, greenish sac.
2. Lift the stomach and locate the pancreas, the light-colored, diffuse gland lying in the mesentery between the stomach and the small intestine. The pancreas has a duct that empties into the duodenum of the small intestine.
3. What is the function of the gallbladder? _____
4. What is the function of the pancreas? _____

   _____

Large Intestine

1. Locate the distal (far) end of the small intestine, which joins the large intestine posteriorly, in the left side of the abdominal cavity (right side in humans). At this junction, note the **cecum,** a blind pouch.
2. Compare the large intestine of your pig to Figure 19.7. The organ does not have the same appearance in humans.
3. Follow the main portion of the large intestine, known as the **colon,** as it runs from the point of juncture with the small intestine into a tight coil (spiral colon), then out of the coil anteriorly, then posteriorly again along the midline of the dorsal wall of the abdominal cavity. In the pelvic region, the **rectum** is the last portion of the large intestine. The rectum leads to the **anus.**
4. The large intestine is a part of what system? _____
5. What is the function of the large intestine? _____
6. Trace the path of food from the mouth to the anus. _____

   _____

Storage of Pigs

1. Before leaving the laboratory, place your pig in the plastic bag provided.
2. Expel excess air from the bag, and tie it shut.
3. Write your *name* and *section* on the tag provided, and attach it to the bag. Your instructor will indicate where the bags are to be stored until the next laboratory period.
4. Clean the dissecting tray and tools, and return them to their proper location.
5. Wipe off your goggles.
6. Wash your hands.

19.7 Human Anatomy

Humans and pigs are both mammals, and their organs are similar. A human torso model shows the exact location of the organs in human beings (Fig. 19.7). You should learn to associate each human organ with its particular system. Four systems are color-coded in Figure 19.7.

Figure 19.7 Human internal organs.
The dotted lines indicate the full shape of an organ that is partially covered by another organ.

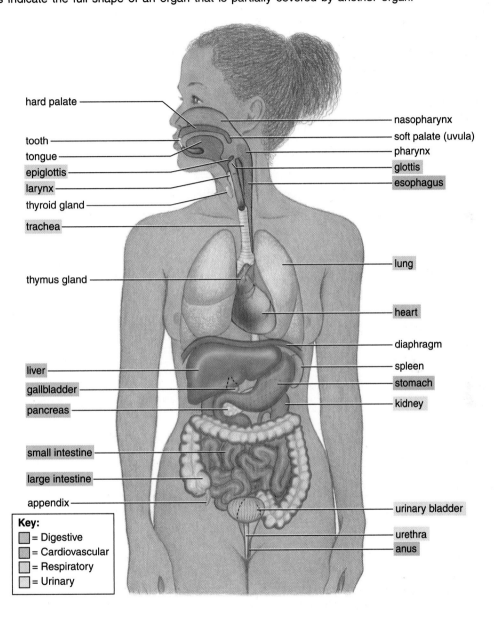

Observation: Human Torso

1. Examine a human torso model, and using Figure 19.7 as a guide, locate the same organs you have just dissected in the fetal pig.
2. In your studies so far, have you seen any major differences between pig internal anatomy and human internal anatomy? _____

1. What two features indicate that a pig is a mammal? _____

_____

2. Put the following organs in logical order: lungs, nasal passages, nasopharynx, trachea, bronchi, glottis.

_____

_____

3. What difficulty would probably arise if a person were born without an epiglottis?

_____

_____

4. The embryonic coelom may be associated with what two cavities studied in this laboratory?

_____

5. Name two principal organs in the thoracic cavity, and give a function for each.

_____

_____

6. What difficulty would arise if a person were born without a thymus gland?

_____

_____

7. Name the largest organ in the abdominal cavity and list several functions. _____

_____

_____

8. A large portion of the abdominal cavity is taken up with digestive organs. What are they?

_____

_____

9. Why is it proper to associate the gallbladder with the liver?

_____

_____

10. Where would you find the pancreas?

_____

_____

20

Basic Mammalian Anatomy II

Learning Outcomes

20.1 Urinary System
- Locate and identify the organs of the urinary system. 262–263
- State a function for the organs of the urinary system. 263

20.2 Male Reproductive System
- Locate and identify the organs of the male reproductive system. 264–266
- State a function for the organs of the male reproductive system. 264–266
- Compare the male pig reproductive system with that of the human male. 266

20.3 Female Reproductive System
- Locate and identify the organs of the female reproductive system. 267–268
- State a function for the organs of the female reproductive system. 267–268
- Compare the female pig reproductive system with that of the human female. 269

20.4 Anatomy of Testis and Ovary
- Identify a cross-section slide of the testis and seminiferous tubules, including sperm. 270
- Identify a slide of the ovary and the follicles, including an oocyte. 271

20.5 Thoracic and Abdominal Organs
- Identify and locate the individual organs of the respiratory system. State a function for each organ. 272
- Identify and locate the individual organs of the digestive system. State a function for each organ. 273
- Identify and locate the chambers of the heart and the vessels connected to these chambers. 273–274
- Name and locate the valves of the heart, and trace the path of blood through the heart. 275
- Name the major vessels of the pulmonary and systemic circuits of the cardiovascular system, and trace the path of blood to various organs. 276–277

Introduction

The **urinary system** and the **reproductive system** are so closely associated in mammals that they are often considered together as the **urogenital system.** They are particularly associated in males, where certain structures function in both systems. In this laboratory, we will focus first on dissecting the urinary and reproductive systems in the fetal pig. We will then compare the anatomy of the reproductive systems in pigs with those in humans.

In mammalian reproductive systems, the testes (sing., testis) are the male gonads, and the ovaries (sing., ovary) are the female gonads. The testes produce sperm, and the ovaries produce oocytes that become eggs. Examining prepared slides in this laboratory will allow you to observe the location of spermatogenesis in the testis and oogenesis in the ovary.

Finally, we will examine parts of the respiratory, digestive, and cardiovascular systems in the fetal pig. You will view the organs of the respiratory system in some detail; remove and examine the heart, stomach, and intestine; and view the exposed hepatic portal system.

Caution: Exercise caution when handling dissecting instruments. Wear protective latex gloves when handling preserved animal organs.

Figure 20.1 Urinary system of the fetal pig.

In (a) females and (b) males, urine is made by the kidneys, transported to the bladder by the ureters, stored in the bladder, and then excreted from the body through the urethra.

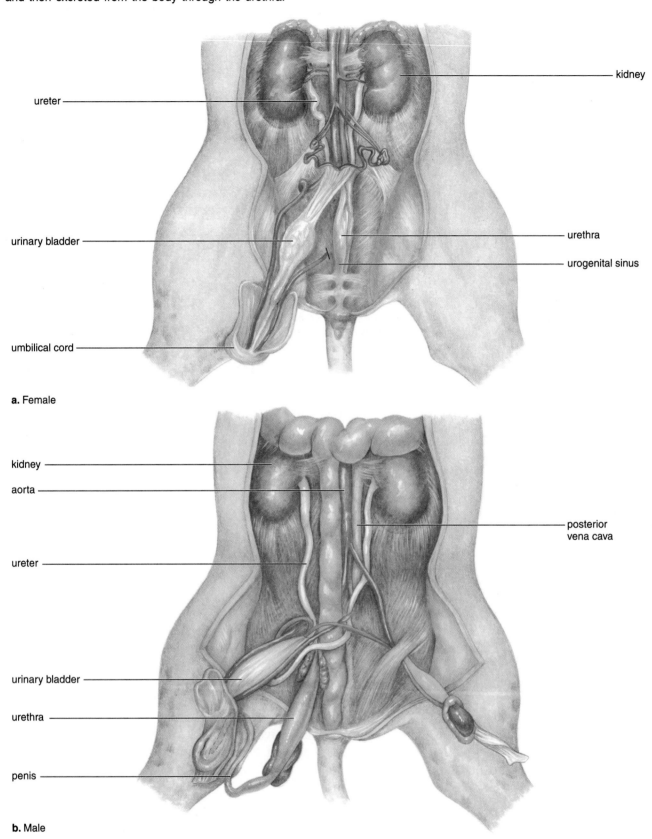

ureter

kidney

urinary bladder

urethra

urogenital sinus

umbilical cord

a. Female

kidney

aorta

posterior
vena cava

ureter

urinary bladder

urethra

penis

b. Male

20.1 Urinary System

The urinary system consists of the **kidneys,** which produce urine; the **ureters,** which transport urine to the **urinary bladder,** where urine is stored; and the **urethra,** which transports urine to the outside. In males, the urethra also transports sperm during ejaculation.

During the upcoming dissection, compare the urinary system structures of both sexes of fetal pigs. Later in this laboratory period, exchange specimens with a neighboring team for a more thorough inspection.

Observation: Urinary System in Pigs

1. The large, paired kidneys (Fig. 20.1) are reddish organs covered by **peritoneum,** a membrane that anchors them to the dorsal wall of the abdominal cavity, sometimes called the **peritoneal cavity.** Clean the peritoneum away from one of the kidneys, and study it more closely.

2. Using a razor blade or scalpel, carefully section one of the kidneys in place, cutting it lengthwise (Fig. 20.2). At the center of the medial portion of the kidney is an irregular, cavity-like reservoir, the **renal pelvis.** The outermost portion of the kidney (the **renal cortex**) shows many small striations perpendicular to the outer surface. This region and the more even-textured **renal medulla** region inward from it are composed of **nephrons** (excretory tubules).

3. Locate the **ureters,** which leave the kidneys and run posteriorly under the peritoneum.

4. Clean the peritoneum away, and follow a ureter to the **urinary bladder,** which normally lies in the posterior ventral portion of the abdominal cavity. The urinary bladder is on the inner surface of the flap of tissue to which the umbilical cord was attached.

5. The **urethra,** which arises from the bladder posteriorly, runs parallel to the rectum. Follow the urethra until it passes from view into the ring formed by the pelvic girdle.

6. Trace the path of urine. _____

_____

Figure 20.2 Anatomy of the kidney.
A kidney has a renal cortex, renal medulla, renal pelvis, and microscopic tubules called nephrons.

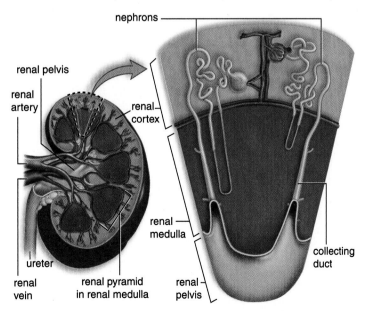

20.2 Male Reproductive System

The **male reproductive system** consists of the **testes** (sing., **testis**), which produce sperm, and the **epididymides** (sing., **epididymis**), which store sperm before they enter the **vasa deferentia** (sing., **vas deferens**). Just prior to ejaculation, sperm leave the vasa deferentia and enter the **urethra,** located in the penis. The **penis** is the male organ of sexual intercourse. **Seminal vesicles,** the **prostate gland,** and the **bulbourethral glands** (Cowper's glands)

Table 20.1	Male Reproductive Organs and Functions
Organ	**Function**
Testis	Produces sperm and sex hormones
Epididymis	Stores sperm as they mature
Vas deferens	Conducts and stores sperm
Seminal vesicle	Contributes secretions to semen
Prostate gland	Contributes secretions to semen
Urethra	Conducts sperm
Bulbourethral glands	Contribute secretions to semen
Penis	Organ of copulation

add fluid to semen after sperm reach the urethra. Table 20.1 summarizes the male reproductive organs.

The testes begin their development in the abdominal cavity, just anterior and dorsal to the kidneys. Before birth, however, they gradually descend into paired **scrotal sacs** within the scrotum, suspended anterior to the anus. Each scrotal sac is connected to the body cavity by an **inguinal canal,** the opening of which can be found in your pig. The passage of the testes from the body cavity into the scrotal sacs is called the descent of the testes and it occurs in human males. The testes in most of the male fetal pigs being dissected will probably be partially or fully descended.

Observation: Male Reproductive System in Pigs

Inguinal Canal, Testis, Epididymis, and Vas Deferens

1. Locate the opening of the left inguinal canal, which leads to the left scrotal sac (Fig. 20.3).
2. Expose the canal and sac by making an incision through the skin and muscle layers from a point over this opening back to the left scrotal sac.
3. Open the sac, and find the testis. Note the much-coiled tubule—the epididymis—that lies alongside the testis. This is continuous with the vas deferens, which passes back toward the abdominal cavity.
4. Trace a vas deferens as it loops over an umbilical artery and ureter and unites with the urethra dorsally at the posterior end of the urinary bladder.

Penis, Urethra, and Accessory Glands

1. Cut through the ventral skin surface just posterior to the umbilical cord. This will expose the rather undeveloped penis, which extends from this point posteriorly toward the anus. The central duct of the penis is the urethra.
2. Lay the penis to one side, and then cut down through the ventral midline, laying the legs wide apart in the process (Fig. 20.4). The cut will pass between muscles and through pelvic cartilage (bone has not developed yet). Do not cut any of the ducts or tracts in the region.
3. You will now see the urethra passing ventrally above the rectum. It is somewhat heavier in the male due to certain accessory glands:
   a. Bulbourethral glands (Cowper's glands), about 1 cm in diameter, lie laterally and well back toward the anal opening.
   b. The prostate gland, about 4 mm across and 3 mm thick, is located on the dorsal surface of the urethra, just posterior to the juncture of the bladder with the urethra. It is often difficult to locate and is not shown in Figures 20.3 and 20.4.
   c. Small, paired seminal vesicles may be seen on either side of the prostate gland.

Figure 20.3 Male reproductive system of the fetal pig.

In males, the urinary system and the reproductive system are joined. The vasa deferentia (sing., vas deferens) enter the urethra, which also carries urine.

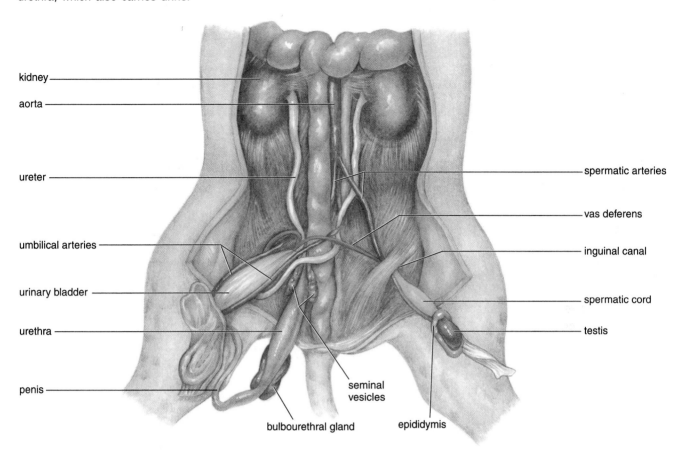

kidney

aorta

ureter

umbilical arteries

urinary bladder

urethra

penis

bulbourethral gland

seminal vesicles

epididymis

spermatic arteries

vas deferens

inguinal canal

spermatic cord

testis

Figure 20.4 Photograph of the male reproductive system of the fetal pig.

Compare the diagram in Figure 20.3 to this photograph to help identify the structures of the male urogenital system.

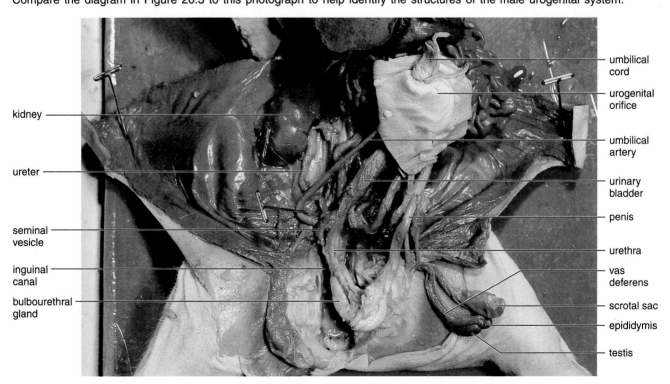

kidney

ureter

seminal vesicle

inguinal canal

bulbourethral gland

umbilical cord

urogenital orifice

umbilical artery

urinary bladder

penis

urethra

vas deferens

scrotal sac

epididymis

testis

4. Trace the urethra as it leaves the bladder. It proceeds posteriorly, but when it nears the posterior end of the body, it turns rather abruptly anterioventrally and runs forward just under the skin of the midventral body wall, where you have just dissected it. This latter portion of the urethra is, then, within the penis.

5. Now you should also be able to see the entrance of the vasa deferentia into the urethra. If necessary, dissect these structures free from surrounding tissue, and expose the point of entrance of these ducts into the urethra near the location of the prostate gland. In males, the urethra transports sperm, as well as urinary wastes from the bladder.

6. Trace the path of sperm in the male. _____

_____

Comparison of Male Fetal Pig and Human Male

Use Figure 20.5 to help you compare the male pig reproductive system with the human male reproductive system. Complete Table 20.2, which compares the location of the penis in these two mammals.

Table 20.2	Location of Penis in Male Fetal Pig and Human Male	
	Fetal Pig	**Human**
Penis		

Figure 20.5 Human male urogenital system.
In the fetal pig, but not in the human male, the penis lies beneath the skin and exits at the urogenital opening.

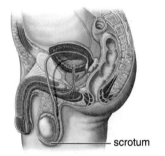

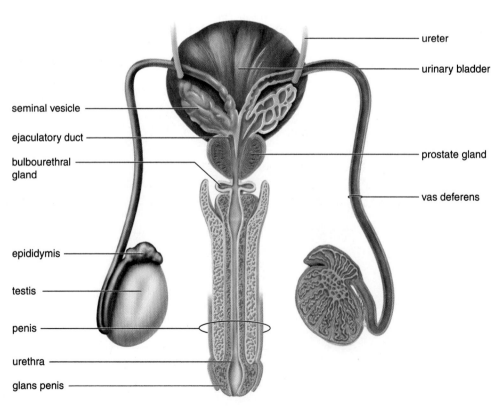

20.3 Female Reproductive System

The **female reproductive system** (Table 20.3) consists of the **ovaries,** which produce eggs, and the **oviducts,** which transport eggs to the **uterus,** where development occurs. In the fetal pig, the uterus does not form a single organ, as in humans, but is partially divided into external structures called **uterine horns,** which connect with the oviduct. The **vagina** is the birth canal and the female organ of sexual intercourse.

Table 20.3	Female Reproductive Organs and Functions	
Organ	**Function**
Ovary	Produces egg and sex hormones
Oviduct (fallopian tube)	Conducts egg toward uterus
Uterus	Houses developing fetus
Vagina	Receives penis during copulation and serves as birth canal

Observation: Female Reproductive System in Pigs

Ovaries and Oviducts

1. Locate the paired ovaries, small bodies suspended from the peritoneal wall in mesenteries, posterior to the kidneys (Figs. 20.6 and 20.7).
2. Closely examine one ovary. Note the small, short, coiled oviduct, sometimes called the **fallopian tube.** The oviduct does not attach directly to the ovary but ends in a funnel-shaped structure with fingerlike processes (fimbriae) that partially encloses the ovary.

Uterine Horns

1. Locate the **uterine horns.** (Do not confuse the uterine horns with the oviducts; the latter are much smaller and are found very close to the ovaries.)
2. Find the median body of the uterus located at the joined posterior ends of the uterine horns.

Vagina

1. Separate the hindlimbs of your specimen, and cut down along the midventral line. The cut will pass through muscle and the cartilaginous pelvic girdle. With your fingers, spread the cut edges apart, and use blunt dissecting instruments to separate connective tissue.
2. Note three ducts passing from the body cavity to the animal's posterior surface. One of these is the urethra, which leaves the bladder and passes into the **urogenital sinus.** The urethra is a part of the urinary system. The most dorsal of the three ducts is the **rectum,** which passes to its own opening, the **anus.** The rectum and anus are, of course, part of the digestive system, not the reproductive system.
3. Find the vagina, located dorsally to the urethra. The vagina is the organ of copulation and is the birth canal. Anteriorly, it connects to the uterus, and posteriorly it enters the urogenital sinus. This sinus is absent in adult humans and several other female mammals.

Figure 20.6 Female reproductive system of the fetal pig.

In the adult female, the urinary system and the reproductive system are separate. In the fetus, the vagina joins the urethra just before the urogenital sinus.

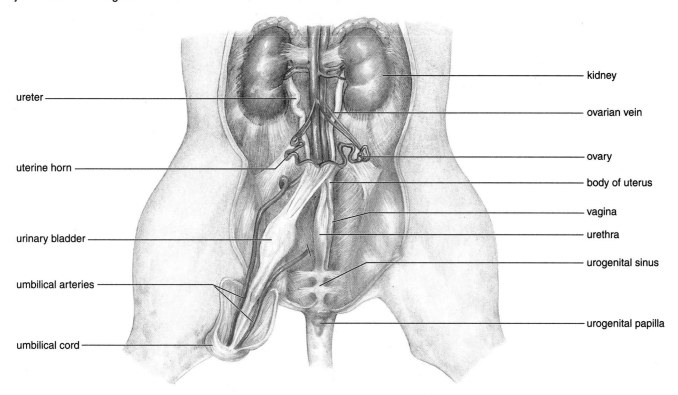

ureter

uterine horn

urinary bladder

umbilical arteries

umbilical cord

kidney

ovarian vein

ovary

body of uterus

vagina

urethra

urogenital sinus

urogenital papilla

Figure 20.7 Photograph of the female reproductive system of the fetal pig.

Compare the diagram in Figure 20.6 with this photograph to help identify the structures of the female urinary and reproductive systems.

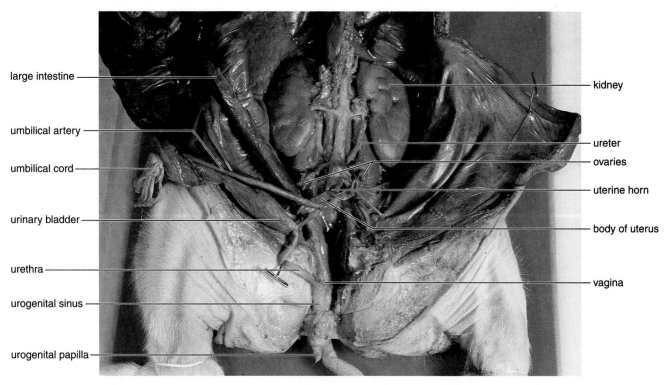

large intestine

umbilical artery

umbilical cord

urinary bladder

urethra

urogenital sinus

urogenital papilla

kidney

ureter

ovaries

uterine horn

body of uterus

vagina

Comparison of Female Fetal Pig with Human Female

Use Figure 20.8 to compare the female pig reproductive system with the human female reproductive system. Complete Table 20.4, which compares the appearance of the oviducts and the uterus, as well as the presence or absence of a urogenital sinus in these two mammals.

Figure 20.8 **Human female reproductive system.**
Especially compare the anatomy of the oviducts in humans with that of the uterine horns in a pig. In a pig, the fetuses develop in the uterine horns; in a human female, the fetus develops in the body of the uterus.

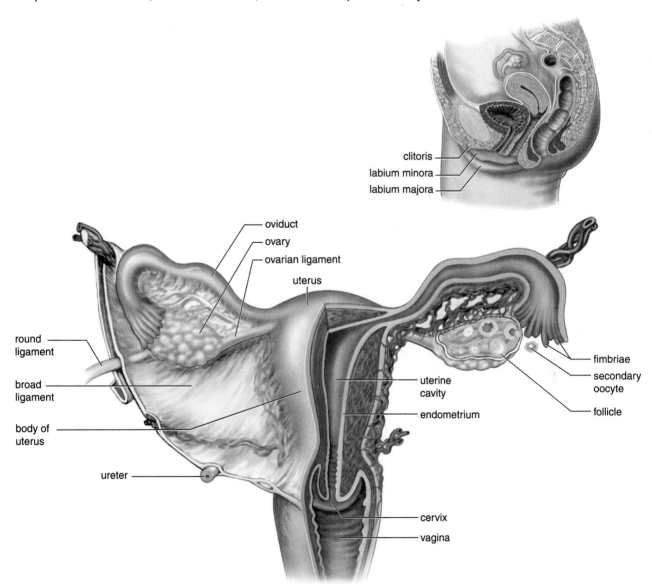

Table 20.4	Comparison of Female Fetal Pig with Human Female	
	Fetal Pig	**Human**
Oviducts		
Uterus		
Urogenital sinus		

20.4 Anatomy of Testis and Ovary

Recall that the testes produce sperm (the male gametes) and that the ovaries produce oocytes (the female gametes). A testis contains **seminiferous tubules,** where sperm formation takes place, and **interstitial cells** scattered in the spaces between seminiferous tubules. Interstitial cells produce the male sex hormone testosterone. Also present are the sertoli cells, which support and nourish the developing sperm cells. An ovary contains **follicles** in various stages of maturation. Ovarian follicles produce the female sex hormones estrogen and progesterone. One or more follicles complete maturation during each cycle and produce an oocyte.

Observation: Testis and Ovary

Testis

1. Examine a prepared slide of the testis. Note under low power the many circular structures—the seminiferous tubules.
2. Switch to high power, and observe one tubule in particular. With the help of Figure 20.9, find mature sperm, which look like thin, fine, dark lines, in the middle of the tubule. Interstitial cells are between the tubules.

Ovary

1. Examine a prepared slide of an ovary, and refer to Figure 20.10 for help in identifying the structures. The female gonad contains an inner core of loose, fibrous tissue. The outer part contains the follicles that produce eggs.

Figure 20.9 **Photomicrograph of a seminiferous tubule.**
A testis contains seminiferous tubules separated by interstitial cells. The tubules produce sperm, and the interstitial cells produce male sex hormones.

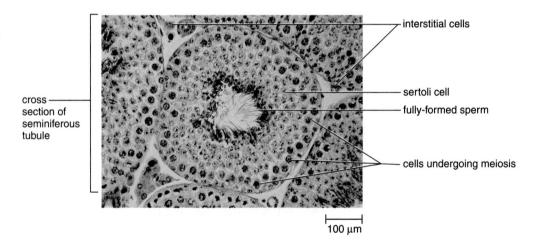

Figure 20.10 Photomicrograph of ovarian tissue.

An ovary contains follicles in different stages of maturity. A secondary follicle contains a secondary oocyte, which will burst from the ovary during ovulation. A follicle also produces the female sex hormones.

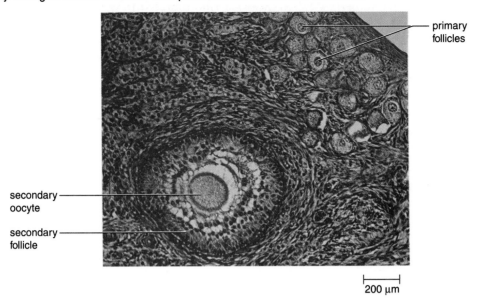

primary follicles

secondary oocyte

secondary follicle

200 μm

2. Locate a **primary follicle,** which appears as a circle of cells surrounding a somewhat larger cell.
3. Find a **secondary follicle,** and switch to high power. Note the **secondary oocyte,** which becomes an egg, surrounded by numerous cells, to one side of the liquid-filled follicle.
4. Also look for a large, fluid-filled vesicular (Graafian) follicle, which contains a mature secondary oocyte to one side. This follicle will be next to the outer surface of the ovary because it is the type of follicle that releases the egg during ovulation.
5. Also look for the remains of the corpus luteum, which will look like scar tissue. The corpus luteum develops after the vesicular follicle has released its egg, and then later it deteriorates. Not all slides will contain a vesicular follicle and corpus luteum because they may not have been present when the slide was made.

Comparison of Reproductive Systems

Complete Table 20.5 to describe the differences between the male and female mammalian reproductive systems.

Table 20.5	Comparison of Human Male and Female Reproductive Systems	
	Male	**Female**
Gonad		
Duct from gonad		
Structure connected to gonad by duct		
Copulatory organ		

20.5 Thoracic and Abdominal Organs

You will now have the opportunity to examine more carefully the organs of the respiratory and digestive systems you merely observed in Laboratory 19 (pp. 254–258). You will also study the path of blood through the heart in both the pulmonary and systemic systems.

Respitory System

The respiratory system contains the lungs and those structures that conduct air to and from the lungs (Fig. 20.11). What are the two ways by which air can enter the trachea? _____
The lungs contain alveoli, air sacs surrounded by capillaries. Here, gas exchange by diffusion brings oxygen into the blood and takes carbon dioxide out of the blood.

Observation: Organs of the Respiratory System

1. Open the pig's mouth, insert your blunt probe into the **glottis,** and explore the pathway of air in normal breathing by carefully working the probe down through the **larynx** to the level of the bronchi.
2. The air and food pathways cross in the **pharynx** (see Fig. 19.3a). If you wish, you can slit open the larynx along its midline and observe the small, paired, lateral flaps inside, known as **vocal cords.** These are not yet well-developed in this fetal animal.
3. Carefully cut the trachea crosswise just below the larynx. Do not cut the esophagus, which lies just dorsal to the trachea. With the removal of the heart, you will see the trachea as it divides into two bronchi.
4. Lift out the entire portion of the respiratory system you have just freed: trachea, bronchi, bronchioles, and lungs (Fig. 20.11). Place them in a small container of water. Holding the trachea with your forceps, gently but firmly stroke the lung repeatedly with the blunt wooden base of one of your probes. If you work carefully, the alveolar tissue will be fragmented and rubbed away, leaving the branching system of air tubes and blood vessels.

Figure 20.11 **Dorsal view of lungs and associated air tubes.**
This drawing also shows the heart and associated blood vessels.

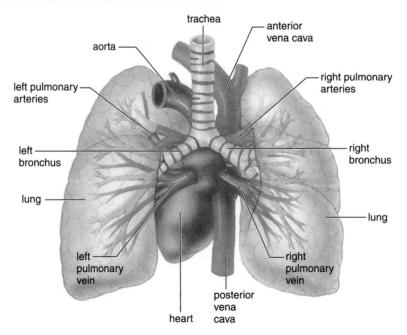

Organs of the Digestive System

The digestive system contains the organs you observed in Laboratory 19 (pp. 256–258) in which food is ingested, digested, absorbed, and prepared for elimination. Using the terms large intestine, mouth, small intestine, esophagus, stomach, and anus, state the path of food from its entrance into the body to the elimination of the remains of digestion:

_____

Observation: Organs of the Digestive System

1. Now you have a better view of the esophagus. Open the mouth again, and insert a blunt probe into the **esophagus** (see Fig. 19.3). Then trace the esophagus to the stomach.
2. Open one side of the **stomach,** and examine its interior surface. Does it appear smooth or rough? _____
3. At the lower end of the stomach, find the **pyloric sphincter,** the muscle that surrounds the entrance to the upper region of the small intestine, the **duodenum.** The pyloric sphincter regulates the entrance of material into the duodenum from the stomach.
4. Find the gallbladder, which is a small greenish sac dorsal to the liver you also observed on page 259. Dissect more carefully now to find the **bile duct system** (see Fig. 21.2) that conducts the bile to the duodenum.
5. Locate the rest of the **small intestine** and the **large intestine** (colon and rectum) first observed on page 259. To complete your study, sever the coiled small intestine just below the duodenum and sever the colon at the point where it joins the rectum.
6. Carefully cut the mesenteries holding the coils of the small intestine so that it can be laid out in a straight line. As defined on page 256, mesenteries are double-layered sheets of membrane that project from the body wall and support the organs.
7. Measure and record in meters the length of the intestinal tract. _____ meters
8. Considering the function of the small intestine (see p. 256), why would such a great length be beneficial to the body? _____

_____
9. Slit open a short portion of the intestines, and note the corrugated texture of the interior lining.

Cardiovascular System

The **cardiovascular system** includes the heart and two major circular pathways: the **pulmonary circuit** to and from the lungs and the **systemic circuit** to and from the body's organs.

Heart

The mammalian heart has a **right** and **left atrium** and a **right** and **left ventricle** (Fig. 20.12). *To tell the left from the right side, mentally position the heart so that it corresponds to your own body.* Contraction of the heart pumps the blood through the heart and out into the arteries. The right side of the heart sends blood through the smaller pulmonary circuit, and the left side of the heart sends blood through the much larger systemic circuit.

Heart Model

1. Study a heart model (Fig. 20.12) or a preserved heart, and identify the four chambers of the heart: right atrium, right ventricle, left atrium, left ventricle.

2. Which ventricle is more muscular? _____ Why is this appropriate? _____
    _____

3. Identify the vessels connected to the heart. Locate the:
    a. **Pulmonary trunk,** which leaves the ventral side of the heart from the top of the right ventricle and then passes forward diagonally before branching into the right and left pulmonary arteries.
    b. **Aorta,** which arises from the anterior end of the left ventricle, just dorsal to the origin of the pulmonary trunk. The aorta soon bends to the animal's left as the **aortic arch.**
    c. **Venae cavae.** The anterior (superior) and posterior (inferior) venae cavae enter the right atrium. They bring blood from the head and body, respectively, to the heart.
    d. **Pulmonary veins,** which return blood from the lungs to the left atrium.

Figure 20.12 **External view of mammalian heart.**
Externally, notice the coronary arteries and cardiac veins that serve the heart itself.

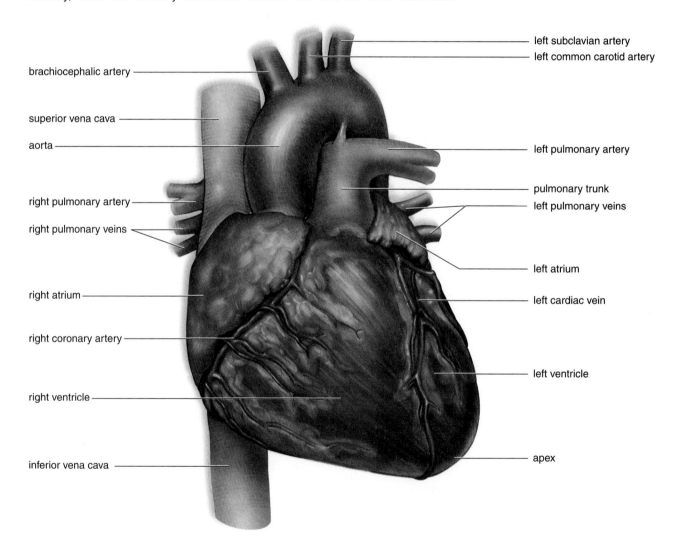

4. With the aid of Fig. 20.13, locate the valves of the heart. Remove the ventral half of the heart model or the ventral half of the heart. Identify the:
   a. **Right atrioventricular (tricuspid) valve** and the **left atrioventricular (bicuspid, mitral) valve.**
   b. **Pulmonary semilunar valve** (in the base of the pulmonary trunk).
   c. **Aortic semilunar valve** (in the base of the aorta).
   d. **Chordae tendineae** that hold the atrioventricular valves in place while the heart contracts. These extend from the papillary muscles.

Fetal Pig Dissection

1. As you remove the heart of the pig by cutting the vessels attached to it, carefully note the pulmonary trunk and pulmonary arteries (Fig. 20.13). Note also the pulmonary veins that leave the lungs and enter the left atrium.
2. Why are the pulmonary veins colored red in Figures 20.11 and 20.13? _____
   _____
3. If you wish, section the heart frontally, separating the ventral from the dorsal side.
4. Identify the four chambers of the heart. Remnants of valves between the chambers can be seen as thin sheets of whitish tissue attached to fine, white, tendinous strands.
5. With your blunt probe, find an oval opening **(foramen ovale)** in the wall between the two atrial chambers. This is a shunt that allows blood to bypass lung circulation prior to birth.

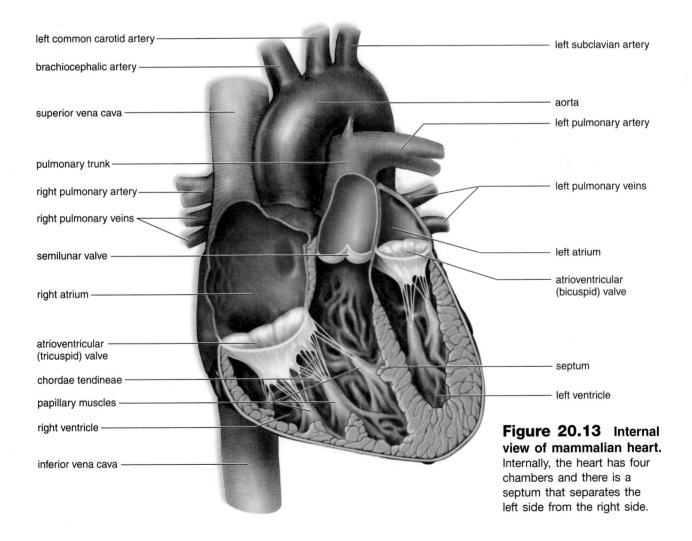

Figure 20.13 Internal view of mammalian heart. Internally, the heart has four chambers and there is a septum that separates the left side from the right side.

Tracing the Path of Blood Through the Heart

To demonstrate that O_2-poor blood is kept separate from O_2-rich blood, trace the path of blood from the right side of the heart to the aorta by filling in the blanks that follow. In the adult mammal, blood passes through the lungs to go from the right side of the heart to the left.

From Venae Cavae

_____

_____ valve

_____

_____ valve

_____

_____

To Lungs

From Lungs

_____

_____

_____ valve

_____

_____ valve

To Aorta

Pulmonary and Systemic Circuits

Blood that leaves the heart enters one of two sets of blood vessels: the pulmonary circuit, which takes blood from the heart to the lungs and from the lungs to the heart, and the systemic circuit, which takes blood from the heart to the body proper and from the body proper to the heart.

Pulmonary Circuit

1. Trace the path of blood in the pulmonary circuit:

    Right ventricle of the heart

    _____

    _____

    Lungs

    _____

    Left atrium of the heart

2. Which of these blood vessels contains O_2-rich blood? _____

Systemic Circuit

With the help of Figure 20.14, review the major systemic blood vessels and complete Table 20.6.

Table 20.6	Major Blood Vessels in the Systemic Circuit	
Body Part	**Artery**	**Vein**
Head		
Front legs in pig (arms in humans)		
Kidney		
Hind legs		

3. With the help of Figure 20.14 and Table 20.6, trace the path of blood in the systemic circuit from the heart to the kidneys and from the kidneys to the heart.

    Left ventricle

    _____

    _____

    Kidneys

    _____

    _____

    Right atrium

Figure 20.14 **Overview of the fetal pig arteries and veins.**
Arteries and veins are found in all parts of the body. In this drawing, only veins are labeled on the left, and only arteries are labeled on the right.

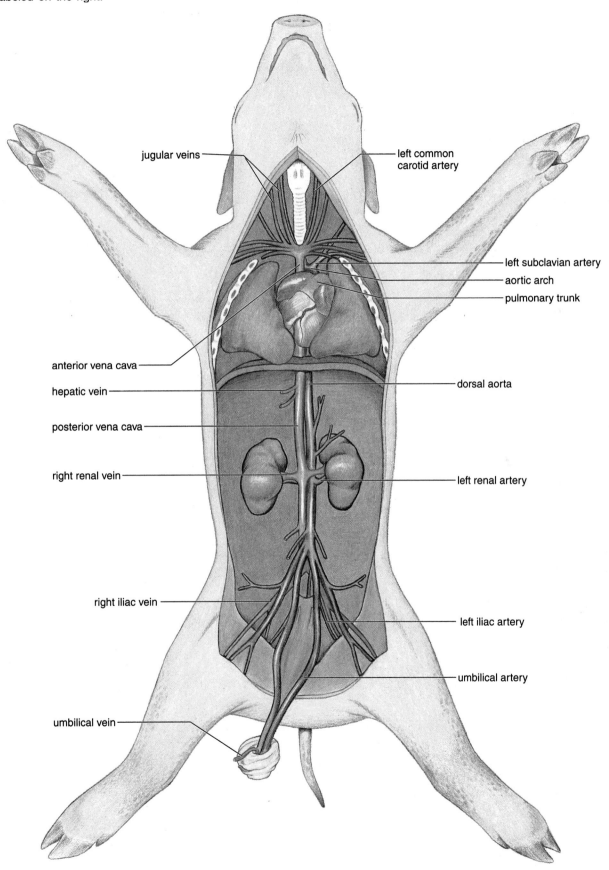

jugular veins

left common carotid artery

left subclavian artery

aortic arch

pulmonary trunk

anterior vena cava

dorsal aorta

hepatic vein

posterior vena cava

right renal vein

left renal artery

right iliac vein

left iliac artery

umbilical artery

umbilical vein

Storage of Pigs

1. Before leaving the laboratory, place your pig in the plastic bag provided.
2. Expel excess air from the bag, and tie it shut.
3. Write your *name* and *section* on the tag provided, and attach it to the bag. Your instructor will indicate where the bags are to be stored until the next laboratory period.
4. Clean the dissecting tray and tools, and return them to their proper location.
5. Wipe off your goggles.
6. Wash your hands.

Laboratory Review 20

1. Which organ is used in the male to carry urine or sperm? At which juncture does the reproductive system join the urinary system?

   _____

2. In which sex of a fetal pig is there a urogenital sinus? _____

3. Which organ in males produces sperm, and which organ in females produces eggs? _____

   _____

4. How and when do sperm acquire access to an egg in mammals? _____

   _____

5. What are the four chambers of the mammalian heart?

   _____

   _____

6. Contrast the functions of the right and left sides of the heart.

   _____

   _____

7. Trace the path of blood from the left ventricle to the kidneys and back to the right atrium.

   _____

   _____

   _____

8. What is a portal system? _____

   _____

9. Trace the path of blood from the mesenteric arteries to the posterior vena cava.

   _____

   _____

10. Put the following organs in logical order: stomach, large intestine, small intestine, pharynx, mouth, esophagus, anus.

    _____

    _____

21

Physiology of the Digestive System

Learning Outcomes

Introduction

Digestion is the process by which food is broken down by enzymes through **hydrolysis.** That is, water is added to a large molecule, which splits into smaller soluble molecules that can be absorbed into the bloodstream. For example, water is added until proteins are broken down into amino acids, starch is broken down into glucose, and fats are broken down into glycerol and fatty acids.

Enzymes are necessary for digestion, just as they are required for other chemical reactions in the body. The experimental procedures in this laboratory will demonstrate that **pepsin** speeds up the hydrolysis of protein, that **pancreatic lipase** acts on fat, and that **pancreatic amylase** hydrolyzes starch. Enzymes are very specific and usually participate in only one type of reaction. According to the Induced Fit model of enzymatic action, an enzyme binds loosely with its specific substrate and undergoes a subtle change in shape to improve its catalytic efficiency. Enzymes must maintain an appropriate shape or configuration to take part in a reaction. Enzymes have an optimum pH that allows them to maintain their usual three-dimensional shape. Enzymatic reactions speed up with increased temperature but are destroyed by excessive heat or boiling.

The Experimental Procedures in this laboratory each contain a **control,** a sample that goes through all the steps of the experiment except the one being tested (the **experimental variable**). If all of the chemicals (materials) are performing as expected, the results of this control should be negative. In experiments such as the ones in this laboratory, however, many factors, such as the purity of the chemical reagents, the age of the reagent, and the conditions of storage, can affect the outcome. In addition, organic compounds can deteriorate upon storage, particularly if they are stored under the wrong conditions.

21.1 Protein Digestion by Pepsin

Certain foods, such as meat and egg whites, are rich in protein. Egg whites contain albumin, which is the protein used in this exercise. Protein is digested by **pepsin** in the stomach (Fig. 21.1), a process described by the following reaction:

$$\text{protein + water} \xrightarrow{\text{pepsin (enzyme)}} \text{peptides}$$

The stomach has a very low pH. Does this indicate that pepsin works effectively in an acidic or basic environment? _____

Test for Protein Digestion

Biuret reagent is used to test for protein digestion. If digestion has not occurred, biuret reagent turns purple, indicating that protein is present. If digestion has occurred, biuret reagent turns pinkish-purple, indicating that peptides are present.

Caution:	Biuret reagent contains a strong solution of sodium or potassium hydroxide. Exercise care in using this reagent, and follow your instructor's directions for disposing of these tubes. If any biuret reagent should spill on your skin, rinse immediately with clear water. Eye protection should be worn.

Experimental Procedure: Protein Digestion

With a wax pencil, number four test tubes, and mark at the 2 cm, 4 cm, 6 cm, and 8 cm levels. Fill all tubes to the 2 cm mark with *albumin solution.*

Tube 1   Fill to the 4 cm mark with *pepsin solution,* and to the 6 cm mark with *0.2% HCl.* Swirl to mix, and incubate at 37°C. After 1½ hours, fill to the 8 cm mark with *biuret reagent.* Record your results in Table 21.1.

Tube 2   Fill to the 4 cm mark with *pepsin solution,* and to the 6 cm mark with *0.2% HCl.* Swirl to mix, and keep at room temperature. After 1½ hours, fill to the 8 cm mark with *biuret reagent.* Record your results in Table 21.1.

Tube 3   Fill to the 4 cm mark with *pepsin solution,* and to the 6 cm mark with *water.* Swirl to mix, and incubate at 37°C. After 1½ hours, fill to the 8 cm mark with *biuret reagent.* Record your results in Table 21.1.

Tube 4   Fill to the 6 cm mark with *water.* Swirl to mix, and incubate at 37°C. After 1½ hours, fill to the 8 cm mark with *biuret reagent.* Record your results in Table 21.1.

Figure 21.1 **Digestion of protein.**
Pepsin, produced by the gastric glands of the stomach, helps digest protein.

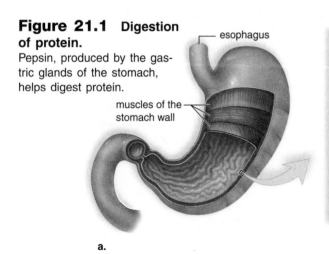

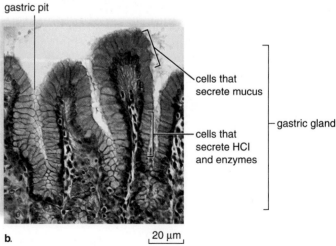

Table 21.1	Protein Digestion by Pepsin			
Tube	Contents	Temperature	Results of Test	Explanation
1	Albumin Pepsin HCl			
2	Albumin Pepsin HCl			
3	Albumin Pepsin Water			
4	Albumin Water			

Conclusions

- Explain your results in Table 21.1 by giving a reason why digestion did or did not occur.
- Which tube was the control? _____
  Explain. _____
- If this control tube had given a positive result for protein digestion, what could you conclude about this experiment? _____

Requirements for Digestion

Explain in Table 21.2 how each of the requirements listed influences effective digestion.

Table 21.2	Requirements for Digestion
Requirement	Explanation
Specific enzyme	
Warm temperature	
Time	
Specific pH	
Fat emulsifier	

21.2 Fat Digestion by Pancreatic Lipase

Lipids include fats, such as butterfat, and oils (sunflower, corn, olive, and canola). Lipids are digested by **pancreatic lipase** in the small intestine, a process described by the following two reactions:

$$(1) \qquad \text{fat} \xrightarrow{\text{bile (emulsifier)}} \text{fat droplets}$$

$$(2) \qquad \text{fat droplets + water} \xrightarrow{\text{lipase (enzyme)}} \text{glycerol + fatty acids}$$

The first reaction is not enzymatic. It is an emulsification reaction in which fat is physically dispersed by the emulsifier (bile) into small droplets. The small droplets provide a greater surface area for enzyme attack. Lipids are hydrophobic and therefore insoluble, so they are hydrolyzed slowly unless an emulsifier is used.

Given the second reaction, would the pH of the solution be lower before or after the reaction?

_____

Hint: Remember that an acid decreases pH and a base increases pH.

Figure 21.2 shows the anatomical relationship between the liver, gallbladder, pancreas, and duodenum. Bile is made in the liver and stored in the gallbladder. When bile is needed, it passes through ducts to reach the duodenum, which also receives pancreatic lipase from the pancreas by way of a duct. Now fat digestion can occur.

Test for Fat Digestion

In the test for fat digestion, you will be using a pH indicator, which changes color as the solution in the test tube goes from basic conditions to acidic conditions. Phenol red is a pH indicator that is red in basic solutions and yellow in acidic solutions. Bile salts will be used as an emulsifier.

Experimental Procedure: Fat Digestion

With a wax pencil, number three clean test tubes, and mark at the 1 cm, 3 cm, and 5 cm levels. Fill all the tubes to the 1 cm mark with *vegetable oil,* and to the 3 cm mark with *phenol red.*

Tube 1   Fill to the 5 cm mark with *pancreatic lipase solution.* Add a pinch of *bile salts.* Swirl to mix, and record the initial color in Table 21.3. Incubate at 37°C, and check every 20 minutes. Record the length of time for any color change. Why is a color change expected? _____

_____

Tube 2   Fill to the 5 cm mark with *pancreatic lipase solution.* Swirl to mix, and record the initial color in Table 21.3. Incubate at 37°C, and check every 20 minutes. Record the length of time for any color change.

Tube 3   Fill to the 5 cm mark with *water.* Swirl to mix, and record the initial color in Table 21.3. Incubate at 37°C, and check every 20 minutes. Record the length of time for any color change.

Table 21.3   Fat Digestion by Pancreatic Lipase

Tube	Contents	Total Time	Color Change		Explanation
			Initial	*Final*	
1	Vegetable oil Phenol red Pancreatic lipase Bile salts				
2	Vegetable oil Phenol red Pancreatic lipase				
3	Vegetable oil Phenol red Water				

Conclusions

- Explain your results in Table 21.3 by giving a reason why digestion did or did not occur.
- What role did bile play in this experiment? _____

  _____

- What role did phenol red play in this experiment? _____
- Which test tube in this experiment could be considered a control? _____

  _____

Figure 21.2 **Emulsification and digestion of fat.**
Bile from the liver (stored in gallbladder) enters small intestine, where lipase in pancreatic juice from the pancreas digests fat.

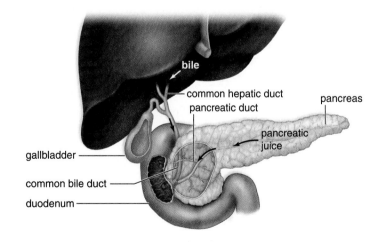

bile
common hepatic duct
pancreatic duct
pancreas
gallbladder
pancreatic juice
common bile duct
duodenum

21.3 Starch Digestion by Pancreatic Amylase

Starch is present in bakery products and in potatoes, rice, and corn. Starch is digested by **pancreatic amylase** in the small intestine, a process described by the following reaction:

$$\text{starch} + \text{water} \xrightarrow{\text{amylase (enzyme)}} \text{maltose}$$

The placement of an enzyme on top of a reaction arrow indicates that the enzyme is not consumed or changed in the process. Enzymes are recycled. Look for this type of notation throughout this chapter.

1. Why is this reaction called a *hydrolytic reaction?* _____

   _____

2. If digestion does *not* occur, which will be present—starch or maltose? _____

3. If digestion *does* occur, which will be present—starch or maltose? _____

Tests for Starch Digestion

You will be using two tests for starch digestion:

1. If digestion has not taken place, the iodine test for starch will be positive. If starch is present, a blue-black color immediately appears after a few drops of iodine are added to the test tube.
2. If digestion has taken place, a test for sugar (maltose) will be positive. To test for sugar, add an equal amount of Benedict's reagent to the test tube. Place the tube in a boiling water bath for 2–5 minutes, and note any color changes. A color change of blue ⟶ green ⟶ yellow ⟶ orange ⟶ red indicates the presence of maltose. Boiling the test tube is necessary for the Benedict's reagent to react. To which category of organic compounds (lipid, carbohydrate, or

   protein) do enzymes such as amylase belong? _____

   What happens when enzymes are boiled? _____

Experimental Procedure: Starch Digestion

Preparation

1. With a wax pencil, number eight clean test tubes, and mark at the 1 cm and 2 cm levels.
2. Fill tubes 1 through 6 to the 1 cm mark with *pancreatic amylase solution.* Fill tubes 7 and 8 to the 1 cm mark with *water.*
3. Fill tubes 3 and 4 to the 2 cm mark with *starch suspension,* and allow them to stand at room temperature for 30 minutes.
4. Place tubes 5 and 6 in a boiling water bath for 10 minutes. After boiling, fill to the 2 cm mark with *starch suspension,* and allow the tubes to stand for 20 minutes.
5. Fill tubes 7 and 8 to the 2 cm mark with *starch suspension.* Allow the tubes to stand for 30 minutes.

   Tube 1   Fill to the 2 cm mark with *starch suspension,* and test for starch immediately, using the iodine test described previously. Record your results in Table 21.4.

   Tube 2   Fill to the 2 cm mark with *starch suspension,* and test for sugar immediately, using *Benedict's reagent* described earlier, which requires boiling. Record your results in Table 21.4.

Why do you expect tube 1 to have a positive test for starch and tube 2 to have a negative test for

sugar? _____

Record your explanation in Table 21.4.

> **Caution:** **Benedict's reagent** Use protective eyewear when performing this experiment. Benedict's reagent is highly corrosive. Exercise care in using this chemical and boiling water bath. If any should spill on your skin, wash the area with mild soap and water. Follow your instructor's directions for disposal of this chemical.

Tubes 3, 5, and 7   After 30 minutes, test for starch using the iodine test. Record your results.

Tubes 4, 6, and 8   After 30 minutes, test for sugar using the Benedict's test. Record your results in Table 21.4.

Why do you expect tube 3 to have a negative test for starch and tube 4 to have a positive test for sugar? _____.

Why do you expect tube 5 to have a positive test for starch and tube 6 to have a negative test for sugar? _____.

Why do you expect tube 7 to have a positive test for starch and tube 8 to have a negative test for sugar? _____.

Record your explanations for your results in Table 21.4.

Table 21.4	Starch Digestion by Amylase				
Tube	Contents	Time	Type of Test	Results	Explanation
1	Pancreatic amylase Starch				
2	Pancreatic amylase Starch				
3	Pancreatic amylase Starch				
4	Pancreatic amylase Starch				
5	Pancreatic amylase, boiled Starch				
6	Pancreatic amylase, boiled Starch				
7	Water Starch				
8	Water Starch				

Physiology of the Digestive System **Laboratory 21**

Conclusions

- This experiment demonstrated that, in order for an enzymatic reaction to occur, an active _____ must be present, and _____ must pass to allow the reaction to occur.
- Which test tubes served as a control in this experiment? _____
Explain. _____

Laboratory Review 21

1. Why would you not expect amylase to digest protein?
   _____

2. Enzymes perform better at room temperature than when they are boiled. Explain.
   _____

3. Relate the expectation of more product per length of time to the fact that enzymes are used over and over.
   _____
   _____

4. Why do enzymes work better at their optimum pH?
   _____

5. Why is an emulsifier needed for the lipase experiment but not for the pepsin and amylase experiments?
   _____

6. Which of the following two combinations is most likely to result in digestion?
   a. Pepsin, protein, water, body temperature
   b. Pepsin, protein, hydrochloric acid (HCl), body temperature
   Explain. _____

7. Which of the following two combinations is most likely to result in digestion?
   a. Amylase, starch, water, body temperature, testing immediately
   b. Amylase, starch, water, body temperature, waiting 30 minutes
   Explain. _____
   _____

8. Relate the composition of fat to the test used for fat digestion.
   _____
   _____

9. Given that, in this laboratory, you tested for the action of digestive enzymes on their substrates, what substance would be missing from a control sample?
   _____

22

Physiology of the Excretory Organs

Learning Outcomes

22.1 Lungs
- Describe the anatomy of the lungs and their role in homeostasis. 288–289
- Describe the microscopic anatomy of the lungs and the role of the alveoli in gas exchange. 288–289

22.2 Liver
- Trace the path of blood from the intestines, through the liver, and to the heart. 290
- Describe the anatomy of the liver and its role in homeostasis. 290–292
- Compare and explain the glucose level in the mesenteric artery, the hepatic portal vein, and the hepatic vein after and before eating. 292–293

22.3 Kidneys
- Describe the anatomy of the kidneys and trace the path of blood about a nephron. 294–296
- State the three steps in urine formation and how they relate to the parts of a nephron. 296–297
- Predict whether substances will normally be in the filtrate and/or urine, and explain. 297–298
- Describe the role of the kidneys in homeostasis and the benefits of urinalysis. 298–299

22.4 Capillary Exchange in the Tissues
- Describe the exchange of molecules across a capillary wall and the mechanisms involved in this exchange. 300

Introduction

Homeostasis refers to the dynamic equilibrium of the body's internal environment. The internal environment of vertebrates, including humans, consists of blood and tissue fluid. The cells take nutrients from tissue fluid and return their waste molecules to it. Tissue fluid, in turn, exchanges molecules with the blood. This is called capillary exchange. All internal organs contribute to homeostasis, but this laboratory specifically examines the contributions of the lungs, liver, and kidneys (Fig. 22.1).

Figure 22.1 **Contributions of organs to homeostasis.**
The lungs contribute to homeostasis because they carry on gas exchange. The kidneys excrete nitrogenous wastes, and the liver, in association with the digestive tract, adds nutrients to the blood.

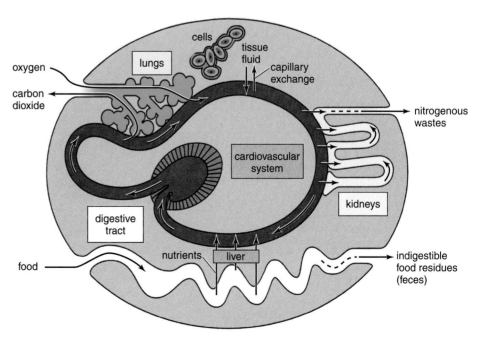

22.1 Lungs

Air moves from the nasal passages to the trachea, bronchi, bronchioles, and finally, lungs. The right and left lungs lie in the thoracic cavity on either side of the heart.

Lung Structure

A **lung** is a spongy organ consisting of irregularly shaped air spaces called **alveoli** (sing., **alveolus**). The alveoli are lined with a single layer of squamous epithelium and are supported by a mesh of fine, elastic fibers. The alveoli are surrounded by a rich network of tiny blood vessels called **pulmonary capillaries.**

Observation: Lung Structure

1. Observe a prepared slide of a stained section of a lung. In stained slides, the nuclei of the cells forming the thin alveolar walls appear purple or dark blue (Fig. 22.2*a*).
2. Look for areas with groups of red- or orange-colored, disc-shaped **erythrocytes** (red blood cells). When these appear in strings, you are looking at capillary vessels in side view.
3. In some part of the slide, you may even observe an artery. Thicker, circular or oval structures with a lumen (cavity) are cross sections of **bronchioles,** tubular pathways through which air reaches the air spaces.

Lung Function

Oxygen concentration in the air in alveoli is *greater* than in the blood in pulmonary capillaries. By the same token, carbon dioxide concentration in the air in alveoli is *less* than in the blood in pulmonary capillaries. Gas exchange in the lungs takes place by diffusion as gases move along a **concentration gradient** from greater to lesser concentration.

During gas exchange in the lungs, carbon dioxide (CO_2) leaves the blood and enters the alveoli, and oxygen (O_2) leaves the alveoli and enters the blood. Label Figure 22.2*b* to show gas exchange in the lungs. State one way the lungs contribute to homeostasis. _____

Carbon Dioxide Transport and Release

Carbon dioxide is carried in the blood as bicarbonate ions (HCO_3^-):

$$CO_2 + H_2O \longrightarrow \underset{\text{carbonic acid}}{H_2CO_3} \longrightarrow \underset{\text{bicarbonate ion}}{HCO_3^-} + H^+$$

1. H^+ increases the acidity of the blood. Is blood more acidic when it is carrying carbon dioxide? _____ Explain. _____

   _____

2. As carbon dioxide leaves the blood, the following reaction is driven to the right:

   $$HCO_3^- + H^+ \longrightarrow H_2CO_3 \longrightarrow H_2O + CO_2$$

   Is blood less acidic when the carbon dioxide exits? _____ Explain. _____

   _____

3. State another way the lungs contribute to homeostasis. _____

   _____

Figure 22.2 Gas exchange in the lungs.

a. A photomicrograph shows that the lungs contain many air sacs called alveoli. The alveoli are surrounded by blood capillaries. **b.** During gas exchange, carbon dioxide leaves the blood and enters the alveoli; oxygen leaves the alveoli and enters the blood. Label the arrows with O_2 and CO_2 to show gas exchange in the lungs.

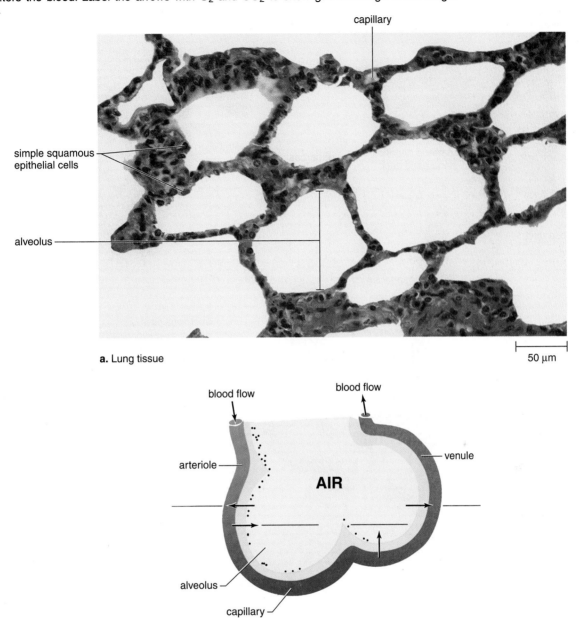

a. Lung tissue

50 μm

b. Alveolus

Figure 22.3 Anatomy of the liver.

a. The liver, a large organ in the abdominal cavity, plays a primary role in homeostasis. **b.** Each lobule is served by branches of the hepatic artery that bring O_2-rich blood to the liver and by a branch of the hepatic portal vein that brings nutrients to the liver from the intestines. The central vein of each lobule takes blood to the hepatic vein, which enters the posterior vena cava. Bile canals enter bile ducts taking bile away from each lobule for storage in the gallbladder. **c.** The hepatic portal system consists of the hepatic portal vein and hepatic vein.

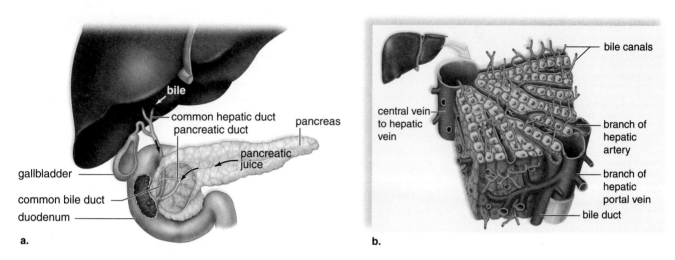

a.

b.

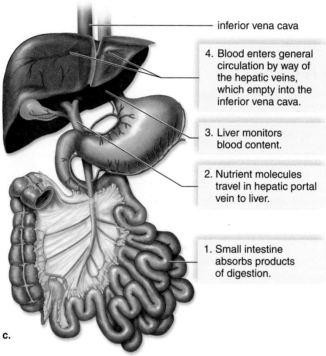

c.

inferior vena cava

4. Blood enters general circulation by way of the hepatic veins, which empty into the inferior vena cava.

3. Liver monitors blood content.

2. Nutrient molecules travel in hepatic portal vein to liver.

1. Small intestine absorbs products of digestion.

22.2 Liver

The **liver,** the largest internal organ in the body, lies mainly in the upper-right quadrant of the abdominal cavity under the diaphragm.

Liver Structure

The liver has two main **lobes;** the right lobe is larger than the left lobe (Fig. 22.3a). The lobes are further divided into **lobules,** which contain the cells of the liver, called **hepatic cells.** Small blood vessels that transport blood out of the liver into the hepatic vein are at the center of each lobule. Between the lobules are three structures: (1) a branch of the hepatic artery, (2) a branch of the hepatic portal vein, and (3) a bile duct to collect bile (Fig. 22.3b). Refer to Figure 22.3c for an expanded view of circulation between the liver and other organs.

Observation: Liver Structure

Study a model of the liver, and identify the following:

1. **Right and left lobes:** The right lobe is larger than the left lobe.
2. **Lobules:** Each lobe has many lobules.
3. **Hepatic cells:** Each lobule has many cells.
4. **Hepatic vein:** The blood vessel that transports O_2-poor blood out of the liver to the inferior vena cava.
5. **Branch of hepatic artery:** The blood vessel that transports O_2-rich blood to the liver.
6. **Branch of hepatic portal vein:** The blood vessel that transports blood containing nutrients from the intestine to the liver.
7. **Bile duct:** The passageway for bile going to the gallbladder.

Liver Function

The liver has many functions in homeostasis. It produces **urea,** the primary nitrogenous end product of humans. In general, the liver is the gatekeeper of the blood—it regulates blood composition. For example, it stores glucose as glycogen and then releases glucose to keep the blood glucose concentration at about 0.1%.

Urea Formation

The liver removes amino groups ($-NH_2$) from amino acids and converts them to urea, a relatively nontoxic nitrogenous end product.

1. In the chemical formula for urea that follows, circle the portions that would have come from amino groups:

$$NH_2 - \overset{\overset{\displaystyle O}{\displaystyle \|}}{C} - NH_2$$

2. State one way the liver contributes to homeostasis. _____

_____

Regulation of Blood Glucose Level

After you eat, the liver stores excess glucose as glycogen. In between eating, glycogen is broken down by liver cells to produce glucose, and this glucose enters the bloodstream. The hormone insulin, made by the pancreas, promotes the uptake and storage of glucose in the liver. The hormone glucagon promotes the breakdown of glycogen and the release of glucose.

1. In the following equation, write the phrases *after eating* and *before eating* above or below the appropriate arrow:

$$\text{glucose} \longleftarrow \text{glycogen}$$

2. Now add the words *insulin* and *glucagon* to the appropriate arrow in the equation. If glucose is excreted in the urine, instead of being stored, the individual has the medical condition called **diabetes mellitus,** commonly known simply as diabetes. In type 1 diabetes, the pancreas is no longer making insulin; in type 2 diabetes, the plasma membrane receptors are unable to bind properly to insulin. In type 1 diabetes, but not type 2, ketones (strong organic acids), which are a breakdown product of fat metabolism, also appear in the urine.

3. State another way in which the liver contributes to homeostasis. _____

Caution: **Benedict's reagent** Use protective eyewear when performing this experiment. Benedict's reagent is highly corrosive. Exercise care in using this chemical and boiling water bath. If any should spill on your skin, wash the area with mild soap and water. Follow your instructor's directions for disposal of this chemical.

Experimental Procedure: Blood Glucose Level After Eating

Study Figure 22.3c and note the function of the hepatic portal vein and the hepatic vein. Simulated serum samples have been prepared to correspond to these blood vessels in a person who ate a short time ago:

A_1: Serum from a mesenteric artery. The **mesenteric arteries** take blood from the aorta to the intestine.

B_1: Serum from the hepatic portal vein, which lies between the intestine and the liver.

C_1: Serum from the hepatic vein, which takes blood from the liver to the inferior vena cava.

1. With a wax pencil, label three test tubes A_1, B_1, and C_1, and mark them at 1 cm and 2 cm.
2. Fill test tube A_1 to the 1 cm mark with *serum A_1* and to the 2 cm mark with *Benedict's reagent.*
3. Fill test tube B_1 to the 1 cm mark with *serum B_1* and to the 2 cm mark with *Benedict's reagent.*
4. Fill test tube C_1 to the 1 cm mark with *serum C_1* and to the 2 cm mark with *Benedict's reagent.*
5. Place all three test tubes in the water bath *at the same time.* Heat the tubes in the same boiling water bath for five minutes.
6. Note the order in which the tubes show a color change, and record your results in Table 22.1. The tube that shows color first has the most glucose, and so forth. The color change sequence of Benedict's reagent that indicates an increasing sugar (glucose) concentration is

$$\text{blue (negative)} \longrightarrow \text{green} \longrightarrow \text{yellow} \longrightarrow \text{orange} \longrightarrow \text{red}$$

Table 22.1	Blood Glucose Level After Eating	
Test Tubes (in Order of Color Change)		Source of Serum

Conclusions

- Which blood vessel—a mesenteric artery, the hepatic portal vein, or the hepatic vein—contains the most glucose *after eating*? _____
- Why do you suppose that the hepatic vein does not contain as much glucose as the hepatic portal vein *after eating*? _____

Experimental Procedure: Blood Glucose Level Before Eating

Simulated serum samples have been prepared to correspond to these blood vessels in a person who has not eaten for some time:

A_2: Serum from a mesenteric artery
B_2: Serum from the hepatic portal vein
C_2: Serum from the hepatic vein

1. With a wax pencil, label three test tubes A_2, B_2, and C_2, and mark them at 1 cm and 2 cm.
2. Fill test tube A_2 to the 1 cm mark with *serum A_2* and to the 2 cm mark with *Benedict's reagent*.
3. Fill test tube B_2 to the 1 cm mark with *serum B_2* and to the 2 cm mark with *Benedict's reagent*.
4. Fill test tube C_2 to the 1 cm mark with *serum C_2* and to the 2 cm mark with *Benedict's reagent*.
5. Heat the tubes in the same boiling water bath for five minutes.
6. Note the order in which the tubes show a color change, and record your results in Table 22.2. The tube that shows color first contains the most glucose, and so forth. The color change sequence of Benedict's reagent that indicates an increasing concentration of sugar (glucose) is shown in Table 3.4 on page 28.

Table 22.2	Blood Glucose Level Before Eating	
Test Tubes (in Order of Color Change)		Source of Serum

Conclusions

- Which blood vessel—a mesenteric artery, the hepatic portal vein, or the hepatic vein—contains the most glucose *before eating*? _____
- Why do you suppose that the hepatic vein *before eating* contains more glucose than the hepatic portal vein? _____

22.3 Kidneys

The **kidneys** are bean-shaped organs that lie along the dorsal wall of the abdominal cavity.

Kidney Structure

Figure 22.4 shows the structure of a kidney, macroscopic and microscopic. The macroscopic structure of a kidney is due to the placement of over 1 million **nephrons.** Nephrons are tubules that do the work of producing urine.

Figure 22.4 Longitudinal section of a kidney.
a. The kidneys are served by the renal artery and renal vein. **b.** Macroscopically, a kidney has three parts: renal cortex, renal medulla, and renal pelvis. **c.** Microscopically, each kidney contains over a million nephrons.

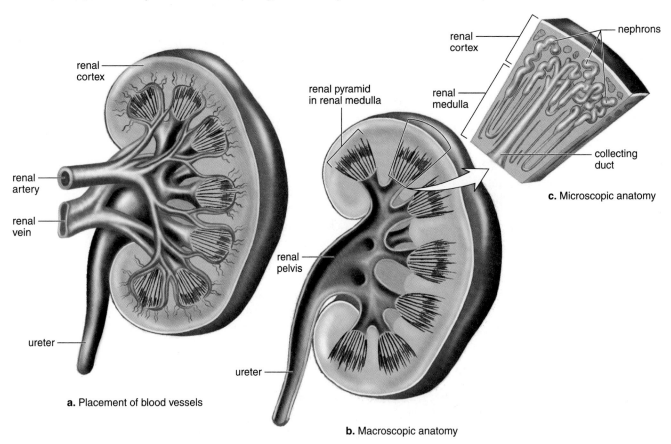

Observation: Kidney Model

Study a model of a kidney, and with the help of Figure 22.4, locate the following:

1. **Renal cortex:** a granular region
2. **Renal medulla:** contains the renal pyramids
3. **Renal pelvis:** where urine collects

Nephron Structure and Circulation

Figure 22.5 shows that the **afferent arteriole** enters the **glomerulus,** situated within the cup-shaped **glomerular capsule** (Bowman's capsule). The **efferent arteriole** leaves the glomerular capsule and enters the **peritubular capillary network** that surrounds the **proximal convoluted tubule,** the **loop of the nephron** (loop of Henle), and the **distal convoluted tubule.** Distal convoluted tubules from several nephrons enter one collecting duct.

Macroscopic and microscopic studies of kidney anatomy show that the glomerular capsule and convoluted tubules are in the renal cortex, while the loop of the nephron and the collecting ducts are in the renal medulla, accounting for the striated appearance of the renal pyramids. The collecting ducts enter the renal pelvis.

Figure 22.5 **Structure of a nephron and its blood supply.**
As the blood moves through the blood vessels about a nephron, substances exit and/or enter the blood from portions of the nephron.

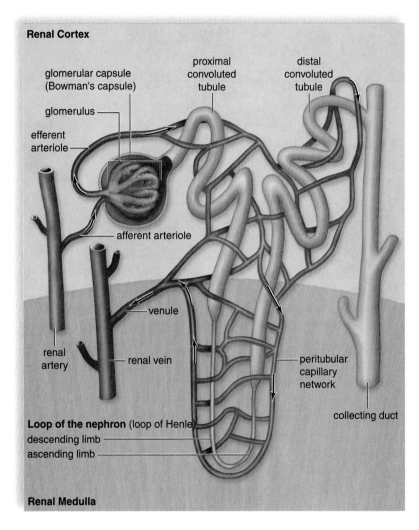

1. With the help of Figure 22.5, list the parts of a nephron, and tell whether they are located in the renal cortex or the renal medulla (assume that the nephron has a long loop of the nephron).

_____

_____

_____

_____

2. With the help of Figure 22.5 and Table 22.3, trace the path of blood toward, around, and away from an individual nephron. _____

_____

_____

Table 22.3	Circulation of Blood About a Nephron
Name of Structure	**Significance**
Afferent arteriole	Brings arteriolar blood toward glomerular capsule
Glomerulus	Capillary tuft enveloped by glomerular capsule
Efferent arteriole	Takes arteriolar blood away from glomerular capsule
Peritubular capillary network	Capillary bed that envelops the rest of the nephron
Veins	Take venous blood away from the nephron

Kidney Function

The kidneys contribute to homeostasis by excreting nitrogenous wastes and by regulating blood volume, blood pressure, and pH.

Urine formation requires three steps:

1. **Glomerular filtration** requires the movement of molecules outward from the glomerulus to the inside of the glomerular capsule. Label the arrow in Figure 22.6 that marks the location of glomerular filtration.
2. **Tubular reabsorption** requires the movement of molecules primarily from the proximal convoluted tubule to the peritubular capillary network. Label the arrow in Figure 22.6 that refers to tubular reabsorption.
3. **Tubular secretion** requires the movement of molecules from the peritubular capillary network to the nephron. Label the arrow in Figure 22.6 that refers to tubular secretion.

The two parts not considered in this discussion—the loop of the nephron and the collecting duct—are both active in water reabsorption. Regulation of water reabsorption maintains blood volume at the proper level.

Figure 22.6 Urine formation.
The three steps in urine formation are glomerular filtration, tubular reabsorption, and tubular secretion. Label the arrows as directed above.

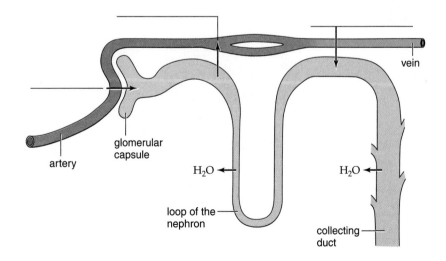

Glomerular Filtration

Blood entering the glomerulus contains cells, proteins, glucose, amino acids, salts, urea, and water. Blood pressure causes small molecules of glucose, amino acids, salts, urea, and water to exit the blood and enter the glomerular capsule. The fluid in the glomerular capsule is called the **filtrate.**

1. In the list that follows, draw an arrow from left to right for all those molecules that leave the glomerulus and enter the glomerular capsule:

    Glomerulus   **Glomerular Capsule (Filtrate)**
    Cells
    Proteins
    Glucose
    Amino acids
    Salts
    Urea
    Water

2. What substances are too large to leave the glomerulus and enter the glomerular capsule?

    _____

    These substances remain in the blood.

Tubular Reabsorption

When the filtrate enters the proximal convoluted tubule, it contains water, glucose, amino acids, urea, and salts. Enough water and salts are passively reabsorbed to maintain blood volume and pH.

1. What would happen to blood volume and blood pressure if water were not reabsorbed?

    _____

    _____

2. The cells that line the proximal convoluted tubule are also engaged in active transport and usually completely reabsorb nutrients (glucose and amino acids). What would happen to cells if the body

    lost all its nutrients by way of the kidneys? _____

    _____

3. In the list that follows, draw an arrow from left to right for all those molecules passively reabsorbed into the blood. Use darker arrows for those actively reabsorbed.

    Proximal Convoluted Tubule   **Peritubular Capillary**
    Water
    Glucose
    Amino acids
    Urea
    Salts

4. What molecule is reabsorbed the least? _____ Urea contributes to making the renal medulla hypertonic to the fluid within the collecting duct.

Tubular Secretion

During tubular secretion, certain substances—for example, penicillin and histamine—are actively secreted from the peritubular capillary into the fluid of the tubule. Also, hydrogen ions and ammonia are secreted as required.

Summary of Urine Formation

For each substance listed at the left in Table 22.4, place an X in the appropriate column(s) to indicate where you expect the substance to be present.

Table 22.4	Urine Constituents		
Substance	In Blood of Glomerulus	In Filtrate	In Urine
Protein (albumin)			
Glucose			
Urea			
Water			

Answer the following questions.

1. What molecule is reabsorbed from the collecting duct so that urine becomes hypertonic?

   _____

2. Based on Table 22.4, state two ways the kidneys contribute to homeostasis. _____

   _____

3. Which organ—the lung, liver, or kidney—makes urea? _____

4. Which organ excretes urine? _____

5. If the blood is more acidic than normal, what pH do you suppose the urine will be? _____

6. If the blood is more basic than normal, what pH do you suppose the urine will be? _____

7. State another way the kidneys contribute to homeostasis. _____

   _____

Urinalysis

Urinalysis can help diagnose a patient's illness. The procedure is easily performed with a Chemstrip test strip, which has indicator spots that produce specific color reactions when certain substances are present in the urine.

Experimental Procedure: Urinalysis

Suppose a patient complains of excessive thirst and urination, loss of weight despite an intake of sweets, and feelings of being tired and run-down. A urinalysis has been ordered, and you are to test the urine. (In this case, you will be testing simulated urine, just as you tested simulated blood sera earlier in this lab.)

1. Review "Regulation of Blood Glucose Level," step 2, on page 292. Obtain a Chemstrip urine test strip (Fig. 22.7) that tests for leukocytes, pH, protein, glucose, ketones, and blood, which are noted in the "Tests For" column of the figure.

2. The color key on the diagnostic color chart or on the Chemstrip vial label will explain what the color changes mean in terms of the pH level and amount of each substance present in the urine sample. You will use these color blocks to read the results of your test.
3. Obtain a "specimen container of the patient's urine."
4. Briefly (no longer than 1 second) dip the test strip into the urine. Ensure that the chemically impregnated patches on the test strip are totally immersed.
5. Draw the edge of the strip along the rim of the specimen container to remove excess urine.
6. Turn the test strip on its side, and tap once on a piece of absorbent paper to remove any remaining urine and to prevent the possible mixing of chemicals.
7. After 60 seconds, read the tests as follows: *Hold the strip close to the color blocks on the vial label and match carefully*, ensuring that the strip is properly oriented to the color blocks.

Figure 22.7 Urinalysis test.
A Chemstrip test strip can help determine illness in a patient by detecting substances in the urine.

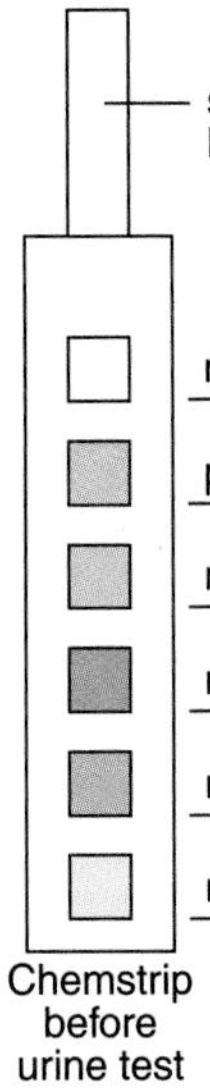

	Normal	Tests For:	Results
	negative	leukocytes	_____
	pH 5	pH	_____
	negative	protein	_____
	normal	glucose	_____
	negative	ketones	_____
	negative	blood	_____

strip handle

Chemstrip before urine test

Conclusions

• According to your results, what condition might the patient have? _____ Explain. _____ _____ _____

• Given that the patient's blood contains excess glucose, why is the patient suffering from excessive thirst and urination? _____

• Since neither the liver nor the body cells are taking up glucose, why is the patient tired? _____ _____

• The metabolism of fat can explain the low pH of the urine. Why? _____

22.4 Capillary Exchange in the Tissues

Tissue fluid is continually created and refreshed at the capillaries when certain molecules leave the blood and others are picked up by the blood.

1. In Figure 22.8, write *oxygen* and *glucose* at the appropriate arrow. Write *wastes* and *carbon dioxide* next to the appropriate arrow.

2. What type of pressure causes water to exit from the arterial side of the capillary? _____

3. What type of pressure causes water to enter the venous side of the capillary? _____

Figure 22.8 Capillary exchange.
A capillary, illustrating the exchange that takes place across a capillary wall.

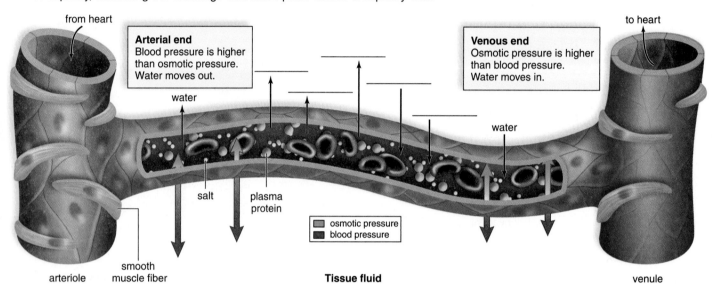

Conclusions

• What is the purpose of the systemic circuit (p. 276)? _____

_____

• Why are cells always in need of glucose and oxygen? _____

_____

• Why are cells always producing carbon dioxide? _____

_____

Summary of Homeostasis

As noted at the beginning of this laboratory, homeostasis is the dynamic equilibrium of the body's internal environment, the blood and tissue fluid surrounding tissue cells. The lungs and kidneys have boundaries that interact with the external environment to refresh blood. The liver also regulates blood content. Fill in the following table to show the activities of the lungs.

Processes	Lungs
Gas exchange	O_2 enters blood; CO_2 leaves blood
pH maintenance	a.
Glucose level	———
Waste removal	b.
Blood volume	———

Fill in the following table to show the activities of the liver.

Processes	Liver
Gas exchange	———
pH maintenance	———
Glucose level	c.
Waste removal	d.
Blood volume	———

Fill in the following table to show the activities of the kidneys.

Processes	Kidneys
Gas exchange	———
pH maintenance	e.
Glucose level	f.
Waste removal	g.
Blood volume	h.

Conclusion

- Which of these organs contributes most to homeostasis? _____

1. What is homeostasis?

   _____

2. What is the role of the lungs in homeostasis?

   _____

   _____

3. What is the role of the liver in maintaining the glucose concentration of the blood?

   _____

   _____

4. Trace the path of blood from the intestinal capillaries to the inferior vena cava.

   _____

   _____

5. Which of the blood vessels in number 4 would you expect to have a high glucose content immediately
   after eating? _____ Explain. _____

   _____

   _____

6. List two ways in which the kidneys aid homeostasis.

   _____

   _____

7. List the three steps in urine formation. Associate each step with a particular part of the nephron.

   Steps   **Part of Nephron**

   a. _____   _____

   b. _____   _____

   c. _____   _____

8. With regard to urine formation, name a substance found in both the filtrate and the urine.
   _____ Explain. _____

   _____

9. With regard to urine formation, name a substance found in the filtrate and not in the urine.
   _____ Explain. _____

   _____

10. During urine formation, reabsorption meets the needs of the body. Explain.

    _____

    _____

11. After a urinalysis test, what would an indicator of glucose signify?

    _____

12. Explain how the capillaries help to maintain homeostasis.

    _____

23

Physiology of the Nervous System

Learning Outcomes

Introduction

The vertebrate nervous system consists of the brain, spinal cord, and nerves. Sensory receptors detect changes in environmental stimuli, and nerve impulses move along sensory nerve fibers to the brain and the spinal cord. The brain and spinal cord sum up the data before sending impulses via motor neurons to effectors (muscles and glands) so a response to stimuli is possible. Nervous tissue consists of neurons; whereas the brain and spinal cord contain all parts of neurons, nerves contain only axons (Fig. 23.1).

Figure 23.1 **Motor neuron anatomy.**
Neurons are cells specialized to conduct nerve impulses.

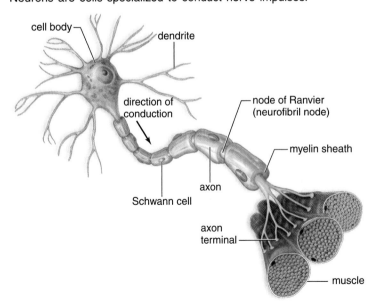

23.1 Animal Nervous Systems

The brain is the enlarged, anterior end of the nerve cord. In vertebrates, the nerve cord is called the spinal cord. The brain contains parts and centers that receive input from, and can command other regions of, the nervous system.

Sheep Brain

The mammalian brain has many parts, and the sheep brain is often used to study the mammalian brain.

Observation: Sheep Brain

Examine the exterior and a transverse (longitudinal) section of a preserved sheep brain or a model of the human brain, and with the help of Figure 23.2, identify the:

1. **Ventricles:** Interconnecting spaces that produce and serve as a reservoir for cerebrospinal fluid, which cushions the brain. Toward the anterior, note the large first lateral ventricle (on one of your transverse sections) and the second lateral ventricle (on your other transverse section). Trace a lateral ventricle to the third and then the fourth ventricles.
2. **Medulla oblongata** (or simply **medulla**): The most posterior portion of the brain stem. It controls internal organs; for example, cardiac and breathing control centers are present in the medulla. Nerve impulses pass from the spinal cord through the medulla to higher brain regions.
3. **Pons:** The ventral, bulblike enlargement on the brain stem. It serves as a passageway for nerve impulses running between the medulla and the higher brain regions.
4. **Midbrain:** Anterior to the pons, the midbrain serves as a relay station for sensory input and motor output. It also contains a reflex center for eye muscles.
5. **Diencephalon:** The portion of the brain where the third ventricle is located. The hypothalamus and thalamus also are located here.
6. **Hypothalamus:** Forms the floor of the third ventricle and contains control centers for appetite, body temperature, and water balance. Its primary function is homeostasis. The hypothalamus also has centers for pleasure, reproductive behavior, hostility, and pain.
7. **Thalamus:** Two connected lobes located in the roof of the third ventricle. The thalamus is the highest portion of the brain to receive sensory impulses before the cerebrum. It is believed to control which of these impulses is passed on to the cerebrum. For this reason, the thalamus sometimes is called the "gatekeeper to the cerebrum."
8. **Cerebellum:** Located just posterior to the cerebrum as you observe the brain dorsally, the cerebellum's two lobes make it appear rather like a butterfly. In cross section, the cerebellum has an internal pattern that looks like a tree. The cerebellum coordinates equilibrium and motor activity to produce smooth movements.
9. **Cerebrum:** The most developed area of the brain and responsible for higher mental capabilities. The cerebrum is divided into the right and left **cerebral hemispheres,** joined by the **corpus callosum,** a broad sheet of white matter. The outer portion of the cerebrum is highly convoluted and divided into the following surface lobes:
   a. **Frontal lobe:** Controls motor functions and permits voluntary muscle control. It also is responsible for abilities to think, problem solve, and speak.
   b. **Parietal lobe:** Receives information from sensory receptors located in the skin. It also helps in the understanding of speech. A groove called the **central sulcus** separates the frontal lobe from the parietal lobe.
   c. **Occipital lobe:** Interprets visual input and also combines visual images with other sensory experiences. The optic nerves split and enter opposite sides of the brain at the optic chiasma.
   d. **Temporal lobe:** Has sensory areas for hearing and smelling. The olfactory bulb contains nerve fibers that communicate with the olfactory cells in the nasal passages and take nerve impulses to the temporal lobe.

Figure 23.2 The sheep brain.

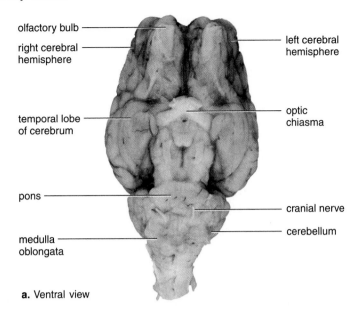

olfactory bulb

right cerebral hemisphere

temporal lobe of cerebrum

pons

medulla oblongata

left cerebral hemisphere

optic chiasma

cranial nerve

cerebellum

a. Ventral view

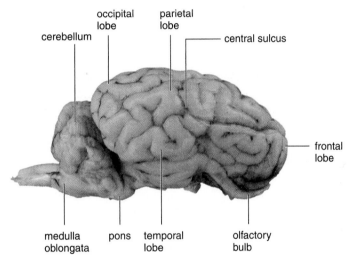

occipital lobe

parietal lobe

cerebellum

central sulcus

frontal lobe

medulla oblongata

pons

temporal lobe

olfactory bulb

b. Lateral view

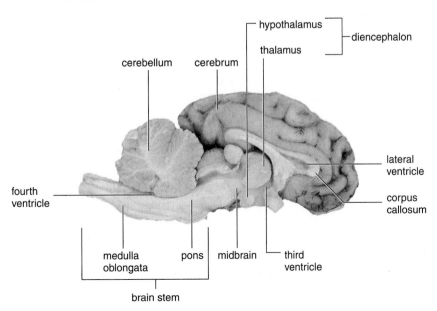

hypothalamus

diencephalon

thalamus

cerebellum

cerebrum

lateral ventricle

fourth ventricle

corpus callosum

medulla oblongata

pons

midbrain

third ventricle

brain stem

c. Longitudinal cut

Comparison of Vertebrate Brains

The vertebrate brain has a forebrain, midbrain, and hindbrain. In the earliest vertebrates, the forebrain was largely a center for sense of smell, the midbrain was a center for the sense of vision, and the hindbrain was a center for the sense of hearing and balance. How the functions of these parts changed to accommodate the lifestyles of different vertebrates can be traced.

Observation: Comparison of Vertebrate Brains

1. Examine the brain of a fish, amphibian, bird, and mammal (Fig. 23.3).
2. Identify the following three areas, and record your observations in Table 23.1.
   a. **Forebrain:** Contains olfactory bulb and cerebrum.
   b. **Midbrain:** Contains optic lobe (and other structures).
   c. **Hindbrain:** Contains cerebellum (and other structures).

In fishes and amphibians, the midbrain is often the most prominent region of the brain because it contains higher control centers. In these animals, the cerebrum largely has an olfactory function. Fishes, birds, and mammals have a well-developed cerebellum, which may be associated with these animals' agility. In reptiles, birds, and mammals, the cerebrum becomes increasingly complex. This may be associated with the cerebrum's increasing control over the rest of the brain and the evolution of areas responsible for thought and reasoning.

Figure 23.3 Comparative vertebrate brains.
Vertebrate brains differ in particular by the comparative sizes of the cerebrum, optic lobe, and cerebellum.

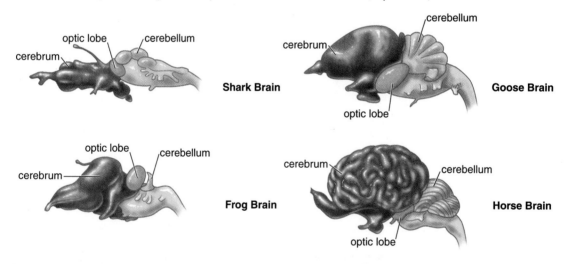

Table 23.1	Comparison of Vertebrate Brains		
Vertebrate	**Area of Brain**		
	Forebrain	*Midbrain*	*Hindbrain*
Fish (shark)			
Amphibian (frog)			
Bird (goose)			
Mammal (horse)			

Anatomy of Spinal Nerves and Spinal Cord

In humans, pairs of spinal nerves are connected to the spinal cord, which lies in the middorsal region of the body protected by the vertebral column. Each spinal nerve contains long fibers of sensory neurons and long fibers of motor neurons. In Figure 23.4, locate the following:

1. **Sensory nerve fiber** takes nerve impulses from a sensory receptor to the spinal cord. The cell body of a sensory neuron is in the dorsal-root ganglion. Why is this neuron called a sensory neuron? _____
_____

2. **Interneuron,** which lies completely within the spinal cord. Some interneurons have long fibers and take nerve impulses to and from the brain. The neuron in Figure 23.4 transmits nerve impulses from the sensory neuron to the motor neuron. Why is this neuron called an interneuron? _____
_____

3. **Motor nerve fiber** takes nerve impulses from the spinal cord to an effector—in this case, a muscle. Muscle contraction often allows a response to a stimulus. Why is this neuron called a motor neuron? _____

Figure 23.4 **Spinal nerves and spinal cord.**
The arrows mark the path of nerve impulses from a sensory receptor to an effector.

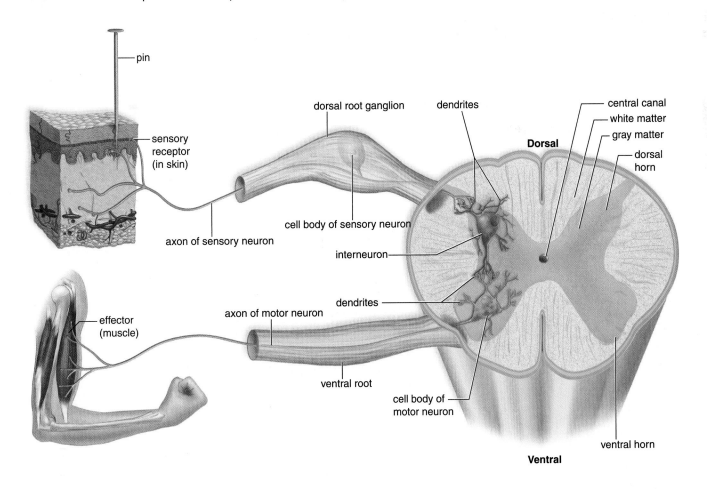

In humans, the spinal cord is a part of the central nervous system and is protected by the vertebral column running down the middle of the back.

1. Examine a prepared slide of a cross section of the spinal cord under the lowest magnification possible. For example, some microscopes are equipped with a very short scanning objective that enlarges about 3.5×, with a total magnification of 35×. If neither of these alternatives is available, then observe the slide against a white background with the naked eye.
2. Identify the following with the help of Figure 23.5:
   a. **Gray matter:** A central, butterfly-shaped area composed of masses of short nerve fibers, interneurons, and motor neuron cell bodies.
   b. **White matter:** Masses of long fibers that lie outside the gray matter and carry impulses up and down the spinal cord. In living animals, white matter appears white because an insulating myelin sheath surrounds long fibers.

Figure 23.5 **Cross section of the spinal cord.**
The diagram in **(a)** labels the photomicrograph shown in **(b)**. Note the butterfly shape of the gray matter.

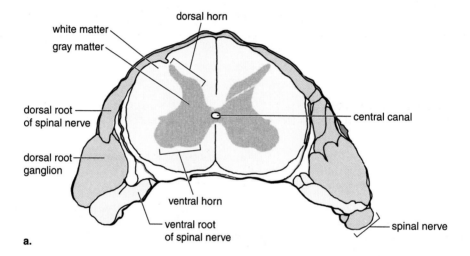

a.

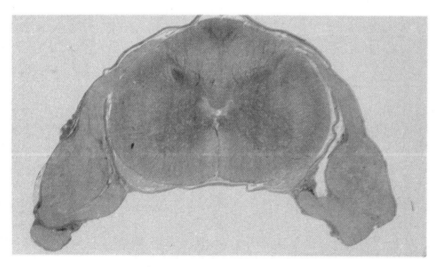

b.

Spinal Reflexes

A **reflex** is an involuntary and predictable response to a given stimulus. When you touch a sharp tack, you immediately withdraw your hand (see Fig. 23.4). When a spinal reflex occurs, a sensory receptor is stimulated and generates nerve impulses that pass along the three neurons mentioned earlier—the sensory neurons, interneurons, and motor neurons—until the effector responds.

Experimental Procedure: Spinal Reflexes

Although many reflexes occur in the body, only tendon reflexes are investigated in this Experimental Procedure. Two easily tested tendon reflexes involve the **Achilles** and **patellar tendons.** When these tendons are mechanically stimulated by tapping with a reflex hammer (Fig. 23.6), or in this case, with a meter stick, sensory neurons in the tendon transmit impulses through the reflex arc to the spinal cord and then back through motor neurons to the muscle attached to the stimulated tendon. The muscle contracts and tugs on the tendon, causing movement of a bone opposite the joint.

Ankle (Achilles) Reflex

1. Have the subject sit on the table so that the leg hangs free.
2. Tap the subject's Achilles tendon at the ankle with a meter stick.
3. Which way does the foot move? Does it extend (move away from the knee) or flex (move

   toward the knee)? _____

Figure 23.6 Two human reflexes.
The quick response when either the **(a)** Achilles tendon or the **(b)** patellar tendon is stimulated by tapping with a rubber hammer indicates that a reflex has occurred.

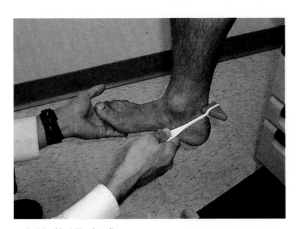

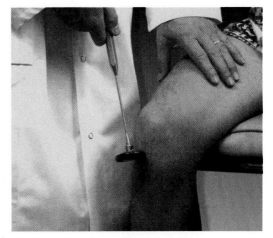

a. Ankle (Achilles) reflex

b. Knee-jerk (patellar) reflex

Knee-Jerk (Patellar) Reflex

1. Have the subject sit on the table so that the leg hangs free.
2. Sharply tap the patellar tendon just below the patella (kneecap) with a meter stick.
3. In this relaxed state, does the leg flex (move toward the buttocks) or extend (move away from

   the buttocks)? _____

23.2 Animal Eyes

The eye is a special sense organ for detecting light rays in the environment.

Anatomy of Invertebrate Eyes

Arthropods have **compound eyes** composed of many independent visual units, called ommatidia, each of which has its own photoreceptor cells (Fig. 23.7). Each unit "sees" a separate portion of the object. How well the brain combines this information is not known.

A squid has a camera type of eye. A single lens focuses an image of the visual field on the photoreceptors, packed closely together (Fig. 23.8).

Figure 23.7 Compound eye of a fly.
Each visual unit of a compound eye has a cornea and lens that focus light onto photoreceptor cells. These cells generate nerve impulses transmitted to the brain, where interpretation produces a mosaic image.

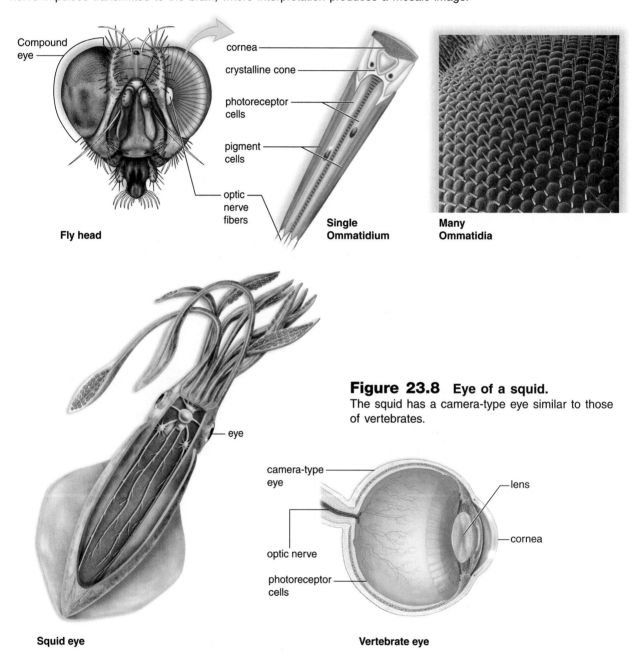

Compound eye

Fly head

cornea
crystalline cone
photoreceptor cells
pigment cells
optic nerve fibers

Single Ommatidium

Many Ommatidia

eye

Squid eye

Figure 23.8 Eye of a squid.
The squid has a camera-type eye similar to those of vertebrates.

camera-type eye
lens
cornea
optic nerve
photoreceptor cells

Vertebrate eye

1. Examine the demonstration slide of a compound eye set up under a binocular dissecting microscope or examine a model of a compound eye.
2. Examine the eyes of any other invertebrates on display.

Anatomy of the Human Eye

The anatomy of the human eye allows light rays to enter the eye and strike the **rod cells** and **cone cells,** the photoreceptors for sight. The rods and cones generate nerve impulses that go to the brain via the optic nerve.

Observation: Human Eye

1. Examine a human eye model, and identify the structures listed in Table 23.2 and depicted in Figure 23.9.
2. Trace the path of light from outside the eye to the retina.

_____

_____

3. Specifically, what are the receptors for sight, and where are they located in the eye?

_____

4. What structure takes nerve impulses to the brain from the rod cells and cone cells?

_____

Figure 23.9 **Anatomy of the human eye.**
The sensory receptors for vision are the rod cells and cone cells present in the retina of the eye.

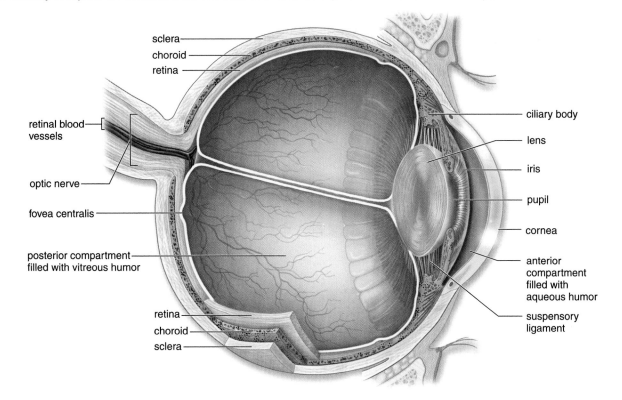

Table 23.2 Parts of the Human Eye

Part	Location	Function
Sclera	Outer layer of the eye	Protects and supports eyeball
Cornea	Transparent portion of sclera	Refracts light rays
Choroid	Middle layer of the eye	Absorbs stray light rays
Retina	Inner layer of the eye	Contains receptors for sight
Rod cells	In retina	Make black-and-white vision possible
Cone cells	Concentrated in fovea centralis	Make color vision possible
Fovea centralis	Special region of retina	Makes acute vision possible
Lens	Interior of eye between cavities	Refracts and focuses light rays
Ciliary body	Extension from choroid	Holds lens in place; functions in accommodation
Iris	More anterior extension of choroid	Regulates light entrance
Pupil	Opening in middle of iris	Admits light
Humors (aqueous and vitreous)	Fluid media in anterior and posterior compartments, respectively, of eye	Transmit and refract light rays; support the eyeball
Optic nerve	Extension from posterior of eye	Transmits impulses to brain

Physiology of the Human Eye

When we look at an object, light rays pass through the cornea, and as they do, they bend and converge. Further bending occurs as the rays pass through the lens. The shape of the lens changes according to whether we are looking at a distant or near object (Fig. 23.10). When looking at a near object, the lens rounds so that the light rays are bent a great deal to bring all rays to a focus on the retina. You will be measuring the ability of your lens to round up (accommodate) for viewing a near object.

Where the optic nerve pierces the wall of the eyeball and, therefore, the retina, there are no rod cells and cone cells, and therefore vision is impossible. You will be finding the blind spot in an Experimental Procedure.

Figure 23.10 Focusing of the human eye.
a. When focusing on a distant object, the ciliary muscle (in the ciliary body) is relaxed and the lens is flattened. **b.** When focusing on a near object, the ciliary muscle contracts and the lens rounds up. This is called accommodation.

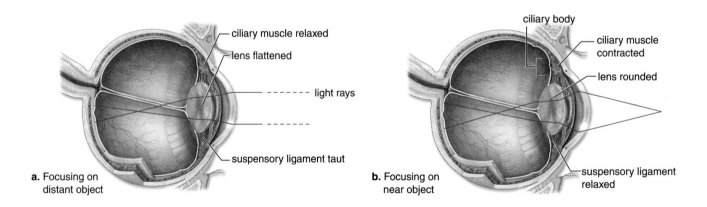

a. Focusing on distant object

b. Focusing on near object

Figure 23.11 Accommodation.

When testing the ability of your eyes to accommodate in order to see a near object, always keep the pencil in this position.

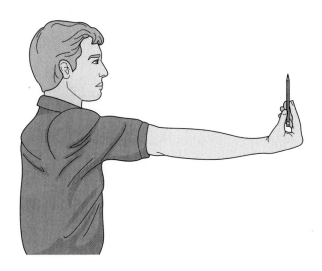

Experimental Procedure: Accommodation of the Eye

When the eye accommodates to see objects at different distances, the shape of the lens changes. The lens shape is controlled by the ciliary muscles attached to it. When you are looking at a distant object, the lens is in a flattened state. When you are looking at a closer object, the lens becomes more rounded. The elasticity of the lens determines how well the eye can accommodate. Lens elasticity decreases with increasing age, a condition called **presbyopia.** Because of presbyopia, many older people need bifocals to see near objects.

This Experimental Procedure requires a laboratory partner. It tests accommodation of either your left or right eye.

1. Hold a pencil upright by the eraser and at arm's length in front of whichever of your eyes you are testing (Fig. 23.11).
2. Close the opposite eye.
3. Move the pencil from arm's length toward your eye.
4. Focus on the end of the pencil.
5. Move the pencil toward you until the end is out of focus. Measure the distance (in centimeters) between the pencil and your eye. _____ cm
6. At what distance can your eye no longer accommodate for distance? _____ cm
7. If you wear glasses, repeat this experiment without your glasses, and note the accommodation distance of your eye without glasses. _____ cm (Contact lens wearers need not make these determinations, and they should write the words *contact lens* in this blank.)
8. The "younger" lens can easily accommodate for closer distances. The nearest point at which the end of the pencil can be clearly seen is called the **near point.** The more elastic the lens, the "younger" the eye (Table 23.3). How "old" is the eye you tested? _____

Table 23.3	Near Point and Age Correlation					
Age (Years)	10	20	30	40	50	60
Near Point (cm)	9	10	13	18	50	83

Figure 23.12 Blind spot.

This dark circle (or cross) will disappear at one location because there are no rod cells or cone cells at each eye's blind spot, where vision does not occur.

Experimental Procedure: Blind Spot of the Eye

The **blind spot** occurs where the optic nerves exit the retina (see Fig. 23.9). No vision is possible at this location because of the absence of rods and cones.

This Experimental Procedure requires a laboratory partner. Figure 23.12 shows a small circle and a cross several centimeters apart.

Left Eye

1. Hold Figure 23.12 approximately 30 cm from your eyes. If you wear glasses, keep them on.
2. Close your right eye.
3. Stare only at the cross with your left eye. You should also be able to see the circle in the same field of vision. Slowly move the paper toward you until the circle disappears.
4. Repeat the procedure as many times as needed to find the blind spot.
5. Then slowly move the paper closer to your eyes until the circle reappears. Because only your left eye is open, you have found the blind spot of your left eye.
6. Measure the distance (with your partner's help) from your eye to the paper when the circle

   first disappeared. _____ cm, left eye

Right Eye

1. Hold Figure 23.12 approximately 30 cm from your eyes. If you wear glasses, keep them on.
2. Close your left eye.
3. Stare only at the circle with your right eye. You should also be able to see the cross in the same field of vision. Slowly move the paper toward you until the cross disappears.
4. Repeat the procedure as many times as needed to find the blind spot.
5. Then slowly move the paper closer to your eyes until the cross reappears. Because only your right eye is open, you have found the blind spot of your right eye.
6. Measure the distance (with your partner's help) from your eye to the paper when the cross

   first disappeared. _____ cm, right eye

23.3 Animal Ears

Ears contain specialized receptors for detecting sound waves in the environment. They also often function as organs of balance.

Anatomy of Invertebrate Ears

Among invertebrates, only certain arthropod groups—crustaceans, spiders, and insects—have receptors for detecting sound waves. The invertebrate ear usually has a simple design: a pair of air pockets enclosed by a tympanum that passes sound waves to sensory neurons.

Observation: An Invertebrate Ear

Examine the preserved grasshopper on display, and with the help of Figure 23.13*a*, locate the tympanum. The tympanum covers an internal air sac that allows it to vibrate when struck by sound waves. Sensory neurons attached to the tympanum are stimulated directly by the vibration.

Anatomy of the Human Ear

The tetrapod vertebrate ear may have evolved from the fishes. Fishes have a lateral line, a series of hair cells with cilia embedded in a mass of gelatinous material that detects water currents and pressure waves from nearby objects (Fig. 23.13*b*). Fishes also have **otic vesicles** that mainly function as organs of balance. In contrast to the "ear" of fishes, the anatomy of the human ear allows sound waves to be received, amplified, and detected by receptors. The human ear also contains receptors for balance.

Figure 23.13 Evolution of the human ear.
a. A few invertebrates have ears such as that of the grasshopper. **b.** The human ear has more in common with that of the lateral line of fishes because this apparatus contains hair cells whose cilia are embedded in a gelatinous cupula.

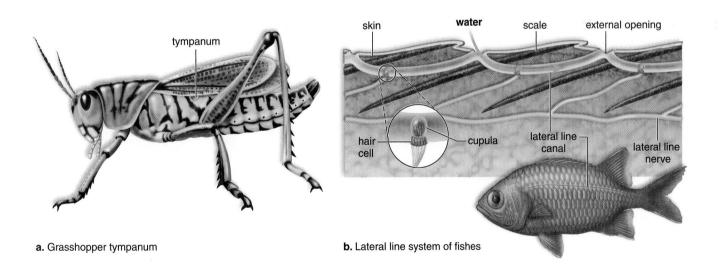

a. Grasshopper tympanum

b. Lateral line system of fishes

Examine a human ear model, and identify the structures depicted in Figure 23.14 and listed in Table 23.4.

Figure 23.14 Anatomy of the human ear.

The outer ear extends from the pinna to the tympanic membrane. The middle ear extends from the tympanic membrane to the oval window. The inner ear encompasses the semicircular canals, the vestibule, and the cochlea.

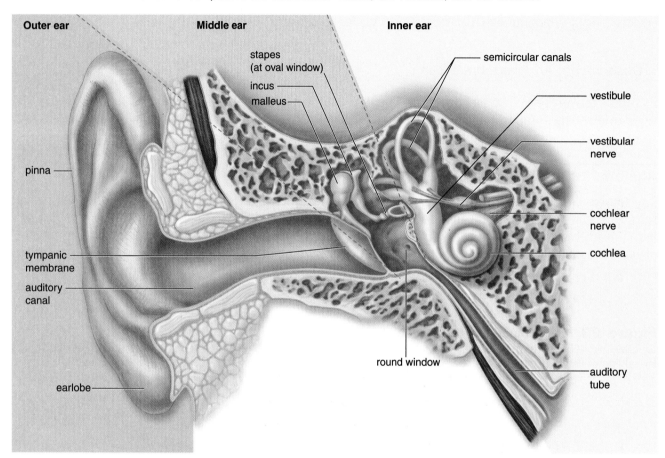

Table 23.4		Parts of the Human Ear		
Part	**Medium**	**Function**	**Mechanoreceptor**	
Outer ear	Air			
Pinna		Collects sound waves	—	
Auditory canal		Filters air	—	
Middle ear	Air			
Tympanic membrane and ossicles		Amplify sound waves	—	
Auditory tube		Equalizes air pressure	—	
Inner ear	Fluid			
Semicircular canals		Rotational equilibrium	Stereocilia embedded in cupula	
Vestibule (contains utricle and saccule)		Gravitational equilibrium	Stereocilia embedded in otolithic membrane	
Cochlea (spiral organ)		Hearing	Stereocilia embedded in tectorial membrane	

Physiology of the Human Ear

The process of hearing begins when sound waves enter the auditory canal. The sound waves are detected by the **tympanic membrane** and amplified by the **ossicles** (bones). The last ossicle, the stapes, strikes the oval window of the cochlea. As a result of the movement of the fluid within the cochlea, the hair cells in the organ of Corti are stimulated. The hair cells are sensory receptors that cause nerve impulses to be conducted in the cochlear (auditory) nerve to the brain.

Experimental Procedure: Locating Sound

Humans locate the direction of sound according to how fast it is detected by either or both ears. A difference in the hearing ability of the two ears can lead to a mistaken judgment about the direction of sound. Both you and a laboratory partner should perform this Experimental Procedure on each other. Enter the data for *your* ears, not your partner's ears, in your laboratory manual.

1. Ask the subject to be seated, with eyes closed.
2. Then strike a tuning fork or rap two spoons together at the five locations listed in number 4. Use a random order.
3. Ask the subject to give the exact location of the sound in relation to his or her head.
4. Record the subject's perceptions when the sound is

   a. directly below and behind the head _____

   _____

   b. directly behind the head _____

   c. directly above the head _____

   _____

   d. directly in front of the face _____

   _____

   e. to the side of the head _____

   _____

5. Is there an apparent difference in hearing between your two ears? _____

   _____

23.4 The Senses of Human Skin

The receptors in human skin are sensitive to touch, pain, temperature, and pressure. There are individual receptors for each of these various stimuli.

Observation: Human Skin

With the help of a model of human skin (see Fig. 25.7 on p. 503 in main book), locate the following areas or structures, and describe the location of each:

1. Subcutaneous layer _____

2. Adipose tissue _____

3. Dermis _____

4. Epidermis _____

5. Hair follicle and hair _____

6. Oil gland _____

7. Sweat gland _____

8. Sensory receptors _____

Sense of Touch

The dermis of the skin contains touch receptors whose concentration differs in various parts of the body.

Experimental Procedure: Sense of Touch

You will need a laboratory partner to perform this Experimental Procedure. Enter *your* data, not the data of your partner, in your laboratory manual.

1. Ask the subject to be seated, with eyes closed.
2. Then test the subject's ability to discriminate between the two points of a hairpin or a pair of scissors at the four locations listed in number 5.
3. Hold the points of the hairpin or scissors on the given skin area, with both of the points simultaneously and gently touching the subject.
4. Ask the subject whether the experience involves one or two touch sensations.
5. Record the shortest distance between the hairpin or scissor points for a two-point discrimination in the following areas:

   a. Forearm _____ mm

   b. Back of the neck _____ mm

   c. Index finger _____ mm

   d. Back of the hand _____ mm

6. Which of these areas apparently contains the greatest density of touch receptors? _____

   Why is this useful? _____

   _____

Sense of Heat and Cold

Temperature receptors respond to a change in temperature.

Experimental Procedure: Sense of Heat and Cold

1. Obtain three 1,000-ml beakers, and fill one with *ice water,* one with *tap water* at room temperature, and one with *warm water* (45–50°C).
2. Immerse your left hand in the ice water beaker and your right hand in the warm water beaker for 30 seconds.
3. Then place both hands in the beaker with room-temperature tap water.
4. Record the sensation in the right and left hands.

   a. Right hand _____

   b. Left hand _____

23.5 Animal Chemoreceptors

Taste buds located on the tongue and olfactory receptors located in the nose are **chemoreceptors** in humans. Most other animals also have receptors that respond to chemicals in the same manner. For example, flies have chemoreceptors on their legs. When flies are ready to feed, chemoreceptors signal the lowering of the **proboscis,** a tubelike feeding structure (Fig. 23.15).

Experimental Procedure: Chemoreceptors

1. Obtain a Styrofoam block that contains rods with flies attached.
2. Obtain a chemplate, and fill a series of wells with *sugar solutions* of increasing concentrations.
3. Fill one well with water.
4. Give the fly a drink of water before you begin, since a thirsty fly will react the same to water as a hungry fly will to food (sugar).
5. Now hold the fly so that its forelegs come into contact with the sugar solutions, as seen in Figure 23.15*b,* starting with the least concentrated. *Do not allow the fly to drink.*
6. Rinse the fly's legs between tests, using a wash bottle. (Collect the water in a beaker.)

   Place an X on the appropriate line whenever your fly lowers its proboscis to feed.

Sugar	0.05 M	0.1 M	0.2 M	0.4 M	0.8 M
Sucrose	_____	_____	_____	_____	_____

7. At what sugar concentration does the fly attempt to feed by lowering its proboscis? _____

Figure 23.15 Chemoreception in flies.

Flies lower the proboscis to feed only after the chemoreceptors on their feet respond to an edible substance.

a. Proboscis retracted

b. Fly attached to rod lowered toward food until feet touch the surface

c. Proboscis fully extended

1. Describe the cerebrum of the human brain, and state a function.

   _____

   _____

2. The brain stem includes the medulla oblongata, the pons, and the midbrain. Explain the expression *brain stem* as an anatomical term. _____

   _____

3. Describe the location of the gray/white matter of the spinal cord, and give a function for each.

   _____

   _____

   _____

4. State, in order from receptor to effector, the neurons associated with a spinal reflex. _____

   _____

5. Trace the path of light in the human eye—from the exterior to the retina and then from retinal nerve impulses to the brain. _____

   _____

6. Contrast the eye of an arthropod with the eye of a squid and human.

   _____

   _____

7. If you move an illustration that contains a dark circle and a dark cross toward an eye, one or the other may disappear. Give an explanation for this.

   _____

   _____

8. Trace the path of sound waves in the human ear—from the tympanic membrane to the receptors for hearing.

   _____

   _____

9. Compare the manner in which a grasshopper "hears" to the way a human hears.

   _____

   _____

10. Name several structures located in the dermis of the skin.

    _____

    _____

24

Animal Development

Learning Outcomes

24.1 Early Embryonic Stages
- Identify and compare the morula, blastula, and gastrula in the sea star and frog. 322–325
- Describe the development of the neural tube in the frog. 325

24.2 Germ Layers
- Name the three germ layers and the major organs that develop from each. 326
- Describe the significance of the process of induction during development. 326

24.3 Chick Development
- Name and contrast the functions of the extraembryonic membranes in mammals and chicks. 326
- Describe, in general, how chick embryos develop. 326–334

24.4 Human Development
- Compare the development of humans to that of the other animals studied. 332–337
- Define embryonic and fetal development, and identify human models as either embryos or fetuses. 335–337

Introduction

The early development of animals is quite similar, regardless of the species. The fertilized egg, or zygote, undergoes successive divisions by cleavage, forming a mulberry-shaped ball of cells called a morula and then a hollow ball of cells called a blastula. The fluid-filled cavity of the blastula is the blastocoel. Later, some of the surface cells fold inward, or invaginate, so that the embryo has an outer and inner layer of cells. Then, a middle layer arises between the outer and inner layer of cells. The embryo is now called a gastrula. In particular, the presence of yolk (nutrient material) influences how the three layers of the gastrula come about.

The layers of the gastrula are called the **germ layers** because they give rise to all the organs of the body: (1) The outer layer, called the **ectoderm,** forms the nervous system and the skin plus accessory structures (hair, nails, scales, feathers, etc.); (2) the inner layer, called the **endoderm,** forms the lining of the digestive system and respiratory system; and (3) the middle layer, called the **mesoderm**, gives rise to the cardiovascular, muscular, reproductive, and skeletal systems and to connective tissue.

Development requires growth, differentiation, and morphogenesis.

Growth occurs when cells divide, get larger, and divide again.

Differentiation occurs when cells become specialized in structure and function. For example, a muscle cell looks and acts quite differently from a nerve cell.

Morphogenesis occurs when body parts become shaped and patterned into a certain form. For example, your arm and leg are very different, even though they contain the same types of tissues.

24.1 Early Embryonic Stages

Successive division of the **zygote** (fertilized egg) results in two-cell, four-cell, eight-cell, and sixteen-cell stages and, finally, in a many-celled stage called a **morula.** The morula becomes a **blastula,** and the blastula becomes a **gastrula.**

Sea Star Development

Echinoderms (e.g., sea stars, sea urchins) are useful for illustrating the stages of early development for multicellular animals. Sea stars develop in an aquatic environment and develop quickly into a larva capable of feeding itself. Explain why you would not expect a sea star's egg to be heavily laden with yolk. _____

_____

Observation: Sea Star Embryos

Examine whole-mount microscope slides of stained sea star embryos at the stages of development shown in Figure 24.1. Try to identify the following:

1. **Unfertilized egg:** Observe the large nucleus and the darkly staining nucleolus. The plasma membrane surrounds the cytoplasm, which contains a small amount of yolk. After fertilization, a fertilization membrane, which prevents the entrance of other sperm, can be seen outside the plasma membrane, and the distinct nucleus disappears.
2. **Cleavage:** Successive division of the embryo results in a two-cell, four-cell, eight-cell, sixteen-cell and, finally, a many-celled stage. Cleavage occurs without an accompanying increase in size.
3. **Morula:** The morula is a ball of cells about the same size as the original zygote. Explain why the morula and zygote are about the same size. _____

_____

4. **Blastula:** The large number of embryonic cells of the morula arrange themselves into a blastula, a single-layered ball with a fluid-filled cavity, called the **blastocoel,** in the middle.
5. **Early gastrula:** The cells of the blastula fold inward to form a two-layered gastrula. The cavity produced by the infolded layer of cells is the **archenteron,** or primitive gut, which has an opening to the outside called the **blastopore.** The outer layer of cells is the **ectoderm,** and the inner layer is the **endoderm.**
6. **Late gastrula:** As development continues, two pouches (the coelomic sacs) form by outpocketing from the endoderm surrounding the gut. These pouches become part of the **coelom** (body cavity), and the walls of the lateral pouches become the third germ layer, the **mesoderm.**

Figure 24.1 Photographs of sea star developmental stages.

All animal embryos go through these same early developmental stages. (**a–f:** Magnification 75×) Courtesy of Carolina Biological Supply Company, Burlington, N.C.

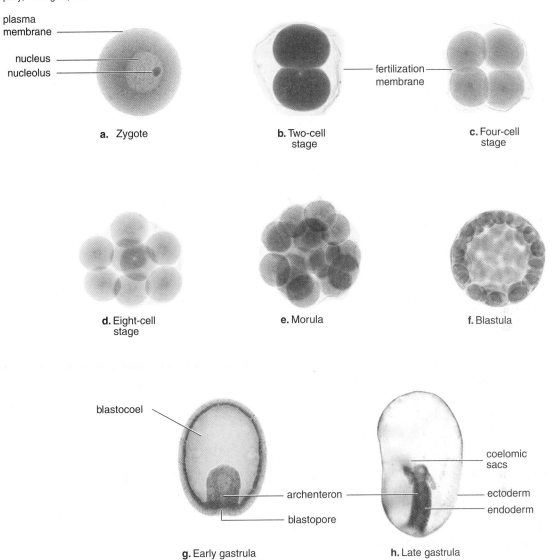

plasma membrane

nucleus

nucleolus

a. Zygote

fertilization membrane

b. Two-cell stage

c. Four-cell stage

d. Eight-cell stage

e. Morula

f. Blastula

blastocoel

archenteron

blastopore

g. Early gastrula

coelomic sacs

ectoderm

endoderm

h. Late gastrula

Frog Development

Like sea star embryos, frog embryos develop in water, but the process takes a bit longer. Frog eggs contain more yolk than sea star eggs.

Observation: Preserved Frog Embryos

Examine preserved frog embryos at various stages of development (Fig. 24.2) in a petri dish, or use models, and identify the following stages:

1. **Fertilized egg:** The eggs are partially pigmented. The black side contains very little yolk (nutrient material) and is called the **animal pole.** The unpigmented, yolky side is called the **vegetal pole.**

   Did the sea star unfertilized egg have an animal pole and a vegetal pole? _____ Explain. _____

2. **Morula:** Cleavage begins at the animal pole (Fig. 24.3*a*). After the first two cell divisions, which run from pole to pole, there are four cells of equal size. Cleavage continues until there is a morula. The cells of the animal pole are smaller and more numerous than those of the vegetal pole. The yolk-laden cells of the vegetal pole are slower to divide than those of the animal pole.

Figure 24.2 Photographs of frog developmental stages.
The presence of yolk alters the appearance of the usual stages somewhat. (**a:** Magnification 25×; **i:** Magnification 22.5×)

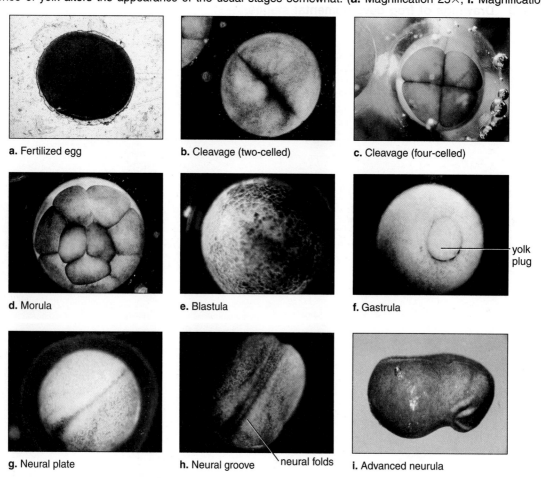

a. Fertilized egg

b. Cleavage (two-celled)

c. Cleavage (four-celled)

d. Morula

e. Blastula

f. Gastrula — yolk plug

g. Neural plate

h. Neural groove — neural folds

i. Advanced neurula

3. **Blastula:** The blastula is a hollow ball of cells. However, the blastocoel (fluid-filled cavity) is found only at the animal pole. The yolk-laden cells of the vegetal pole do not divide rapidly and, therefore, do not help in the formation of the blastocoel.

   Compare the frog blastula to that of the sea star. _____

   _____

4. **Early gastrula:** Gastrulation is recognized by the presence of a crescentic slit. This is the location of the blastopore, where cells are invaginating (Fig. 24.3b). The blastopore later takes on a circular shape, but is plugged by yolk cells that do not invaginate rapidly. In the frog, invagination is accomplished primarily by yolkless cells from the animal pole.

5. **Late gastrula:** The moderate amount of yolk also influences the formation of the mesoderm. This germ layer develops by invagination of cells at the lateral and ventral lips of the blastopore.

   Compare formation of the mesoderm in the frog to that in the sea star. _____

   _____

6. **Neurula:** During neurulation in the frog, two folds of ectoderm grow upward as the neural folds with a groove between them (Fig. 24.3c). The flat layer of ectoderm between them is the **neural plate.** The tube resulting from closure of the folds is the **neural tube,** which will become the spinal (nerve) cord and brain. An examination of the neurula in cross section shows that the nervous system develops directly above the **notochord,** a structure that arises from invaginated cells in the middorsal region.

Figure 24.3 **Drawings of frog developmental stages.**
Compare these series of drawings with the photographs in Figure 24.2.

a. Cleavage

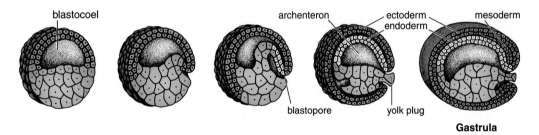

b. Gastrulation

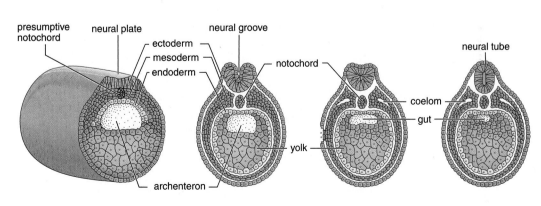

c. Neurulation

Animal Development Laboratory 24 **325**

24.2 Germ Layers

As mentioned previously, early development proceeds to the formation of three germ layers. In Table 24.1, list the three germ layers and the major organs that develop from each in the frog.

Table 24.1	Germ Layer Organization
Germ Layers	**Organs/Systems Associated with Germ Layer**
1.	
2.	
3.	

Induction

The notochord is said to induce the formation of the nervous system. Experiments have shown that, if contact with notochord tissue is prevented, no neural plate is formed. Even more dramatic are experiments in which presumptive (soon-to-be) notochord is transplanted under an area of ectoderm not in the dorsal midline. This ectoderm then is induced to differentiate into neural plate tissue, something it would not normally do.

Induction is believed to be one means by which development is usually orderly. The part of the embryo that induces the formation of an adjacent organ is said to be an **organizer** and is believed to carry out its function by releasing one or more chemical substances.

24.3 Chick Development

Unlike sea stars and frogs, chicks develop on land, and there is no larval stage. Therefore, chick development is markedly different from that of sea stars and frogs because of three features: (1) The embryo is enclosed by extraembryonic membranes, (2) there is a large amount of yolk to sustain development, and (3) there is a hard outer shell.

Extraembryonic Membranes

The embryos of land vertebrates (reptiles, birds, and mammals) are surrounded and covered by membranes that do not become a part of the animal. These **extraembryonic membranes** are the **chorion, amnion, yolk sac,** and **allantois** (Fig. 24.4).

   The chorion is the outermost membrane and, in chicks, it lies just below the porous shell, where it functions in gas exchange. In mammals, such as humans, the chorion forms the fetal portion of the placenta through which nutrients, gases, and wastes are exchanged with the blood of the mother.

   The amnion is a delicate membrane containing amniotic fluid that bathes the embryo. The embryo, suspended in this fluid medium, is thus protected from drying out and from mechanical shocks.

   The yolk sac surrounds the yolk in bird and reptile eggs. In mammals, it is the first site of blood cell formation.

   The allantois serves as a storage area for metabolic waste products in reptiles and birds. In mammals, the allantoic blood vessels are a part of the umbilical cord.

Figure 24.4 Extraembryonic membranes.

The chick and a human have the same extraembryonic membranes, but except for the amnion, they have different functions.

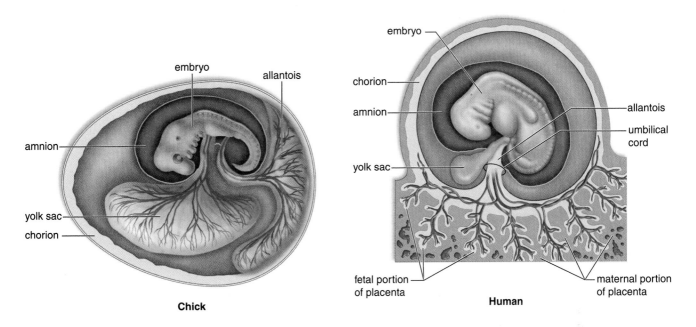

Chick ... **Human**

Observation: Raw Chick Egg

1. Carefully crack an unfertilized chick egg into a 5-inch finger bowl.
2. Examine the contents of the raw egg (Fig. 24.5). Some of the egg white (**albumen**) is denser and thicker than the remainder. These thicker masses are the **chalazae** (sing., **chalaza**), two spirally wound strands of albumen that result from the passage of the egg through the oviduct.
3. On the top of the yolk mass, note the germinal vesicle, a small, white spot containing the clear cytoplasm of the egg cell and the egg nucleus. Only the yolk with its germinal vesicle is the ovum. The various layers of albumen, including the chalazae, the shell membranes, and the shell, are all secreted by the oviduct as the ovum travels through it.

Figure 24.5 Unfertilized chick egg.
Chalazae are two spirally wound strands of albumen.

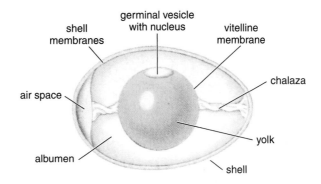

1. Follow the standard procedure (see page 329) for selecting and opening an egg containing a 24-hour chick embryo.
2. Neurulation is occurring anteriorly, and invagination is occurring posteriorly. Refer to Figure 24.6, and identify the following:

   a. **Embryo,** which lies atop the yolk.

   b. **Head fold,** which is the beginning of the embryo.

   c. **Nervous system,** which is in the process of developing and has a **head fold** at the anterior end and **neural folds** toward the posterior end, because the neural tube is still in the process of forming.

   d. **Primitive streak,** an elongated mass of cells. In the chick, the germ layers develop as flat sheets of cells. The mesodermal cells invaginate as the germ layers form and give rise to the early organs.

   e. **Somites,** blocks of developing muscle tissue that differentiate from mesoderm.

Figure 24.6 **Dorsal view of 24-hour chick embryo.**
The head fold is the amnion that will eventually envelop the entire embryo.

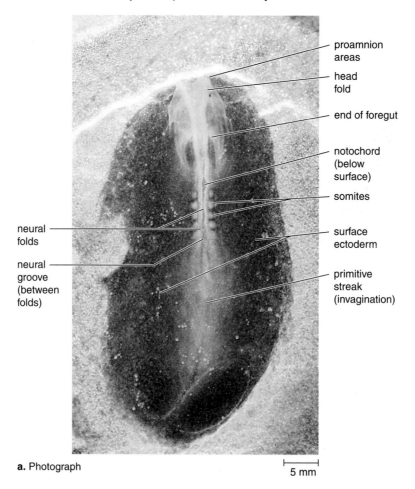

a. Photograph

5 mm

Figure 24.6 Twenty-four-hour chick embryo—*continued*.

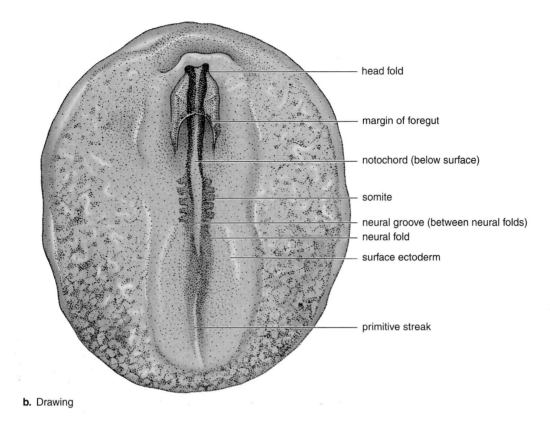

head fold

margin of foregut

notochord (below surface)

somite

neural groove (between neural folds)
neural fold
surface ectoderm

primitive streak

b. Drawing

Observing Live Chick Embryos

Use the following procedure for selecting and opening the eggs of live chick embryos:

1. Choose an egg of the proper age to remove from the incubator, and put a penciled X on the uppermost side. The embryo is just below the shell.
2. Add warmed chicken Ringer solution to a finger bowl until the bowl is about half full. (Chicken Ringer solution is an isotonic salt solution for chick tissue that maintains the living state.) The chicken Ringer solution should not cover the yolk of the egg.
3. On the edge of the dish, gently crack the egg on the side opposite the X.
4. With your thumbs placed over the X, hold the egg in the chicken Ringer solution while you pry it open from below and allow its contents to enter the solution. If you open the egg too slowly or too quickly, the shell may damage the delicate membranes surrounding the embryo.

1. Follow the standard procedure (on the previous page) for selecting and opening an egg containing a 48-hour chick embryo.
2. The embryo has started to twist so that the head region is lying on its side. Refer to Figure 24.7, and identify the following:
   a. **Shape of the embryo,** which has started to bend. The head is now almost touching the heart.
   b. **Heart,** contracting and circulating blood. Can you make out a ventricle, an atrium, and the aortic arches in the region below the head? Later, only one aortic arch will remain.

Figure 24.7 **Forty-eight-hour chick embryo.**
The most prominent organs are labeled.

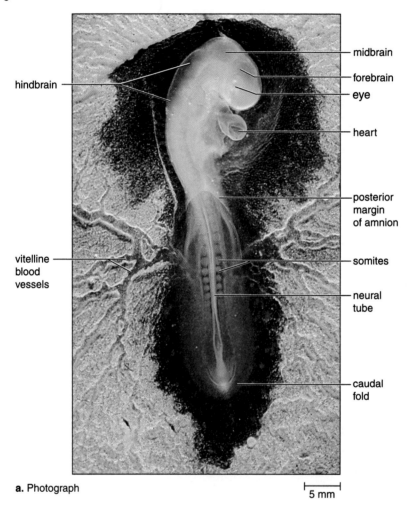

hindbrain

midbrain

forebrain

eye

heart

posterior margin of amnion

vitelline blood vessels

somites

neural tube

caudal fold

a. Photograph

5 mm

c. **Vitelline arteries** and **veins,** which extend over the yolk. The vitelline veins carry nutrients from the yolk to the embryo.

d. **Brain** divided into three regions: forebrain, midbrain, and hindbrain.

e. **Eye,** which has a developing lens.

f. **Margin (edge) of the amnion,** which can be seen above the vitelline arteries.

g. **Somites,** which now number 24 pairs.

h. **Caudal fold** of the amnion. The embryo will be completely enveloped when the head fold and caudal fold meet the margin of the amnion.

Figure 24.7 Forty-eight-hour chick embryo—*continued.*
The most prominent organs are labeled.

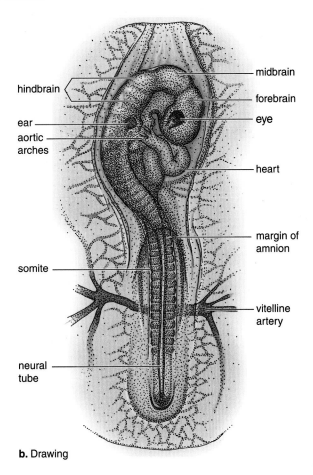

b. Drawing

1. Follow the standard procedure for selecting and opening an egg. For this observation, the egg should contain a 72-hour chick embryo.
2. The chick is now almost completely on its side, and the brain is even more flexed than before. Refer to Figure 24.8, and identify the following:

   a. **Brain,** which has a ↄ shape and is very prominent.

   b. **Eye,** which has a distinctive lens.

Figure 24.8 **Seventy-two-hour chick embryo.**
Sense organs and limb buds have appeared.

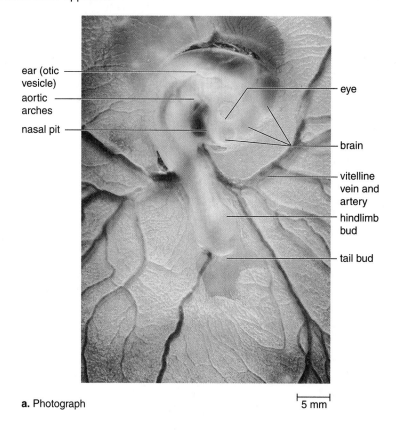

ear (otic vesicle)
aortic arches
nasal pit
eye
brain
vitelline vein and artery
hindlimb bud
tail bud

a. Photograph

5 mm

c. **Ear (otic vesicle),** enlarged and more noticeable than before.

d. **Heart,** which has a distinctive ventricle and atrium and is actively pumping blood. There are pharyngeal pouches between the aortic arches.

e. **Vitelline blood vessels,** which extend over the yolk.

f. **Limb buds,** which become the wings and hindlimbs. The posterior limb buds are easier to see at this point.

g. **Somites,** which now number 36 pairs.

h. **Tail bud,** which marks the end of the embryo.

Figure 24.8 **Seventy-two-hour chick embryo—*continued*.**
Sense organs and limb buds have appeared.

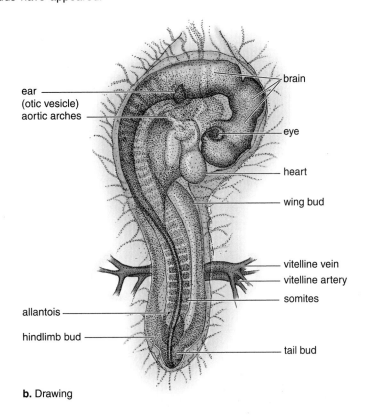

b. Drawing

As a chick embryo continues to grow, various organs differentiate further (Fig. 24.9). The neural tube closes along the entire length of the body and is now called the spinal cord. The allantois, an extraembryonic membrane, is seen as a sac extending from the ventral surface of the hindgut near the tail bud. The digestive system forms specialized regions, and there is both a mouth and an anus. The yolk sac, the extraembryonic membrane that encloses the yolk, is attached to the ventral wall, but when the yolk is used up, the ventral wall closes.

Figure 24.9 **Ninety-six-hour chick embryo.**
Brain regions listed in the key below can now be seen. Courtesy of Carolina Biological Supply, Burlington, N.C.

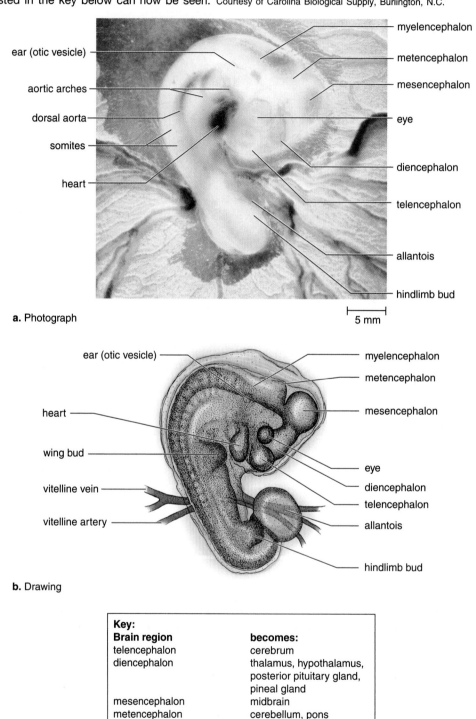

a. Photograph

5 mm

b. Drawing

Key:	
Brain region	**becomes:**
telencephalon	cerebrum
diencephalon	thalamus, hypothalamus, posterior pituitary gland, pineal gland
mesencephalon	midbrain
metencephalon	cerebellum, pons
myelencephalon	medulla oblongata

24.4 Human Development

As illustrated in Figure 24.10, the early stages of human development are quite similar to those of the chick. Differences become marked only as development proceeds.

Figure 24.10 **Comparison of mammalian embryos.**
Successive stages in the embryonic development of chick and human. Early stages are similar; differences become apparent only during later stages.

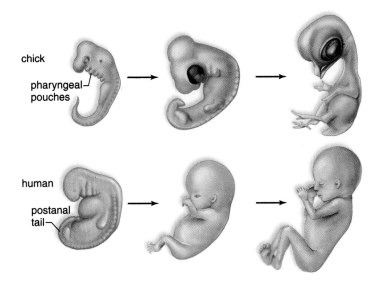

Stages of Human Development

Human development is divided into embryonic development and fetal development. During **embryonic development** (first two months), the germ layers and extraembryonic membranes develop; then the internal organs form (Fig. 24.11). At the end of this period, the embryo has a human appearance.

1. Review frog development on pages 394–395 and state here the stages you would expect to see during early embryonic development in humans:

   _____

   _____

2. Review chick development and decide what two organs will most likely be first to make their appearance during human development.

   _____

3. Figure 24.11 shows in particular the development of the extraembryonic membranes in humans. Circle the labels for the membranes in Figure 24.11.

During **fetal development** (last seven months), the skeleton becomes ossified (bony), reproductive organs form, arms and legs fully develop, and the fetus enlarges in size and gains weight (Fig. 24.12).

Figure 24.11 Human development.

Changes occurring during the third to the fifth week include the development of the extraembryonic membranes and the umbilical cord.

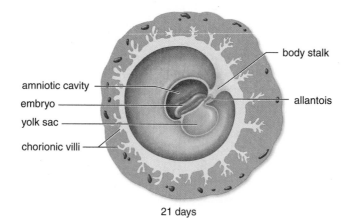

- amniotic cavity
- embryo
- yolk sac
- chorionic villi
- body stalk
- allantois

21 days

- chorion
- amniotic cavity
- amnion
- chorionic villi
- allantois
- yolk sac

25 days

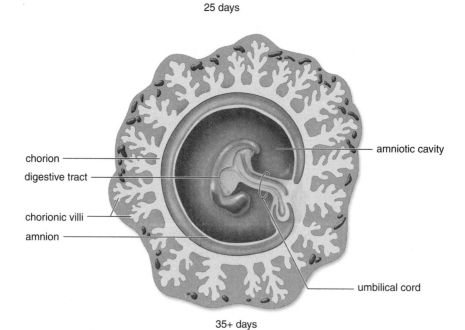

- chorion
- digestive tract
- chorionic villi
- amnion
- amniotic cavity
- umbilical cord

35+ days

Figure 24.12 Human development.

Changes occurring from the fifth week to the eighth month.

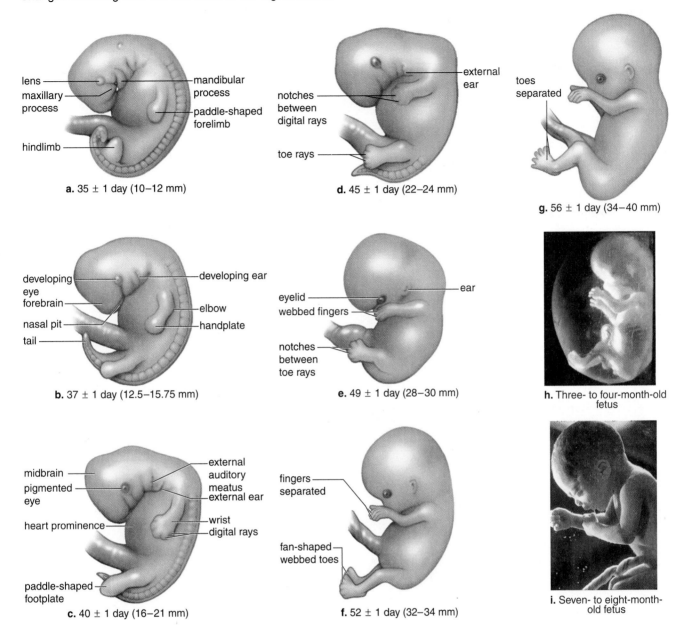

a. 35 ± 1 day (10–12 mm)

- lens
- maxillary process
- hindlimb
- mandibular process
- paddle-shaped forelimb

b. 37 ± 1 day (12.5–15.75 mm)

- developing eye
- forebrain
- nasal pit
- tail
- developing ear
- elbow
- handplate

c. 40 ± 1 day (16–21 mm)

- midbrain
- pigmented eye
- heart prominence
- paddle-shaped footplate
- external auditory meatus
- external ear
- wrist
- digital rays

d. 45 ± 1 day (22–24 mm)

- notches between digital rays
- toe rays
- external ear

e. 49 ± 1 day (28–30 mm)

- eyelid
- webbed fingers
- notches between toe rays
- ear

f. 52 ± 1 day (32–34 mm)

- fingers separated
- fan-shaped webbed toes

g. 56 ± 1 day (34–40 mm)

- toes separated

h. Three- to four-month-old fetus

i. Seven- to eight-month-old fetus

1. Describe how an embryo becomes a morula, a blastula, and a gastrula. _____
    _____
    _____

2. What factor causes a frog's morula, blastula, and gastrula to appear differently from those of a sea star?
    _____
    _____

3. Describe how induction may control development.
    _____

4. Name the four extraembryonic membranes, and state the function of each in birds and mammals.

Extraembryonic Membrane	Function in Birds	Function in Mammals
a.		
b.		
c.		
d.		

5. What three features are quite noticeable in a 24-hour chick embryo? _____
    _____

6. List the two stages of human development, and state a reason for dividing human development into
    these two stages. _____
    _____
    _____
    _____

25

Symbiotic Relationships

Learning Outcomes

Introduction
- Define and differentiate the three types of symbiotic relationships examined in this laboratory. 339

25.1 Mutualism
- Identify root nodules of leguminous plants, and use them as an example of mutualism. 340
- Identify lichens, and use them as examples of mutualism. 341–342
- Identify a prepared slide of *Trichonympha* or *Pyrsonympha,* and use their relationship with termites as an example of mutualism. 342
- Describe the relationship between ants and acacias as an example of mutualism. 343

25.2 Commensalism
- Identify epiphytes, and use them as examples of commensalism. 344

25.3 Parasitism
- Describe the life cycle of *Plasmodium,* and identify a prepared slide of human red blood cells infected with *Plasmodium.* 345–346
- Contrast free-living worms with parasitic worms in six ways. 347
- Identify a tapeworm, and describe its life cycle. 348
- Identify a slide of a fluke and describe the life cycle of *Schistosoma.* 349–350
- Identify a free-living roundworm. 350
- Identify a slide of meat infected with *Trichinella* and a slide of a hookworm. 351–352

Introduction

Symbiosis is a close relationship that may occur when two organisms of different species live together. At least three types of symbiotic relationships have been identified. These types are differentiated by whether and to what extent organisms benefit from the relationship (Table 25.1).

In commensalism and parasitism, the larger organism is called a **host.** Commensalists do not generally harm their host, and even parasites do not

Table 25.1	Relationships Between Symbionts	
Type of Symbiosis	**Symbiont**	**Host**
Mutualism	+	+
Commensalism	+	0
Parasitism	+	−

Key: "+" = benefited; "0" = unaffected; "−" = harmed.

usually cause the host's death because that would also cause the death of the parasite. Host and parasite are expected to coevolve into a stable relationship. If the host does die, it is likely due to an infection that is new to that host.

Not all symbiotic relationships are as clear-cut as Table 25.1 implies. For instance, commensalism and even mutualism may lean toward parasitism if disease, lack of food, or other circumstances change the balance.

25.1 Mutualism

Mutualism is a relationship in which both organisms benefit from the association. Mutualistic relationships often help organisms obtain food or avoid predation. Indeed, mutualistic relationships are often essential to the life of the organism.

Root Nodules of Leguminous Plants

The bacterium *Rhizobium* invades the roots of certain leguminous plants (clover, alfalfa, soybeans), where it forms nodules (Fig. 25.1). Here it obtains carbohydrates from its host and provides the host with fixed forms of nitrogen. *Rhizobium* changes atmospheric nitrogen (approximately 80% of our air) to ammonia and other fixed forms used by the plant to make its own proteins.

Observation: Root Nodules

1. Study the living or preserved specimen of a plant with root nodules.
2. Do the roots run through the center of the nodules, or are the nodules to the side or at the ends of the roots?

_____

3. Examine a prepared slide showing cross sections of nodule cells. The cells are filled with masses of stained bacteria.

Figure 25.1 Root nodule formation.

Steps in the infection of a legume plant by *Rhizobium,* the symbiotic root-nodule-forming bacterium.

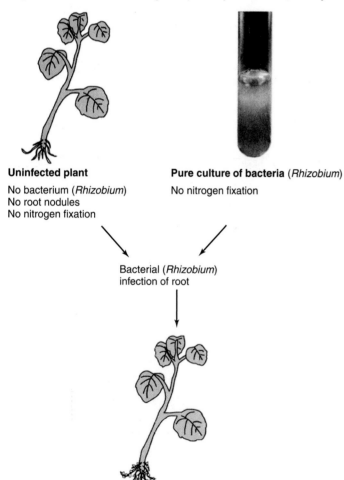

Uninfected plant

No bacterium (*Rhizobium*)
No root nodules
No nitrogen fixation

Pure culture of bacteria (*Rhizobium*)

No nitrogen fixation

Bacterial (*Rhizobium*)
infection of root

Root nodules with bacteria (*Rhizobium*)

Legume root with nodules from nodule-forming bacterium *Rhizobium.*

Figure 25.2 Lichen morphology.

a. The diagram shows the arrangement of algae and fungi in lichens. **b.–d.** The photographs illustrate three different types of lichens.

Lichens

Lichens are composed of a fungus and either an alga or a cyanobacterium growing in intimate association. While the association is often considered mutualistic, the fungus is probably somewhat parasitic on the alga or cyanobacterium.

A lichen called reindeer moss provides food for reindeer and caribou in the Arctic regions. Lichens are also important as soil-forming agents. Lichens can grow on rocks because the fungus sends acids into the rock and dissolves minerals, while the alga or cyanobacterium carries on photosynthesis.

Observation: Lichen Morphology and Diversity

1. Obtain a prepared slide of a lichen, and note the fungus, composed of **hyphae** and photosynthesizing cells, in close proximity within the thallus (body) of the lichen (Fig. 25.2a).

2. Which member of a lichen provides inorganic food? _____

3. Which member provides organic food?

   _____

4. Lichens can be **crustose** (appressed to a substratum), **foliose** (leaflike lobes), or **fruticose** (erect or pendant branching structures) (Fig. 25.2b, c, and d). Examine the lichens on display. Then complete Table 25.2 by listing these lichens by name and growth habit. Also note any other pertinent information about each type of lichen.

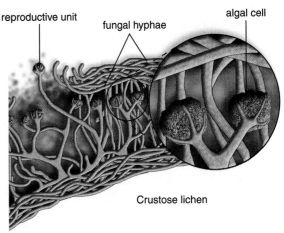

reproductive unit   fungal hyphae   algal cell

Crustose lichen

a. This section of a lichen shows the placement of the photosynthetic cells and the fungal hyphae, which encircle and penetrate them.

b. Crustose lichens are compact.

c. Foliose lichens are leaflike.

d. Fruticose lichens are shrublike.

Table 25.2	Lichen Diversity	
Lichen Name	Growth Habit	Other Observations

Termites and Zooflagellates

As is well known, termites are insects with the capacity to eat wood. Termites, however, cannot actually digest wood. Instead, symbiotic zooflagellates (*Trichonympha* and *Pyrsonympha*) within termites are responsible for wood digestion.

Observation: Symbiotic Zooflagellates in Termites

1. Place a drop of *invertebrate saline solution* on a clean slide.
2. Place a living, wood-eating termite in the saline solution.
3. Holding the termite's head with fine forceps, use a dissecting needle to express material from the termite's abdomen.
4. Remove the termite from the solution.
5. Cover the slide with a coverslip, and examine it microscopically for moving creatures.
6. Carefully adjust the light source to enable you to see the characteristic shapes of the zooflagellates and, perhaps, their numerous flagella.
7. If live specimens are not available, study a prepared slide of *Trichonympha* (Fig. 25.3) or *Pyrsonympha*.
8. In this example of mutualism, how does the termite benefit?

_____

9. How does the zooflagellate benefit? _____

_____

_____

Figure 25.3 *Trichonympha.*
Termites would be unable to digest wood without the zooflagellate *Trichonympha,* which lives in their gut.

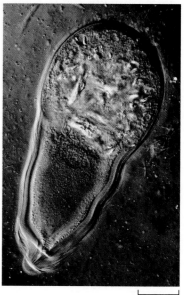

100 µm

Ants and Acacias

In Central America, the bull's horn acacia is adapted to provide a home for ants of the species *Pseudomyrmex ferruginea* (Fig. 25.4). Unlike other acacias, it has swollen thorns with a soft interior where ant larvae can grow and develop. In addition to housing the ants, the bull's horn acacia provides the ants with food. The ants feed from **nectaries** (plant glands that secrete nectar) at the base of the leaves and eat nodules called **beltian bodies** at some of the leaf tips. The ants constantly protect the plant from herbivorous insects because, unlike other ants, they are active 24 hours a day. The ants also girdle any branch (remove a strip of branch) that shades the acacia, which kills the branch and allows the acacia to receive the sunlight it needs to grow taller.

Figure 25.4 Bull's horn acacia.
The bull's horn acacia is adapted to provide a home for the ant *Pseudomyrmex ferruginea,* which promotes the well-being of the plant.

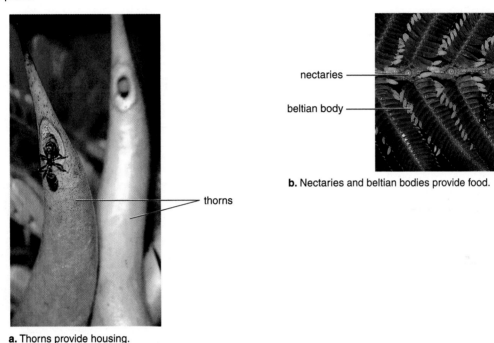

nectaries

beltian body

b. Nectaries and beltian bodies provide food.

thorns

a. Thorns provide housing.

1. If this is a mutualistic relationship, neither partner can live without the other. How would you test the hypothesis that the plant cannot live without the ants? _____

   _____

   _____

2. In 1964, Daniel Janzen removed the ants from a bull's horn acacia, and the mutualism hypothesis was supported. What do you suppose happened to the plant? _____

   _____

   _____

25.2 Commensalism

Commensalism is a relationship between two species in which only one of the species benefits; the other neither benefits nor is harmed. Often, the latter provides a home and/or transportation for the former.

Barnacles, which attach themselves to the backs of whales and to the shells of horseshoe crabs, are provided with both a home and transportation. Remoras are fish that attach themselves to sharks by means of a modified dorsal fin that acts as a suction cup. Remoras obtain a free ride and also feed on the remains of the shark's prey.

Epiphytes

Epiphytes, common in tropical climates, are plants that grow on other plants. However, they obtain their nutrients from rainwater and decaying leaves that collect around their roots, not from the host plant. Various mosses, ferns, bromeliads, and orchids are epiphytes, particularly on trees (Fig. 25.5).

Observation: Live Epiphyte

1. Examine an orchid living as an epiphyte.
2. In what way is an epiphyte living in the upper branches of a tropical tree benefiting from its symbiotic relationship with the tree?

_____

Figure 25.5 Epiphytes.
Epiphytes grow on other plants, but derive their nutrients from leaf wash and organic matter that collect at their roots.

Epiphytic orchid growing on a tree

Epiphyte growing on a fig tree

25.3 Parasitism

Parasitism is a relationship between two species in which one species—the **parasite**—derives nourishment from the other—the **host.** Usually, the parasite does the host some harm, but does not kill it. Why is it better for the parasite if the host is not killed? _____

_____

Viruses are **obligate parasites** because they are incapable of reproducing on their own. When a virus invades a cell, it takes over the cell's metabolic machinery and causes the cell to reproduce more viruses. Some bacteria, protozoans, fungi, and worms are also parasites. These are usually extracellular parasites that obtain nutrients from their hosts.

Plasmodium

Organisms in the genus *Plasmodium* cause malaria in humans. Members of this genus have no means of locomotion and are dispersed by infected mosquitoes. Study Figure 25.6, which depicts the life cycle of this parasite. In Figure 25.7, write the word *human* or *mosquito* on the lines provided.

Given the life cycle of *Plasmodium*, by what method has malaria usually been controlled?

_____

Figure 25.6 **Life cycle of the protist *Plasmodium*.**
Asexual reproduction of *Plasmodium* occurs in humans, while sexual reproduction takes place within the *Anopheles* mosquito.

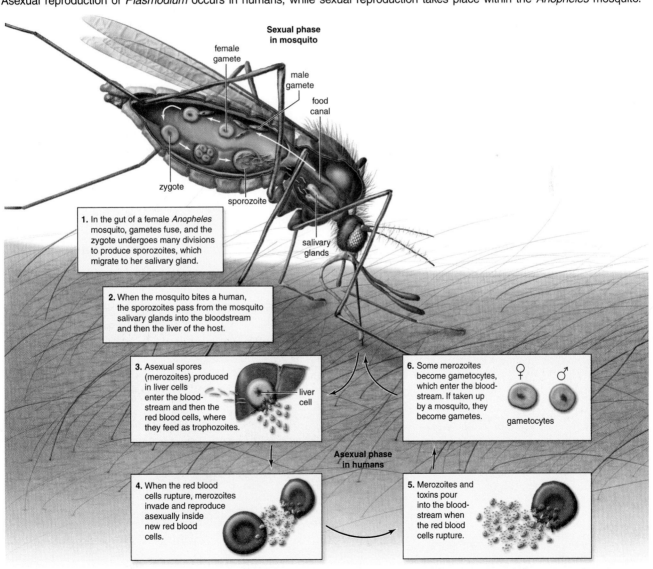

Figure 25.7 Life cycle of *Plasmodium* in diagram.

Complete this representation of the *Plasmodium* life cycle by writing the word *human* or *mosquito* on the lines provided.

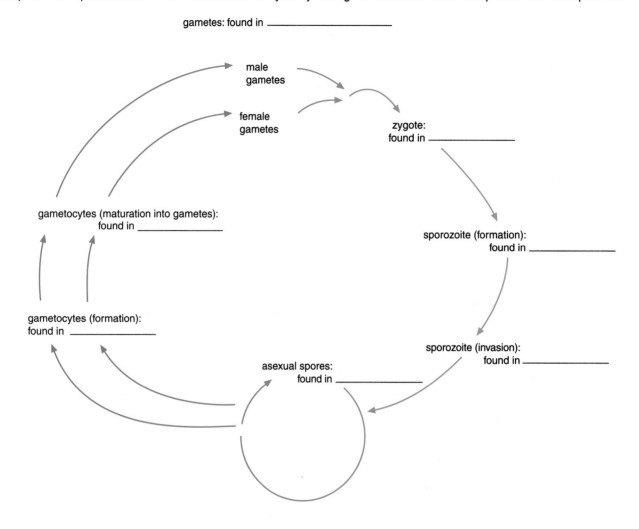

gametes: found in _____

male gametes

female gametes

zygote: found in _____

gametocytes (maturation into gametes): found in _____

sporozoite (formation): found in _____

gametocytes (formation): found in _____

sporozoite (invasion): found in _____

asexual spores: found in _____

Observation: Plasmodium *Infection*

Examine a prepared slide of human red blood cells infected with *Plasmodium,* and identify the infected blood cells (Fig. 25.8).

Figure 25.8 Human red blood cells infected by *Plasmodium.*
Plasmodium forms spores inside human red blood cells.

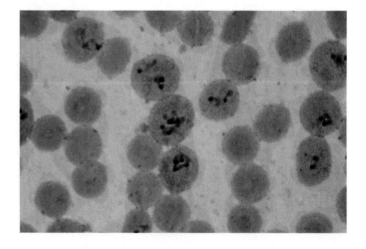

Worms

A variety of organisms, such as flatworms, roundworms, and segmented worms, are referred to simply as "worms," and some are parasites that infect humans. Worms are multicellular animals of considerable size, compared to the parasites examined thus far in this laboratory. All parasitic worms have various modifications suited to their parasitic way of life. One way to emphasize these modifications is to compare the parasite to the free-living animal, as in Table 25.3. The free-living worm needs a well-developed nervous system, sense organs, and good musculature to acquire food and escape predation. Because a parasitic worm takes nourishment from its host, these attributes, as well as a circulatory system, are not as critical. The complicated life cycle of a parasite includes a **secondary host** that carries the immobile parasite from host to host. As you examine the parasitic worms in this section, note how they are modified to be parasites.

Table 25.3	Free-Living Worm Compared to Parasitic Worm
Free-Living Worm	**Parasitic Worm**
1. Well-developed nervous system	1. Reduced nervous system
2. Sense organs, such as eyes	2. Sense organs, such as those of touch
3. Fast-moving, with protective devices	3. Limited locomotion
4. Well-developed muscles	4. Minimal muscle fibers
5. Efficient circulatory system	5. Reduced circulatory system
6. Normal reproduction	6. Complicated life cycle

Flatworms

Among worms, flatworms literally have flat bodies. Tapeworms and flukes are parasitic flatworms.

Tapeworm Anatomy

Tapeworms are parasitic flatworms that live in the intestines of vertebrate animals, including humans (Fig. 25.9). The worms consist of a **scolex** (head), usually with suckers and hooks, and **proglottids** (segments of the body) (Fig. 25.10). Ripe proglottids detach and pass out with the host's feces, scattering fertilized eggs on the ground. If pigs or cattle happen to ingest these, larvae develop and eventually become encysted in muscle, which humans may then eat in poorly cooked or raw meat. Upon digestion, a bladder worm that escapes from the cyst develops into a new tapeworm that attaches to the intestinal wall.

1. How do humans get infected with a tapeworm? _____

_____

2. What is the function of a tapeworm's hooks (if present) and suckers? _____

_____

3. Proglottids mature into "bags of eggs." Given the life cycle of the tapeworm, why might a

tapeworm produce so many eggs? _____

_____

1. Examine a preserved specimen and/or slide of *Taenia pisiformis*, a tapeworm.
2. With the help of Figure 25.10, identify the scolex, with hooks and suckers, and the proglottids.

Figure 25.9 Life cycle of the tapeworm *Taenia*.

The secondary host (pig) is the means by which the worm is dispersed to the primary host (humans).

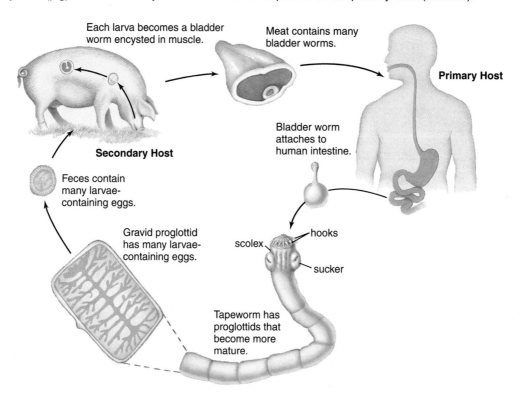

Figure 25.10 Anatomy of *Taenia*.

The adult worm is modified for its parasitic way of life. It consists of a scolex and many proglottids, which become bags of eggs. (**a:** Magnification 35×; **b:** Magnification 7×)

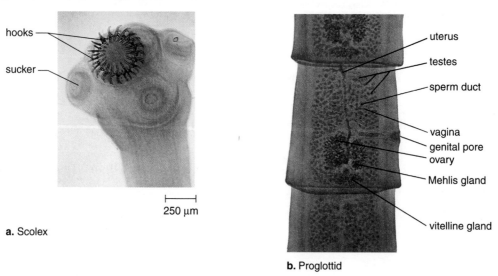

a. Scolex

b. Proglottid

Fluke Diversity and Anatomy

The many different types of flukes are usually designated by the type of organ they inhabit; for example, there are blood, liver, and lung flukes. While the structure may vary slightly, in general the fluke body tends to be oval to elongate with no definite head except for an oral sucker surrounded by sensory papillae at the anterior end. Usually there is at least one other sucker for attachment to the host. Inside, a fluke has a reduced digestive system, a reduced nervous system, and an excretory system. As expected, its reproductive system is well developed.

The human blood fluke, *Schistosoma mansoni*, is a flatworm that has separate sexes. Female worms are round, long, and slender (1.2–2.6 cm in length), while males are shorter and flattened. The males look cylindrical during copulation, however, because their body wall curves inward to form a canal in which the female resides.

Observation: Fluke Diversity and Anatomy

Fluke Diversity
Examine flukes on display in the laboratory, and fill in Table 25.4.

Table 25.4	Flukes
Name of Fluke	**Organ Infected by Fluke**

Fluke Anatomy
1. Examine a prepared slide of the blood fluke such as *Schistosoma* that shows a male and female fluke during copulation.
2. How can you identify the male? _____
3. How can you identify the female? _____

Life Cycle of Schistosoma

Females of the genus *Schistosata* deposit their eggs in small venules, close to the human intestinal passageway (lumen). The eggs have sharp spines and secrete an enzyme that causes enough damage to allow the eggs to escape into the intestine. The eggs are liberated from the human body by way of the feces. In fresh water, the eggs rupture, and the miracidia (ciliated larvae) that escape swim in search of appropriate snail hosts. If successful, they penetrate the snails and then undergo a cycle of development, giving rise to a large number of cercariae (tadpole-shaped larvae). Humans are infected when the cercariae penetrate the skin in water and invade the cardiovascular system. Young flukes mature in the human liver; as adults, they take up residence in blood vessels near the intestines (Fig. 25.11). This disease is called **schistosomiasis.**

Figure 25.11 Life cycle of the blood fluke *Schistosoma.*
This blood fluke has two hosts; humans are the primary host and snails are the secondary host.

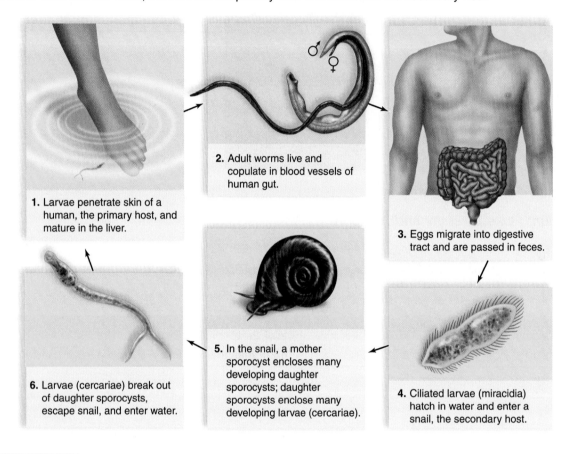

1. Larvae penetrate skin of a human, the primary host, and mature in the liver.

2. Adult worms live and copulate in blood vessels of human gut.

3. Eggs migrate into digestive tract and are passed in feces.

6. Larvae (cercariae) break out of daughter sporocysts, escape snail, and enter water.

5. In the snail, a mother sporocyst encloses many developing daughter sporocysts; daughter sporocysts enclose many developing larvae (cercariae).

4. Ciliated larvae (miracidia) hatch in water and enter a snail, the secondary host.

Roundworms

Roundworms have a smooth outside wall, indicating that they are not segmented. They are usually small (less than 5 cm long) and occur in great numbers in aquatic and terrestrial habitats. Most roundworms are free-living.

Observation: Roundworm Anatomy

1. Remove a small sample from a culture of *Rhabditis,* a free-living roundworm.
2. Place the sample in a drop of water on a slide. Cover the slide with a coverslip, and examine. If the culture is not too old, you should see many eggs and larvae, as well as adult males and females.
3. Describe the characteristic movement of roundworms.

Trichinella

Trichinella is a parasitic roundworm that causes the disease **trichinosis.** When humans eat raw or undercooked pork infected with *Trichinella* cysts, juvenile worms are released in the digestive tract where they penetrate the wall of the small intestine and mature sexually. Male and female worms mate. Females then produce juvenile worms that migrate and form cysts in various human muscles (Fig. 25.12). A human with trichinosis has muscular aches and pains that can lead to death if the respiratory muscles fail.

Figure 25.12 Larvae of the roundworm *Trichinella* embedded in a muscle.
A larva coils in a spiral and is surrounded by a sheath derived from a muscle fiber.

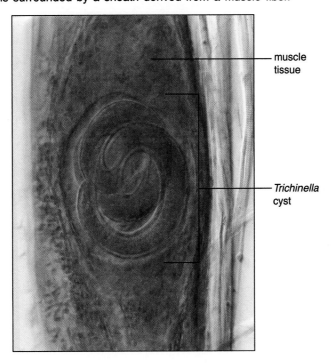

muscle tissue

Trichinella cyst

Observation: Trichinella

1. Examine preserved, infected muscle or a slide of infected muscle, and locate the *Trichinella* cysts, which contain the juvenile worms.

2. How can trichinosis be prevented in humans? ———————————————————————

3. How can pig farmers help to stamp out trichinosis so that humans are not threatened by the disease? ———————————————————————

———————————————————————————————————————

Hookworm

Necator americanus (literally, "the American killer") is a roundworm that causes **hookworm disease** in humans. Before adequate sanitation, this animal was a common cause of death in the United States. Other species infested other regions of the world. Hookworm eggs are passed in human feces, and when deposited on moist, sandy soil, the larvae develop and hatch within 24 to 48 hours. After about seven days, they become infective larvae. Then they extend their bodies into the air and remain waving about in this position until they contact human skin. After penetrating the skin, the adult lives in the intestinal tract of the human host (Fig. 25.13). Given that transmission is from human to human and that the eggs of these organisms pass out with the feces, how can

transmission best be prevented? _____

_____

Figure 25.13   Hookworm attached to an intestinal wall.
The worms attach to the intestinal mucosa and suck the blood and tissue juices of the host.

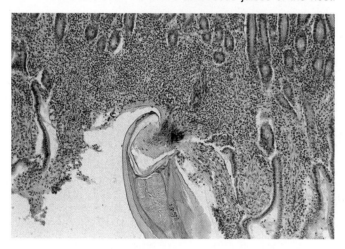

Observation: Hookworm Anatomy

1. Examine a prepared slide of the adult female hookworm, which attains a length of approximately 10 mm.
2. A hookworm has teeth by which it attaches itself to the intestinal wall. Explain the name "hookworm." _____
3. A hookworm sucks blood and tissue juices from the host. Make a list of possible host symptoms, and compare your list with that of a classmate. _____

_____

1. Distinguish between the three types of symbiotic relationships on the basis of potential harm to a host organism.

_____

_____

_____

2. In the process of taking nectar from flowers, bees also transport pollen from flower to flower. What type of symbiotic relationship is this? _____ Explain. _____

_____

3. Explain the composition of a lichen and why this composition is usually considered a mutualistic relationship.

_____

_____

4. Explain why the relationship between an epiphyte and a tree is considered commensalistic.

_____

_____

5. Why was DDT used to try to control the spread of malaria?

_____

_____

6. Would you expect to find eyes in a parasitic animal or in a free-living animal? _____
Explain. _____

_____

7. For each of the following types of organisms, give examples of a parasite:
   a. Protozoan _____
   b. Flatworm _____ and _____
   c. Roundworm _____ and _____
8. Why should meat be cooked thoroughly to avoid a tapeworm infection as well as trichinosis?

_____

_____

9. Tapeworms produce thousands of eggs, which indicates an emphasis on what part of the life cycle?

_____

26

Effects of Pollution on Ecosystems

Learning Outcomes

26.1 Studying the Effects of Pollutants
- Predict the effect of acid deposition on the growth of organisms. 356–358
- Predict the effect of oxygen deprivation on organisms. 356–358
- Predict the effect of thermal pollution on organisms. 358

26.2 Studying the Effects of Cultural Eutrophication
- Predict the effect of cultural eutrophication on food chains so that pollution results. 359–360

Introduction

Pollutants are substances added to the environment, particularly by human activities, that lead to undesirable effects for all living things. In this laboratory, we will examine the effect of two types of pollutants—acid rain and thermal pollution—on the well-being of organisms, and hypothesize how these pollutants might affect an ecological pyramid, such as the one shown in Figure 26.1.

We will also explore how ecosystem interactions can change when excess nutrients are added to a system. Over time, bodies of water undergo a natural enrichment process called **eutrophication.** Eutrophication leads to overgrowth that can eventually cause a pond or lake to fill in and disappear. Sometimes, human activities add excess nutrients to bodies of water, and this is called *cultural eutrophication.* Cultural eutrophication leads to lack of oxygen and death of organisms.

Figure 26.1 Ecological pyramid.
The amount of living matter at each succeeding trophic (feeding) level in an ecosystem decreases. Below are examples of organisms at four trophic levels.

26.1 Studying the Effects of Pollutants

We are going to study the effects of pollution by observing its effects on hay infusion organisms, on seed germination, and on an animal called *Gammarus*.

Study of Hay Infusion Cultures

A hay infusion culture (hay soaked in water) contains various microscopic organisms, such as those depicted in Figure 26.2. What do you predict will happen to organisms in a hay infusion culture when conditions are acidic or when overenrichment occurs? _____

_____

_____

    When nutrients produced by human activities enter a body of water, the algae overpopulate. Zooplankton are unable to significantly reduce the algal population, and bacteria use up all the available oxygen when decomposing them. What do you predict will happen next? _____

_____

Experimental Procedure: Study of Hay Infusion Cultures

The following four hay infusion cultures have been provided:

1. **Control culture:** This hay infusion culture simulates the optimum conditions for normal growth.
2. **Enriched culture:** Same as control culture but with more inorganic nutrients for the growth of algae.
3. **Oxygen-deprived culture:** Same as control culture, but with minimal oxygen to test the effect of lack of oxygen on organisms.
4. **Acidic culture:** Same as control culture except that it is adjusted to pH 4 with sulfuric acid (H_2SO_4). This simulates the effect of acid deposition on organisms.

Figure 26.2 Microorganisms in hay infusion cultures.

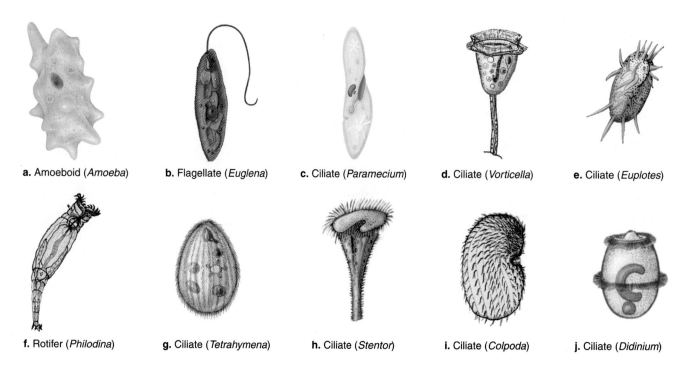

a. Amoeboid (*Amoeba*)   b. Flagellate (*Euglena*)   c. Ciliate (*Paramecium*)   d. Ciliate (*Vorticella*)   e. Ciliate (*Euplotes*)

f. Rotifer (*Philodina*)   g. Ciliate (*Tetrahymena*)   h. Ciliate (*Stentor*)   i. Ciliate (*Colpoda*)   j. Ciliate (*Didinium*)

Examine each of these cultures by preparing a wet mount. Record in Table 26.1 the diversity of life that you observe and the relative quantity of organisms.

Table 26.1	Hay Infusion Cultures		
Wet Mount	Type of Culture	Diversity of Life (List Organisms)	Relative Quantity of Organisms (High, Medium, or Low)
1	Control		
2	Enriched		
3	Oxygen-deprived		
4	Acidic		

Study of Seed Germination

Just like animals, plant seeds depend on proper conditions of temperature, light, and moisture to germinate, grow, and reproduce. Many pollutants affect seeds' ability to germinate.

Experimental Procedure: Study of Seed Germination

Observe these petri dishes, and record your observations in Table 26.2.

1. **Control petri dish:** The seeds in this petri dish have been watered with a control solution having a neutral pH.
2. **Acidic petri dish:** The seeds in this petri dish have been watered with an acidic solution having a pH of 4 to simulate acid rain.

Table 26.2	Seed Germination		
Petri Dish	Type of Solution	pH	Observations
1	Control		
2	Acidic		

Effects of Pollution on Ecosystems Laboratory 26 **357**

Study of *Gammarus*

We will study the effect of thermal pollution and acid rain on a small crustacean called *Gammarus*, found in ponds and streams (Fig. 26.3).

Experimental Procedure: Gammarus

Control Culture

1. Add 25 ml of spring water to a container.

2. Measure the pH of the spring water with a pH meter or litmus paper, and record it here: _____ . Add four *Gammarus* to the container.

3. Observe the behavior of *Gammarus* for 10 to 15 minutes, and then answer the following questions. Retain these *Gammarus* for subsequent exercises.

   • Where do the *Gammarus* spend their time in the container? _____

   _____

   • How do they spend their time? _____

   • What percentage of their time is spent moving? _____

   • Do they use all their legs in swimming? _____

   • Which legs are used in jumping and climbing? _____

   • Do *Gammarus* avoid each other? _____

   • What do *Gammarus* do when they "bump" into each other? _____

Thermal Pollution

1. To simulate thermal pollution, boil 100 ml of *water,* and then allow it to cool to 31°C.

2. Put 25 ml of this 31°C water in a beaker, and add two *Gammarus.*

3. Observe the behavior of *Gammarus,* and answer the following question:

   • What is the effect of thermal pollution on the behavior of *Gammarus?*

   _____

   _____

Acid Pollution

1. Put 25 ml of *acidic spring water* (adjusted to pH 4) in a beaker, and add two *Gammarus.*

2. Observe the animals' behavior, and answer the following questions:

   • What is the difference between the pH of the control culture and the pH of the acidic culture? _____

   • Compare the behavior of *Gammarus* in the control culture to its behavior in the acidic culture. _____

   _____

Figure 26.3 *Gammarus.*
Gammarus is a type of crustacean related to shrimp.

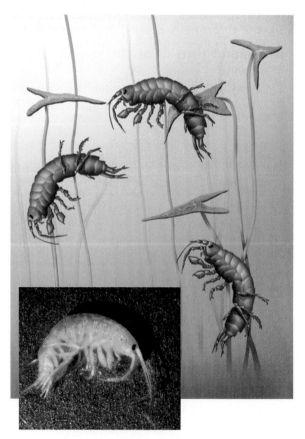

26.2 Studying the Effects of Cultural Eutrophication

Chlorella, the alga used in this study, is considered to be representative of phytoplankton in bodies of fresh water. The protozoan *Daphnia* is a zooplankton that feeds on *Chlorella.* First, you will observe how *Daphnia* feeds, and then you will determine the extent to which *Daphnia* could keep cultural eutrophication from occurring in a hypothetical example. Keep in mind that this case study is an oversimplification of a generally complex problem.

Observation: Daphnia *Feeding*

1. Place a small pool of petroleum jelly in the center of a small petri dish.
2. Use a dropper to take a *Daphnia* from the stock culture, place it on its back (covered by water) in the petroleum jelly, and observe it under the binocular dissecting microscope.
3. Note the clamlike carapace and the legs waving rapidly as the *Daphnia* filters the water.
4. Add a drop of *carmine solution,* and observe how the *Daphnia* filters the "food" from the water and passes it through the gut. The gut is more visible if you push the animal onto its side. In this position, you may also observe the heart beating in the region above the gut and just behind the head.
5. Allow the *Daphnia* to filter-feed for up to 30 minutes, and observe the progress of the carmine particles through the gut. Does the carmine travel completely through the gut in 30 minutes?

_____

Experimental Procedure: Daphnia *Feeding on* Chlorella

This exercise requires the use of a spectrophotometer. Absorbance will be a measure of the algal population level; the greater the number of algal cells, the greater the absorbance. The higher the absorbance, the greater the amount of light absorbed and *not* passed through the solution.

1. Obtain two spectrophotometer tubes (cuvettes) and a Pasteur pipette.
2. Fill one of the cuvettes with distilled water, and use it to zero the spectrophotometer. Save this tube for number 6.
3. Use the Pasteur pipette to fill the second cuvette with *Chlorella.* Gently aspirate and expel the sample several times (without creating bubbles) to give a uniform dispersion of the algae.
4. Add ten hungry *Daphnia* and, following your instructor's directions, immediately measure the absorbance with the spectrophotometer. If a *Daphnia* swims through the beam of light, a strong deflection should occur; do not use any such higher readings—instead, use the lower figure for the absorbance. Record your reading in the first column of Table 26.3.
5. Remove the cuvette with the *Daphnia* to a safe place in a test tube rack. Allow the *Daphnia* to feed for 30 minutes.
6. Rezero the spectrophotometer with the distilled water cuvette.
7. Measure the absorbance of the experimental cuvette again. Record your reading in the second column of Table 26.3, and explain your results in the third column.

Table 26.3	Spectrophotometer Data of *Daphnia* Feeding on *Chlorella*	
Absorbance Before Feeding	**Absorbance After Feeding**	**Explanation**

The following problem will test your understanding of the value of a single species—in this case, *Daphnia*. Please realize that this is an oversimplification of a generally complex problem.

1. Assume that developers want to build condominium units on the shores of Silver Lake. Homeowners in the area have asked the regional council to determine how many units can be built without altering the nature of the lake. As a member of the council, you have been given the following information:

   The current population of *Daphnia,* 10 animals/liter, presently filters 24% of the lake per day, meaning that it removes this percentage of the algal population per day. This is sufficient to keep the lake essentially clear. Predation—the eating of the algae—will allow the *Daphnia* population to increase to no more than 50 animals/liter. Therefore, 50 *Daphnia*/liter will be available for feeding on the increased number of algae that would result from building the condominiums.

   Using this information, complete Table 26.4.

Table 26.4	*Daphnia* Filtering
Number of *Daphnia*/liter	**Percent of Lake Filtered**
10	24%
50	

2. The sewage system of the condominiums will add nutrients to the lake. Phosphorus output will be 1 kg per day for every 10 condominiums. This will cause a 30% increase in the algal population. Using this information, complete Table 26.5.

Table 26.5	Cultural Eutrophication	
Number of Condominiums	**Phosphorus Added**	**Increase in Algal Population**
10	1 kg	30%
20		
30		
40		
50		

Conclusion

- Assume that phosphorus is the only nutrient that will cause an increase in the algal population and that *Daphnia* is the only type of zooplankton available to feed on the algae. How many condominiums would you allow the developer to build? _____

1. What type of population would you expect to be the largest in most ecosystems? _____
   Explain. _____
   _____

2. What causes acid rain? _____

3. What condition does acid deposition result in that can be harmful to organisms?
   _____

4. Name the type of pollution that results when water from rivers and ponds is used for cooling, and
   explain why it has detrimental effects. _____
   _____

5. Use biological magnification to show that pollution affects all living things, including humans.
   _____
   _____
   _____

6. When excess nutrients enter an aquatic ecosystem, pollution can result. Why?
   _____
   _____

7. How does cultural eutrophication show that balance of population sizes in ecosystems is beneficial?
   _____

8. When pollutants enter the environment, they have far-ranging effects. Give an example from this
   laboratory.
   _____
   _____

Metric System

Unit and Abbreviation	Metric Equivalent	Approximate English-to-Metric Equivalents	Units of Temperature
Length			
nanometer (nm)	$= 10^{-9}$ m (10^{-3} μm)		
micrometer (μm)	$= 10^{-6}$ m (10^{-3} mm)		
millimeter (mm)	$= 0.001$ (10^{-3}) m		
centimeter (cm)	$= 0.01$ (10^{-2}) m	1 inch $= 2.54$ cm	
		1 foot $= 30.5$ cm	
meter (m)	$= 100$ (10^{2}) cm	1 foot $= 0.30$ m	
	$= 1,000$ mm	1 yard $= 0.91$ m	
kilometer (km)	$= 1,000$ (10^{3}) m	1 mi $= 1.6$ km	
Weight (mass)			
nanogram (ng)	$= 10^{-9}$ g		
microgram (μg)	$= 10^{-6}$ g		
milligram (mg)	$= 10^{-3}$ g		
gram (g)	$= 1,000$ mg	1 ounce $= 28.3$ g	
		1 pound $= 454$ g	
kilogram (kg)	$= 1,000$ (10^{3}) g	$= 0.45$ kg	
metric ton (t)	$= 1,000$ kg	1 ton $= 0.91$ t	
Volume			
microliter (μl)	$= 10^{-6}$ l (10^{-3} ml)		
milliliter (ml)	$= 10^{-3}$ liter	1 tsp $= 5$ ml	
	$= 1$ cm^3 (cc)	1 fl oz $= 30$ ml	
	$= 1,000$ mm^3		
liter (l)	$= 1,000$ ml	1 pint $= 0.47$ liter	
		1 quart $= 0.95$ liter	
		1 gallon $= 3.79$ liter	
kiloliter (kl)	$= 1,000$ liter		

Units of Temperature scale markings:

°F: 230, 220, 212°–210, 200, 190, 180, 170, 160°–160, 150, 140, 134°/131°–130, 120, 110, 105.8°, 98.6°–100, 90, 80, 70, 56.66°–60, 50, 40, 32°–30, 20, 10, 0, -10, -20, -30, -40

°C: 110, 100 –100°, 90, 80, 70 –71°, 60 –57°, 50, 40 –41°, –37°, 30, 20, 13.7°, 10, 0 –0°, -10, -20, -30, -40

Common Temperatures

°C	°F	
100	212	Water boils at standard temperature and pressure.
71	160	Flash pasteurization of milk
57	134	Highest recorded temperature in the United States, Death Valley, July 10, 1913
41	105.8	Average body temperature of a marathon runner in hot weather
37	98.6	Human body temperature
13.7	56.66	Human survival is still possible at this temperature.
0	32.0	Water freezes at standard temperature and pressure.

To convert temperature scales:

$$°C = \frac{(°F - 32)}{1.8}$$

$$°F = 1.8°C + 32$$

Practical Examination Answer Sheet

1. _____

2. _____

3. _____

4. _____

5. _____

6. _____

7. _____

8. _____

9. _____

10. _____

11. _____

12. _____

13. _____

14. _____

15. _____

16. _____

17. _____

18. _____

19. _____

20. _____

21. _____

22. _____

23. _____

24. _____

25. _____

26. _____

27. _____

28. _____

29. _____

30. _____

31. _____

32. _____

33. _____

34. _____

35. _____

36. _____

37. _____

38. _____

39. _____

40. _____

41. _____

42. _____

43. _____

44. _____

45. _____

46. _____

47. _____

48. _____

49. _____

50. _____

Practical Examination Answer Sheet

1. _____
2. _____
3. _____
4. _____
5. _____
6. _____
7. _____
8. _____
9. _____
10. _____
11. _____
12. _____
13. _____
14. _____
15. _____
16. _____
17. _____
18. _____
19. _____
20. _____
21. _____
22. _____
23. _____
24. _____
25. _____

26. _____
27. _____
28. _____
29. _____
30. _____
31. _____
32. _____
33. _____
34. _____
35. _____
36. _____
37. _____
38. _____
39. _____
40. _____
41. _____
42. _____
43. _____
44. _____
45. _____
46. _____
47. _____
48. _____
49. _____
50. _____

Credits

Photos

Laboratory 1
Figure 1.1: © James Robinson/Animals Animals/Earth Scenes.

Laboratory 2
Figure 2.3a: © Michael Ross/Photo Researchers; **2.3b:** © CNRI/SPL/Photo Researchers; **2.3c:** © Steve Gschmeissner/Photo Researchers; **2.4:** Courtesy of Leica, Inc. Deerfield, IL; **2.5:** Courtesy of Leica, Inc. Deerfield, IL; **2.9:** © T.E. Adams/Visuals Unlimited.

Laboratory 3
Figure 3.3a: © Jeremy Burgess/SPL/Photo Researchers.

Laboratory 4
Figure 4.1: © Ed Reschke; **4.2b:** © Ralph Slepecky/Visuals Unlimited; **4.5a,b:** Courtesy of Ray F. Evert, University of Wisconsin; **4.9a:** © David M. Phillips/Visuals Unlimited; **4.9c:** © David M. Phillips/Visuals Unlimited; **4.9c:** © David M. Phillips/Visuals Unlimited; **4.10a:** © Dwight Kuhn; **4.10b:** © Alfred Owczarzak/Biological Photo Service.

Laboratory 8
Figure 8.2a: © Andrew Syred/Photo Researchers; **8.3a:** © Ed Reschke; **8.3b:** © Ed Reschke; **8.3c Early metaphase:** © Michael Abbey/Photo Researchers; **8.3d:** © Ed Reschke; **8.3e:** © Ed Reschke; **8.3f:** © Ed Reschke; **8.3g:** © Ed Reschke; **8.3h (prophase):** © Robert Calentine/Visuals Unlimited; **8.3i:** © Ed Reschke; **8.3k (metaphase):** © Robert Calentine/Visuals Unlimited; **8.3l (anaphase):** © Robert Calentine/Visuals Unlimited; **8.3m (telophase):** © Jack M. Bostrack/Visuals Unlimited; **8.4a:** © R.G. Kessel and C.Y. Shih, "Scanning Electron Microscopy in Biology: A Students' Atlas on Biological Organization," 1974 Springer-Verlag, New York.; **8.4b:** © R.G. Kessel and C.Y. Shih, "Scanning Electron Microscopy in Biology: A Students' Atlas on Biological Organization," 1974 Springer-Verlag, New York; **8.5:** © B.A. Palevitz & E.H. Newcomb/BPS/Tom Stack & Assoc.; **8.7a:** © Ed Reschke; **8.7b:** © Ed Reschke; **8.7c:** © Ed Reschke; **8.7d:** © Ed Reschke; **8.7e:** © Ed Reschke; **8.7f:** © Ed Reschke; **8.7g:** © Ed Reschke; **8.7h:** © Ed Reschke; **8.7i:** © Ed Reschke; **8.7j:** © Ed Reschke.

Laboratory 9
Figures 9.5a-c and 9.6: © Carolina Biological Supply/PhotoTake.

Laboratory 10
Figure 10.1c: © James King-Holmes/SPL/Photo Researchers; **10.1d,e:** © CNRI/SPL/Photo Researchers; **10.3a-d:** © The McGraw-Hill Companies, Inc./Bob Coyle, photographer.

Laboratory 11
Figure 11.9a,b: © Bill Longcore/Photo Researchers.

Laboratory 13
Figure 13.1: © Ralph Slepecky/Visuals Unlimited; **13.2a:** © Alfred Pasieka/SPL/Photo Researchers, Inc.; **13.2b:** © David Scharf/SPL/Photo Researchers, Inc.; **13.2c:** © Science Source/Photo Researchers, Inc.; **13.3a:** © M.I. Walker/Photo Researchers, Inc.; **13.3b:** © R. Knauft/Science Scource/Photo Researchers, Inc.; **13.3c:** © David Hall, University of London, Kings College/Photo Researchers, Inc.; **13.4c:** Image acquired by Prof. Michael Duszenko, University of Tubingen and Eye of Science, Reutlingen (Germany); **13.6a:** © V. Duran/Visuals Unlimited; **13.6b:** © Cabisco/Visuals Unlimited; **13.7b:** © M.I. Walker/Photo Researchers, Inc.; **13.8c:** © Carolina Biological Supply/Phototake; **13.9a (inset):** © Dr. Anne Smith/SPL/Photo Researchers, Inc.; **13.9b:** © John Cunningham/Visuals Unlimited; **13.10a:** © Biophoto Assoc./Photo Researchers; **13.10b:** © Bill Keogh/Visuals Unlimited; **13.10c:** © Gary R. Robinson/Visuals Unlimited; **13.10d:** © Michael Vivard/Peter Arnold; **13.11:** © Gary T. Cole/Biological Photo Service; **13.12:** © Gary R. Robinson/Visuals Unlimited; **13.13a:** © Carolina Biological Supply/Phototake; **13.13b:** © Carolina Biological Supply/Phototake; **13.15a:** © Everett S. Beneke/Visuals Unlimited; **13.15b:** © John Hadfield/SPL/Photo Researchers, Inc.; **13.15c:** © P. Marazzi/SPL/Photo Researchers, Inc.

Laboratory 14
Figure 14.1: © Stephen P. Lynch; **14.3a Moss:** © John Shaw/Tom Stack & Associates; **14.3a (inset):** © Ed Reschke; **14.3b (fern):** © William E. Ferguson; **14.3b (inset):** Courtesy Graham Kent; **14.3c (conifer):** © Kent Dannen/Photo Researchers; **14.3c (inset):** Courtesy Graham Kent; **14.3d (cherry blossom):** © E. Webber/Visuals Unlimited; **14.3d (inset):** © Richard Shiell/Animals Animals/Earth Scenes; **14.5a:** © Heather Angel/Biofotos; **14.5b:** © Bruce Iverson; **14.6:** © Matt Meadows/Peter Arnold, Inc.; **14.7:** © The McGraw-Hill Companies/Carlyn Iverson, photographer; **14.8a (conifer):** © D. Giannechini/Photo Researchers, Inc.; **14.8b (ginkgo):** © Kingsley Stern; **14.8c (ephedra):** © Virginia Weinland/Photo Researchers, Inc.; **14.8d (conifer):** © D. Giannechini/Photo Researchers, Inc.; **14.9:** © PhotoTake; **14.11a:** © J.R. Waaland/Biological Photo Service; **14.11b:** © Ed Reschke; **14.12:** © Carolina Biological Supply/PhotoTake.

Laboratory 15
Figure 15.1: © A & J Visage/Peter Arnold; **15.3a (snail):** © Bob Coyle/Red Diamond Stock Photo; **15.3b (nudibranch):** © Kenneth W. Fink/Bruce Coleman; **15.3c (octopus):** © Alex Kerstitch/Bruce Coleman; **15.3d (nautilus):** © Douglas Faulkner/Photo Researchers; **15.3e (mussels):** © Rick Harbo; **15.3f (scallop):** Courtesy of Larry S. Roberts; **15.5b:** Photo by Ken Taylor/Wildlife Images; **15.6:** Photo by Ken Taylor/Wildlife Images; **15.7a:** © Kim Taylor/Bruce Coleman; **15.7b:** © James Carmichael/Nature Photographics; **15.7c:** © Natural History Photographic Agency; **15.7d:** © Kjell Sandved/Butterfly Alphabet; **15.8b:** Photo by Ken Taylor/Wildlife Images; **15.9a (scale):** © Science VU/Visuals Unlimited; **15.9b (beetle):** © Kjell Sandved/Bruce Coleman; **15.9c (lacewing):** © Glenn Oliver/Visuals Unlimited; **15.9d (housefly):** © L. West/Bruce Coleman; **15.9e (walking stick):** © Art Wolfe; **15.9f (honeybee):** © John Shaw/Tom Stack & Associates; **15.9g (flea):** © Edward S. Ross; **15.9h (dragonfly):** © John Gerlach/Visuals Unlimited; **15.9i (moth):** © Clevland P. Hickman; **15.9j (grasshopper):** © Gary Meszaros/Visuals Unlimited; **15.9k (leaf hopper):** © Farley Bridges; **15.14a:** © Hal Beral/Visuals Unlimited; **15.14b:** © Patrice/Visuals Unlimited; **15.14c:** © Rod Planck/Tom Stack & Associates; **15.14d:** © Suzanne and Joseph Collins/Photo Researchers, Inc.; **15.14e:** © Robert and Linda Mitchell Photography; **15.14f:** © Craig Lorenz/Photo Researchers, Inc.; **15.15:** © Rod Planck/Photo Researchers, Inc.; **15.16b:** © Carolina Biological Supply/Phototake; **15.17:** © Ken Taylor Wildlife Images; **15.18b:** © Ken Taylor Wildlife Images; **15.19:** © Ken Taylor Wildlife Images.

Laboratory 16
Figure 16.4: Courtesy Ray F. Evert/University of Wisconsin Madison; **16.5a:** © Dr. Robert Calentine/Visuals Unlimited; **16.5b:** © Ed Degginger/Color Pic; **16.6a:** © Ed Reschke; **16.6b:** Courtesy Ray F. Evert/University of Wisconsin Madison; **16.7a:** © Carolina Biological Supply/PhotoTake; **16.7b:** © Kingsley Stern; **16.9b:** © Carolina Biological Supply/PhotoTake.

Laboratory 17

Figure 17.1a: Courtesy of Graham Kent; **17.1b:** © Ed Reschke; **17.2a-c:** Courtesy Dr. Chun-Ming Liu; **17.2d (torpedo):** © Biology Media/Photo Researchers, Inc.; **17.2e (embryo):** Jack Bostrack/Visuals Unlimited; **17.4:** © Ed Reschke; **17.5a (coleoptile):** © James Mauseth; **17.5b (first leaf):** © Barry L. Runk/Grant Heilman, Inc.

Laboratory 18

Figure 18.1a-e: © Ed Reschke; **18.1f:** © The McGraw Hill Companies, Inc./Dennis Strete, photographer; **18.1g,h:** © Ed Reschke; **18.1i:** © National Cancer Institute/Photo Researchers; **18.1j-l:** © Ed Reschke; **18.1m:** © The McGraw Hill Companies, Inc./Dennis Strete, photographer; **p. 234:** © Ed Reschke; **p. 235 (both):** © Ed Reschke; **p. 236:** © Ed Reschke; **p. 237 (left):** © Ed Reschke; **p. 237 (right):** © The McGraw Hill Companies, Inc./Dennis Strete, photographer; **p. 238 (both):** © Ed Reschke; **p. 239:** © Ed Reschke; **18.2a-d:** © Ed Reschke; **18.2e:** © R. Kessel/Visuals Unlimited; **p. 241 (top and bottom):** © Ed Reschke; **p. 242:** © The McGraw Hill Companies, Inc./Dennis Strete, photographer; **p. 245:** © John Cunningham/Visuals Unlimited; **p. 246: (a)** © The McGraw Hill Companies, Inc./Dennis Strete, photographer; **(b):** © National Cancer Institute/Photo Researchers; **(c):** © Ed Reschke; **(d):** © Ed Reschke; **18.3:** © Ed Reschke; **18 4:** © Ed Reschke.

Laboratory 19

Figure 19.3: Photo by Ken Taylor/Wildlife Images; **19.6:** Photo by Ken Taylor/Wildlife Images.

Laboratory 20

Figures 20.4 and 20.7: © The McGraw Hill Companies, Inc./Carlyn Iverson, photographer; **20.9, 20 10:** © Ed Reschke.

Laboratory 21

Figure 21.1b: © Ed Reschke/Peter Arnold.

Laboratory 22

Figure 22.2a: © Dwight Kuhn.

Laboratory 23

Figure 23.2a-c: Courtesy of Dr. J. Timothy Cannon; **23.5b:** © Manfred Kage/Peter Arnold; **23.6a:** Copyright © 2005 The Regents of the University of California. All Rights Reserved. Used by permission; **23.6b:** © PH Gerbier/SPL/Photo Researchers; **23.7:** © S.L. Flegler/Visuals Unlimited.

Laboratory 24

Figure 24.1a-g: © Carolina Biological Supply/PhotoTake; **24.1h:** © Ed Reschke/ Peter Arnold; **24.2a:** © Martin Rotker/ Phototake; **24.2b-h:** © Carolina Biological Supply/PhotoTake; **24.2i:** © Alfred Owczarzak/BPS; **24.6a, 24.7a, 24.8a, 24.9a:** © Carolina Biological Supply/PhotoTake; **24.2h,i:** © Petit format/Photo Researchers.

Laboratory 25

Figure 25.1a: © Kathy Talaro/Visuals Unlimited; **25.1b:** © Dwight Kuhn; **25.2b:** © S.J. Kraseman/Peter Arnold, Inc.; **25.2c:** © Kerry T. Givens/Tom Stack & Assoc.; **25.2d:** © John Shaw/Tom Stack & Assoc.; **25.3:** © M. Abbey/Visuals Unlimited; **25.4a,b:** © Robert and Linda Mitchell Photography; **25.5a:** © Kenneth Fink/Photo Researchers; **25.5b, 25.8:** © Ed Reschke; **25.10a:** © R. Calentine/Visuals Unlimited; **25.10b:** © Carolina Biological Supply/Visuals Unlimited; **p. 349:** © SPL/ Photo Researchers; **25.12:** © Carolina Biological Supply/PhotoTake; **25.13:** © R. Calentine/Visuals Unlimited.

Laboratory 26

Figure 26 3: © NOAA/Visuals Unlimited.

Text and Line Art

Laboratory 1

Laboratory adapted from Kathy Liu, "Eye to Eye with Garden Snails." Reprinted by permission of Kathy Liu, Port Townsend WA. http://www.accessexcellence.org/AE/AEC/AEF/1994/liu_snails.html

Laboratory 15

Figures 15.12 and 15.13: After Carolina Biological Supply.

Laboratory 23

Figure 23.14: BCSC, Colorado Springs, CO, from A. Glenn Richards. The Complementarity of Structure and Function: A Laboratory Block, 1969.

Laboratory 26

Figure 26.3: American Biology Teacher, April/May 1983:221. Drawn by Kristine A. Kohn.

Index

Note: page references followed by *f* and *t* refer to figures and tables respectively

A

Abdomen, crayfish, 189, 189*f*
Abdominal cavity, fetal pig, 252, 255*f*, 256–258, 257*f*
Acacias, symbiosis with ants, 343, 343*f*
Accommodation, in eyes, 312–313, 312*f*, 313*f*
Acetone, 58–59
Achenes, 224, 225*t*
Achilles reflex, 309, 309*f*
Achilles tendon, 309
Acidic group, of protein, 24, 24*f*
Acid rain, effects of, 358
Actin filaments, 241
Action spectrum, photosynthesis, 62*f*
Active site, 49–50, 50*f*
Adaptations, 131
Adductor muscles, clam, 184
Adenine (A), 118*f*, 119
Adipose tissue, 232*f*–233*f*, 238, 240*t*
Adrenal gland, frog, 201*f*, 202*f*
Adult stage, of metamorphosis, 194
Afferent arterioles, 295, 295*f*, 296*t*
Agar, 151
Aggregate fruits, 224
AIDS, and oral thrush, 161, 161*f*
Albumen, 327, 327*f*
Algae, 156–157, 156*f*, 157*f*
  blue-green, 152
  cell structure, 147*t*
  geological history, 135*f*
  in lichens, 341, 341*f*, 342*t*
  nutrition, 147*t*
  in pond water, 154–155, 154*f*
Allantois, 326
  chick embryo, 327*f*, 333*f*, 334, 334*f*
  human embryo, 327*f*, 336*f*
Alleles, 91, 91*f*
  autosomal traits, 111–113, 111*f*
  dominant, 111
  recessive, 111
  X-linked, 113
Alternation of generations, in plants, 165, 165*f*
Alveoli, 272, 288, 289*f*
Amino acids, 24, 24*f*
  in protein synthesis, 124–125, 124*f*, 125*f*
  similarity in living organisms, 143–145
Amino group, of protein, 24, 24*f*

Amniocentesis, 106, 106*f*
Amnion, 326
  in chick embryo, 327*f*
  in human embryo, 327*f*, 336*f*
  margin of, 330*f*, 331, 331*f*
Amniotic cavity, 336*f*
Amoeba, 153*f*, 154*f*, 356*f*
Amphibians
  brain, 306, 306*f*
  metamorphosis, 197
Amylase, 24, 30
Anabaena, 152, 152*f*
Analogous structures, 137
Anaphase
  in meiosis
    anaphase I, 82, 84*f*–85*f*, 86*f*–87*f*
    anaphase II, 83, 84*f*–85*f*
  in mitosis, 76*f*–77*f*, 86*f*–87*f*
Anatomy, comparative, 137–143
Angiosperms, 163*f*, 174–177. *See also* Eudicots; Monocots
  adaptation to land environment, 163, 177
  characteristics of, 177
  evolution of, 164, 164*f*, 165*f*
  external anatomy, 206, 206*f*
  flower
    anatomy, 219–220, 219*f*, 221*f*
    eudicot *vs.* monocot, 219–220, 219*f*, 221*f*
  fruit
    definition of, 224
    in plant life cycle, 174, 174*f*, 219
    types, 224, 225*f*, 225*t*
  geological history, 135*f*
  herbaceous
    characteristics of, 205
    stem, 212–213, 212*f*–213*f*
  leaf. *See* Leaf
  life cycle, 165, 174, 174*f*, 220, 221*f*
  reproduction, 219–228
  root system, 206, 206*f*, 209–211, 209*f*, 210*f*
    anatomy, 209–211, 209*f*, 210*f*
    functions, 205, 209
  seeds. *See* Seed(s)
  shoot system, 206, 206*f*
  stem. *See* Stem, angiosperm
  tissues, 208, 208*t*
  woody
    characteristics, 205
    stem, 214–215, 214*f*, 215*f*
Animal(s)
  chemoreceptors in, 319, 319*f*
  classification of, 181, 181*t*

development. *See* Development
ear. *See* Ear(s)
evolution of, 180*f*
eyes. *See* Eye(s)
geological history, 134*f*
Animal cells
  cell cycle, 74
  cytokinesis in, 79, 79*f*
  meiosis in, 84*f*–85*f*
  mitosis in, 75, 76*f*–77*f*, 78
  osmosis in, 43–44, 43*f*
  structure, 35, 35*t*, 36, 36*f*
Animal pole, 324–325, 325*f*
Ankle, fetal pig, 249*f*
Ankle reflex, 309, 309*f*
Annelids
  evolution of, 180*f*
  geological history, 134*f*
Annual rings, in wood, 215, 215*f*
Annulus, 160*f*, 161
Anopheles mosquito, 345*f*
Ant(s), symbiosis with acacias, 343, 343*f*
Antennae
  crayfish, 189*f*, 190
  grasshopper, 192, 192*f*
Anterior compartment, of eye, 311*f*, 312*t*
Anterior vena cava
  fetal pig, 272*f*, 277*f*
  mammal, 274
Anther, 174*f*, 175, 175*f*, 219, 220, 221*f*
Antheridium
  fern, 168, 168*f*
  moss, 166, 166*f*, 167
Antibiotics
  for bacterial infection, 148
  bacterial resistance to, 149
Antibody(ies), 144
Antibody-antigen reaction, and degree of relatedness between animals, 144–145, 144*f*, 145*f*
Anticodons, 124, 125*f*
Antigens, 144
Antipodals, of angiosperm, 221*f*
Anus
  clam, 184, 185*f*
  crayfish, 189*f*
  fetal pig, 249, 249*f*, 258, 267
  human, 259*f*
  squid, 187*f*
Aorta
  fetal pig, 262*f*, 265*f*, 272*f*
  mammal, 274, 274*f*, 275*f*

Pupation, in fruit flies, 94
Pupil, 311*f*, 312*t*
Pyloric sphincter, of fetal pig, 273
Pyrsonumpha, 342, 342*f*

Q

Quart (qt), 362
Quaternary period, 132*t*, 134*f*, 135*f*

R

R group, of protein, 24, 24*f*
Radicle, 222, 222*f*–223*f*, 226, 226*f*, 227, 227*f*
Radula, of squid, 186
Rays, in wood, 215*f*
Reactants, 49
Reagents
    Benedict's, 28–29, 28*t*, 41, 284
    biuret, 24–25, 24*t*, 280
Receptacle, of angiosperm, 174*f*, 175, 175*f*, 219, 221*f*
Recessive alleles, 111
Rectum
    clam, 185*f*
    fetal pig, 258, 267
    grasshopper, 193*f*
Red blood cells. *See* Erythrocytes
Reflexes, spinal, 309, 309*f*
Reindeer moss, 341
Remoras, 344
Renal arteries, 277*f*, 294*f*
Renal cortex, 263, 263*f*, 294*f*, 295, 295*f*
Renal medulla, 263, 263*f*, 294*f*, 295, 295*f*
Renal pelvis, 263, 263*f*, 294*f*
Renal pyramid, 294*f*
Renal veins, 277*f*, 294*f*
Reproduction, in angiosperms, 219–228
Reproductive system, 261. *See also* Urogenital system
    fetal pig
        female, 267, 267*t*, 268*f*
        male, 264–266, 264*t*, 265*f*
    human
        female, 269*f*
        male, 266*f*
Reptiles, embryo, comparative anatomy, 142*f*
Resolution, of microscope, 13
Respiratory passage, fetal pig, 254
Respiratory system
    fetal pig, 272, 272*f*
    frog, 199
    human, 259*f*
Respirometers, 68–69, 68*f*
Retina, 311*f*, 312*t*
Retinal blood vessels, 311*f*
Rhizobium, 340, 340*f*
Rhizopus, 159
Ribosomal DNA (rRNA), 122
Ribosomes
    of bacteria, 148, 149*f*

in cell anatomy, 34, 34*f*, 35*t*
    in protein synthesis, 122, 124, 125, 125*f*
Right atrioventricular (tricuspid) valve, in mammals, 275, 275*f*
Right atrium, 255*f*, 273, 274*f*, 275*f*
Right ventricle, 255*f*, 273, 274*f*, 275*f*
Ringworm, 161, 161*f*
RNA (ribonucleic acid), structure, 121–122, 121*f*
Rod cells, 311, 312*t*
Roly-poly bugs. *See* Pillbugs
Romalea, 191–193, 192*f*, 193*f*
Root(s)
    eudicot, 207*f*, 209–211, 209*f*, 210*f*
    fibrous, 211, 211*f*
    monocot, 207*f*, 210–211
    prop, 227*f*
    taproots, 211, 211*f*
    tip, anatomy, 209, 209*f*
    tissues, 208
Root apical meristem, 209, 209*f*, 222*f*–223*f*
Root cap, 209, 209*f*
Root ganglion, 307*f*, 308, 308*f*
Root hairs, 209, 209*f*
Root nodules of leguminous plants, 340, 340*f*
Root system, angiosperm, 206, 206*f*, 209–211, 209*f*
    anatomy, 209–211, 209*f*, 210*f*
    functions, 205, 209
Rotating lens mechanism, of binocular dissecting microscope, 15
Rotifers, 356*f*
Rough ER, 35*t*
Round window, 316*f*
Roundworms
    evolution of, 180*f*
    parasitism in, 350–352
rRNA. *See* Ribosomal DNA

S

S stage, 74, 74*f*
Saccule, 316*t*
Salivary amylase, digestion of starch with, 279, 284–285
Salivary glands, of grasshopper, 193, 193*f*
Samaras, 224, 225*t*
Scallops, 182*f*
Scanning electron micrographs (SEM), 12*f*, 13
Scanning electron microscope, 13
Scanning objective, of compound light microscope, 17
Scarlet hood mushroom, 158*f*
Scarlet leafhopper, 191*f*
Schwann cell, 303*f*
Scientific method, 2, 3*f*
Scientific theory, defined, 2, 3*f*
Scissor-tailed flycatcher (*Aves tyrannidae*), 196*f*

Sclera, 311*f*, 312*t*
Sclerenchyma cells, of angiosperm, 208, 208*t*
Scolex, of tapeworm, 347, 348*f*
Scrotal sac, fetal pig, 249*f*, 264, 265*f*
Scrotum, human, 266*f*
Sea stars, development, 322, 323*f*
Secondary follicle, 271, 271*f*
Secondary growth, 205, 209, 214
Secondary host, 347, 350*f*
Secondary oocyte, 269*f*, 271, 271*f*
Secondary phloem, 214, 215
Secondary xylem, 214, 215, 215*f*
Sectioning, of tissue, in slide preparation, 231
Seed(s)
    in angiosperm life cycle, 221*f*
    dispersion of, 219
    eudicot, 207*f*
    evolution of, 164, 164*f*, 165*f*
    germination, 226*f*, 227*f*, 228
        effect of pH on, 357
    monocot, 207*f*, 227, 227*f*
    structure, 226–227, 226*f*, 227*f*
Seed coat, 221*f*, 222*f*–223*f*, 226, 226*f*
Seed cones, 172, 172*f*, 173, 173*f*
Seed plants. *See* Angiosperms; Gymnosperms
Seedless plants, 165–169
Segmentation
    in classification of animals, 181*t*
    evolution of, 180*f*
Segregation, law of, 92
SEM. *See* Scanning electron micrographs
Semicircular canals, 316*f*, 316*t*
Semilunar valve, in mammals, 275, 275*f*
Seminal receptacles
    crayfish, 190
    grasshopper, 193*f*
Seminal vesicle
    fetal pig, 264, 264*t*, 265*f*
    human, 266*f*
Seminiferous tubules, 270, 270*f*
Senses. *See also* Ear(s); Eye(s)
    smell chemoreceptors, 319
    taste chemoreceptors, 319
    temperature receptors, in skin, 319
    touch receptors, in skin, 318
Sensory nerve fiber, 307, 307*f*
Sensory neurons, 307*f*, 309
Sepals, 175, 175*f*, 219, 220, 221*f*
Septum
    cardiac, 275*f*
    in fungal hyphae, 158, 158*f*
Serosa, 244, 244*f*
Sertoli cell, 270*f*
Sex chromosomes
    abnormalities in, 107, 107*t*, 109*f*
    defined, 105
Sex pilus, 34*f*, 148, 149*f*

Suspensory ligament, of eye, 311f, 312f
Swimmerets, of crayfish, 189f, 190
Symbiosis, types, 339, 339t
Symmetry
    bilateral, evolution of, 180f
    in classification of animals, 181t
Synapsis, 82
Syndrome, defined, 105
Synergids, of angiosperm, 221f
Synthesis, enzymatic, 50, 50f
Systemic artery, of frog, 199
Systemic circuit, 276

T

Taenia pisiformis, 348, 348f
Tail bud, of chick embryo, 332f, 333f
Tapeworms, 347–348, 348f
Taproots, 211, 211f
Taste buds, 319
Taste chemoreceptors, 319
Teaspoon (tsp), 362
Teeth
    fetal pig, 250, 250f
    frog, 198, 198f
    human, 259f
Telencephalon, of chick embryo, 334f
Telophase
    in meiosis
        telophase I, 82, 84f–85f, 86f–87f
        telophase II, 83, 84f–85f
    in mitosis, 76f–77f, 86f–87f
Telson, of crayfish, 189f
TEM. *See* Transmission electron
    micrographs
Temperature
    conversion of Celsius to/from
        Fahrenheit, 362
    and enzyme activity, 52, 279
    metric units, 362
Temperature receptors, in skin, 319
Template strand, 123, 123f
Temporal bone, 141f
Temporal lobe, 304, 305f
Tendon reflexes, 309
Tentacles
    cephalopod, 182, 182f
    squid, 186, 187f
Terminal bud, 206, 206f
Terminal bud scales, 214, 214f
Terminal bud scars, 214, 214f
Termites, symbiosis with zooflagellate,
    342, 342f
Tertiary period, 132t, 134f, 135f
Testes
    fetal pig, 264, 264t, 265f, 270, 270f
    frog, 200, 201f
    grasshopper, 193
    human, 266f
    mammal, 261
    tapeworm, 348f
Testosterone, 270
Tetrahymena, 154f, 356f

Thalamus, 304, 305f, 334f
Thermal pollution, effects of, 358
Thoracic cavity, of fetal pig, 252, 254,
    255f, 257f
Thorax
    crayfish, 189
    grasshopper, 191, 192f
Thylakoids, 34, 57, 57f, 152
Thymine (T), 118f, 119
Thymus gland
    fetal pig, 254, 255f
    human, 259f
Thyroid gland
    fetal pig, 251f, 254, 257f
    human, 259f
Tinea, 161, 161f
Tissue(s). *See also* Connective tissue;
        Epithelial tissue; Muscular
        tissue; Nervous tissue
    angiosperms, 208, 208t
    definition of, 231
    evolution of, 180f
    human, types of, 232f–233f
Tissue level of organization, 181t, 231,
    234–243
Tobacco, genetics of seedling color,
    92–93, 92f
Ton, 362
Tongue
    chemoreceptors on, 319
    fetal pig, 250, 250f
    frog, 198, 198f
    human, 259f
Tonicity, 43–45, 43f, 45f
Torpedo stage, of eudicot embryo,
    222f–223f
Tortoiseshell scale, 191f
Touch receptors, in skin, 318
Trachea(e)
    fetal pig, 250, 251, 251f, 254,
        272, 272f
    frog, 198
    grasshopper, 192, 193
    human, 259f
    insect, 191
Transcription, 122, 123, 123f
Transfer RNA (tRNA), 122, 124, 124f,
    125, 125f
Translation, 122, 124–125
Transmission electron micrographs
    (TEM), 12f, 13
Transmission electron microscope,
    13, 13t
Treponema pallidum, 150f
Triassic period, 132t, 134f, 135f, 164f
Trichinella, 351, 351f
Trichinosis, 351, 351f
Trichomes, 216, 216f
Trichonympha, 342, 342f
Tricuspid (right atrioventricular) valve,
    in mammals, 275, 275f
Trilobites, geological history, 134f

Triplet code, of DNA, 124, 143
Trisomies, 105
Trisomy 21. *See* Down syndrome
tRNA. *See* Transfer RNA
Trophozoites, 345f
Trypanosoma, 153f
Tube cell nucleus, of angiosperm, 221f
Tubular reabsorption, 296, 296f, 297
Tubular secretion, 296, 296f, 297
Turgor pressure, 44–45, 45f
Turner syndrome, 107, 107t, 109f
Two-trait genetic crosses, 98–101
Tympanic membrane, 316f, 316t, 317
Tympanum
    frog, 197, 197f
    grasshopper, 192, 192f, 315, 315f

U

Umbilical artery, 255f, 257f, 265f,
    268f, 277f
Umbilical cord
    fetal pig, 248, 249f, 252, 253f, 255f,
        257f, 262f, 265f, 268f
    human embryo, 327f, 336f
    mammal, 248
Umbilical vein, 252, 255f, 277f
Umbo (beak), clam, 183f, 184
Upper epidermis, leaf, 216, 216f
Uracil (U), 121, 121f
Urea, formation, 291
Ureter
    fetal pig, 262f, 263, 265f, 268f
    human, 266f, 269f
    mammal, 294f
Urethra
    fetal pig, 262f, 263, 264, 264t, 265f,
        266, 267, 268f
    human, 259f, 266f
Urinalysis, 298–299, 299f
Urinary bladder
    fetal pig, 255f, 257f, 262f, 263,
        265f, 268f
    frog, 200, 200f, 201f, 202f
    human, 259f, 266f
Urinary system. *See also* Urogenital
        system
    fetal pig, 262f, 263, 267
    human, 259f
    mammal, 261
Urine, production of, 294, 298
Urogenital opening, fetal pig, 249, 249f
Urogenital orifice, fetal pig, 265f
Urogenital papilla, fetal pig, 249,
    249f, 268f
Urogenital sinus, fetal pig, 262f,
    267, 268f
Urogenital system, frog, 200, 201f, 202f
Uropods, crayfish, 189f, 190
U.S. customary units, conversion
        to/from metric units, 362
Uterine cavity, human, 269f
Uterine horns, fetal pig, 267, 268f

Uterus
  fetal pig, 267, 267t, 268f
  human, 269f
  tapeworm, 348f
Utricle, 316t
Uvula
  fetal pig, 250
  human, 259f

V

Vacuoles
  contractile, 153, 153f
  food, 153, 153f
Vagina
  fetal pig, 267, 267t, 268f
  human, 269f
  tapeworm, 348f
Valves, clam, 183
Vas deferens
  fetal pig, 264, 264t, 265f, 266
  human, 266f
Vasa efferentia, frog, 200, 201f
Vascular bundle scars, 214, 214f
Vascular bundles, 207f, 212, 212f, 213f
Vascular cambium, 212f, 214, 215, 215f
Vascular cylinder, 209f, 210f
Vascular plants, 168
Vascular tissue
  of angiosperm, 208, 208t
  plant, evolution of, 164, 164f, 165f
Vegetable, definition of, 224
Vegetal pole, 324–325, 325f
Veins
  cardiac, 274f
  hepatic, 277f, 292–293
  hepatic portal, 290f, 291, 292–293
  iliac, 277f
  jugular, 277f
  of leaf, 207f, 216, 216f
  ovarian, 268f
  pulmonary, 272f, 274, 274f, 275f
  renal, 277f, 294f, 296t
  umbilical, 252, 255f, 277f
  vitelline, 330f, 331, 331f, 332f,
    333f, 334f
Vena cava
  fetal pig, 262f, 272f, 277f
  mammal, 274, 274f, 275f
  squid, 187f
Ventral horn, of spinal cord, 307f,
  308, 308f
Ventral nerve cord, grasshopper, 193f
Ventricles
  brain, 304, 305f

heart
  fetal pig, 255f
  frog, 199
  mammal, 273, 274f, 275f
Venus, 183–184, 184f, 185f, 186t
Vertebrates, 196–202. See also
    Mammals
  anatomy, frog, 197–202, 197f, 198f,
    200f, 201f, 202f
  brains, comparison of, 306, 306f
  characteristics of, 179, 196–202
  diversity of, 196, 196f
  embryos, comparative anatomy,
    142–143, 142f
  forelimbs, comparative anatomy,
    137–138, 138f
  geological history, 134f
  vs. invertebrates, 203
  nervous system
    brain, 306, 306f
    overview, 303
Vesicles, 35t
Vessel elements, of angiosperm, 213f
Vestibular nerve, 316f
Vestibule, 316f, 316t
Villi, intestinal, 244
Viruses, as obligate parasites, 345
Visceral mass
  clam, 184
  mollusc, 182
Visceral muscle, 242
Vitelline arteries and veins, of chick
    embryo, 330f, 331, 331f, 332f,
    333f, 334f
Vitelline gland, tapeworm, 348f
Vitreous humor, 311f, 312t
Vocal cords, of fetal pig, 272
Volume
  measurement of, 10–11
  metric units, 10–11, 362
Volumeters, 60, 61f, 70–71, 71f
Voluntary muscle, 241
Volvox, 156, 157f
Vomerine teeth, of frog, 198, 198f
Vorticella, 154f, 356f

W

Walking stick, 191f
Wet mount, 19, 19f
White blood cells. See Leukocytes
White light
  components of, 60f
  rate of photosynthesis in, 60–61
White matter, 307f, 308, 308f

Whitefish blastula, mitosis in, 78
Wing buds, of chick embryo, 333f, 334f
Wood, 215, 215f
Woody angiosperms
  characteristics, 205
  stem, 214–215, 214f, 215f
Worms
  anatomy or parasitic vs. free-living
    worms, 347, 347f
  parasitism in, 347–352
Wrist, of fetal pig, 249f

X

Xanthophylls, paper chromatography,
  58–59, 58f
X chromosomes, 105, 106
X-linked alleles, 113
X-linked crosses, 102–103
X-linked genes, 105
Xylem, 207f, 208, 208t, 209, 210f, 211,
    212, 212f, 213f
  primary, 215f
  secondary, 214, 215, 215f

Y

Y chromosomes, 105, 106
Yard (yd), 362
Yeast, fermentation in, 67, 68–69, 68f
Yolk, 321, 322, 323f, 324, 325f, 334
Yolk plug, 324f, 325f
Yolk sac
  chick embryo, 334, 327f
  human embryo, 336f, 327f
  vertebrate, 326, 327f

Z

Zone of cell division, 209, 209f
Zone of elongation, 209, 209f
Zone of maturation, 209, 209f
Zooflagellates, symbiosis with termites,
    342, 342f
Zoom mechanism, of binocular
    dissecting microscope, 15
Zygomatic arch, 141f
Zygote
  animal, 321, 322
  fern, 168, 168f
  frog, 324, 324f
  human, 88, 89f
  pine tree, 171, 171f
  plant, 165, 165f
Zygote stage, of eudicot embryo,
    222f–223f